124 Advances in Polymer Science

Polysoaps/Stabilizers/ Nitrogen-15 NMR

With contributions by
M. Andreis, J. Koenig, A. Laschewsky, J. Pospíšil

With 61 Figures and 11 Tables

Springer

ISBN 3-540-58983-X Springer-Verlag Berlin Heidelberg NewYork
ISBN 0-387-58983-X Springer-Verlag NewYork Berlin Heidelberg

© Springer-Verlag Berlin Heidelberg 1995
Library of Congress Catalog Card Number 61-642
Printed in Germany

The use of registered names, trademarks, etc. in this publication does not imply, even in the absence of a specific statement, that such names are exempt from the relevant protective laws and regulations and therefore free for general use.

Typesetting: Macmillan India Ltd., Bangalore-25
SPIN: 10495079   02/3020 - 5 4 3 2 1 0 - Printed on acid-free paper

Editors

Table of Contents

**Molecular Concepts, Self-Organisation and Properties
of Polysoaps**
A. Laschewsky   1

Aromatic and Heterocyclic Amines in Polymer Stabilization
Pospíšil, J   87

Application of Nitrogen-15 NMR to Polymers
M. Andreis, J. L. Koenig   191

Author Index Volumes 101 - 124   239

Subject Index   247

Molecular Concepts, Self-Organisation and Properties of Polysoaps

A. Laschewsky

Université Catholique de Louvain, Dépt. de Chimie, Place L. Pasteur 1,
B-1348 Louvain-la-Neuve

The article reviews water-soluble polymers characterized by surfactant side chains, and related amphiphilic polymers. Various synthetic approaches are presented, and rules for useful molecular architectures are given. Models for the self-organization of such polymers in water are presented comparing them with the micellization of low molecular weight surfactants. Highlighting key properties of aqueous polysoap solutions such as viscosity, surface tension and solubilization power, some structure-property relationships are established. Further, the formation of mesophases and of superstructures in bulk is addressed. Finally, the functionalization of polysoaps, and potential applications are discussed.

1   Introduction   3

2   The Structure of Polysoaps   3
   2.1   Micellar Polymers.   3
   2.2   Synthetic Strategies to Polysoaps   6
   2.3   Molecular Architecture of Polysoaps   9
      2.3.1   General.   9
      2.3.2   Surfactant Side Chains   10
      2.3.3   Steric Requirements of the Polymer Architecture.   12
      2.3.4   The Spacer Concept   15
      2.3.5   Flexibility of the Polymer Backbone   19
   2.4   Functional Polysoaps   20

3   Properties of Polysoaps in Aqueous Solution   22
   3.1   Viscosity   22
   3.2   Surface Activity.   26
   3.3   Solubilization   32
      3.3.1   Solubilization of Probe Molecules   34
      3.3.2   Solubilization Capacity.   37
   3.4   Emulsifying and Dispersing Properties   39
   3.5   Dynamic Properties   40

4   Aggregation in Aqueous Solution   42
   4.1   Micelles.   42
   4.2   Polymeric Micelles   43
      4.2.1   Models   43
      4.2.2   Experimental Data and Aggregation Numbers.   46
   4.3   Lyotropic Liquid Crystals.   49

Advances in Polymer Science, Vol. 124
© Springer-Verlag Berlin Heidelberg 1995

5   Aggregation in the Solid State 50

6   Molecular Weight Effects and the Behaviour of Oligomers 53
　6.1   Defined Oligomeric Surfactants 53
　6.2   Oligomeric Mixtures 58

7   Applications of Polysoaps 59

8   Survey on Polymerizable Surfactants 76

9   Conclusions 76

10 References 76

1 Introduction

Amphiphilic polymers, i.e. polymers bearing hydrophilic and hydrophobic fragments, have been known for a long time. They have attracted much attention in recent years because of their resemblance to biological systems as well as their strong tendency for self-organization in aqueous environments due to the hydrophobic effect [1–3], creating plentiful superstructures [4, 5] on the intermediate level between molecular and macroscopic structures ("mesoscopic scale" [6]). In particular, the subgroups of polymeric lipids and lipid analogs have been investigated in detail with respect to their self-organization in insoluble monolayers, in (Langmuir-Blodgett) multilayers, in vesicles and in myelin structures, thus providing general concepts for their understanding [7–17].

Whereas the above systems which give rise mainly to phase separated superstructures in aqueous systems are now fairly well understood, the situation for more hydrophilic and thus watersoluble amphiphilic polymers is less clear. Such polymers are often referred to as "micellar polymers" because they have properties similar to surfactant micelles [18–23], and thus similar super-structures are implicitely assumed.

There is a considerable practical interest in the micellar polymers based on their many attractive properties and thus their potential uses, e.g. as protective colloids, emulsifiers, surfactants, wetting agents, lubricants, viscosity modifiers, (anti) foaming agents, pharmaceutic and cosmetic formulation ingredients, catalysts etc. (see also Sect. 7). This is reflected by the plethora of compounds in the scientific and patent literature which can be addressed as micellar polymers. However, the number of systematic studies of the molecular structures is surprisingly small, as is the number of investigations of the detailed nature of the hydrophobic aggregation. Therefore our understanding of polysoaps is still limited, despite many advances in recent years, compared to the state of knowledge in the mid-1970s documented in the review of Bekturov and Bakauova [24].

2 The Structure of Polysoaps

2.1 Micellar Polymers

Compared to low molecular weight amphiphiles, the size of polymeric amphiphiles allows for much more diverse arrangements of the hydrophilic and hydrophobic segments, as exemplified in Fig. 1. Accordingly, micellar polymers are characterized by versatile molecular architectures, giving rise to distinct subgroups. Diblock-copolymers with a clear separation of the hydrophilic

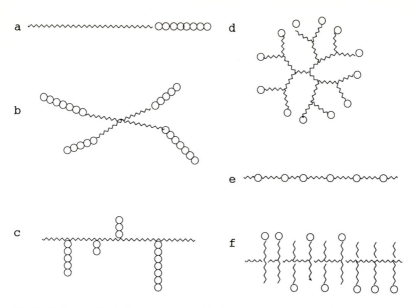

Fig. 1a–f. Types of micellar polymers: **a** block copolymers (macrosurfactants); **b** stars; **c** graft copolymers; **d** dendrimers; **e** segmented block copolymers; **f** polysoaps

("head") and the hydrophobic ("tail") parts come closest to the architecture of standard surfactants; they are often referred to as "macrosurfactants" (Fig. 1a). Representing the best studied micellar polymers at present, they behave in many respects like "oversize" standard surfactants although a unified picture is still missing [25–32]. Hydrophobic aggregation generally takes place by *inter-molecular association*. In contrast, amphiphilic star block-copolymers [32–35] and graft polymers [36–40] should undergo preferentially intramolecular aggregation (Fig. 1b,c). Amphiphilic dendrimers [6, 41–45] represent the extreme case (Fig. 1d). They can form intramolecular hydrophobic aggregates comprising the whole macromolecule. All of these types of micellar polymers are characterized by large, well separated blocks of hydrophilic and hydrophobic groups, and the amphiphilic character is based on the overall macromolecular architecture.

Alternatively, the hydrophilic and hydrophobic groups may be scattered all over the macromolecule. Here, the amphiphilic character of micellar polymers results from the presence of many independent, surfactant-like structural units which are covalently linked. This is realized in polymers which bear a limited number of ionic groups in their otherwise hydrophobic backbones – corresponding to a longitudinal linkage (Fig. 1e) – or in polymers with functional side-chains – corresponding to a lateral linkage of surfactant units (Fig. 1f).

The structural similarities of the individual polymer fragments with low molecular weight surfactants are paralleled by the similarities of two important properties of such polymers and of surfactants: i) high solubilization capacity for hydrophobic molecules implying a "molten character" of the aggregated hydrophobic parts, and ii) low viscosities of aqueous solution due to the hydrophobic aggregation which reduces the hydrodynamic radii dramatically. Thus the term "polysoap" was coined [24, 46–55]. Originally confined to functional side-chain polymers (Fig. 1f), the term is increasingly used as well for segmented block-copolymers [56–60] (Fig. 1e).

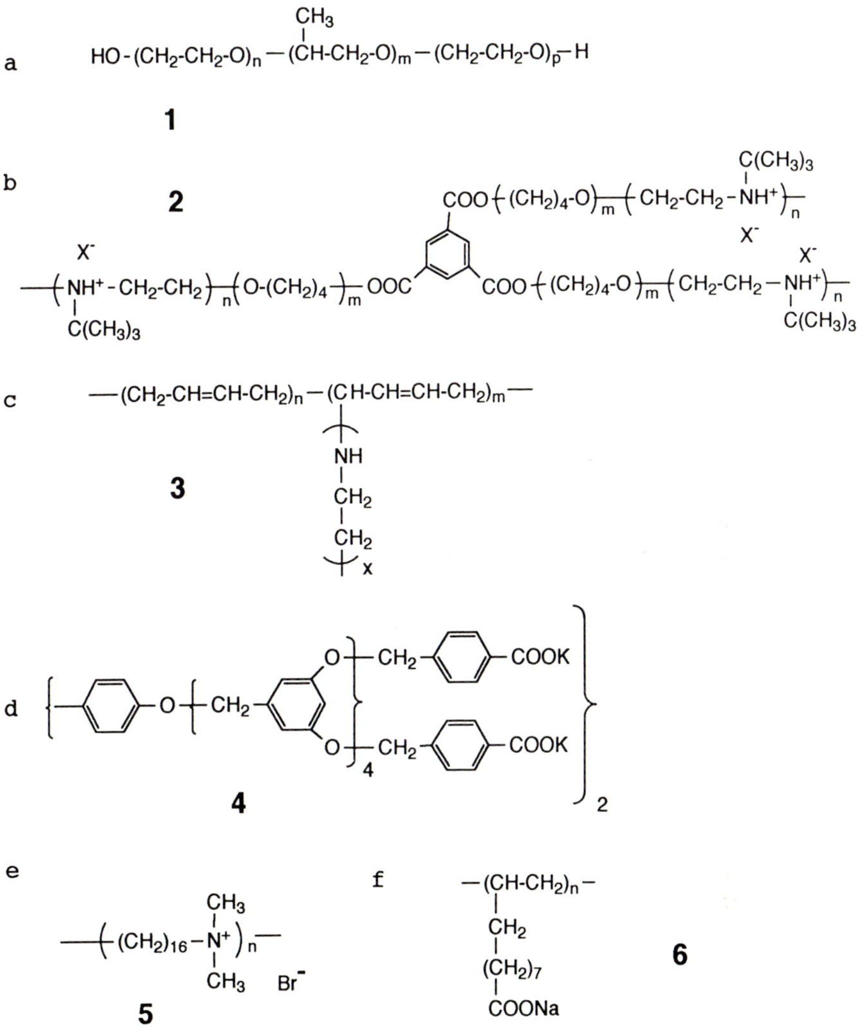

a $\qquad$ HO-(CH$_2$-CH$_2$-O)$_n$ $\overset{\text{CH}_3}{-\text{(CH-CH}_2\text{-O)}_m}$ $-$(CH$_2$-CH$_2$-O)$_p$$-$H

1

b $\qquad$ **2**

$$-\left(\text{NH}^+\text{-CH}_2\text{-CH}_2\right)_n\left(\text{O-(CH}_2)_4\right)_m\text{OOC}$$

$$\text{COO}\left(\text{(CH}_2)_4\text{-O}\right)_m\left(\text{CH}_2\text{-CH}_2\text{-NH}^+\right)_n$$

$$\text{COO}\left(\text{(CH}_2)_4\text{-O}\right)_m\left(\text{CH}_2\text{-CH}_2\text{-NH}^+\right)_n$$

c $\qquad$ $-$(CH$_2$-CH=CH-CH$_2$)$_n$$-$(CH-CH=CH-CH$_2$)$_m$$-$

3 $\qquad$ NH

CH$_2$

CH$_2$ $\}_x$

d

4 $\qquad$ $\}_2$

e $\qquad\qquad$ f $\qquad$ $-$(CH-CH$_2$)$_n$$-$

$$-\left(\text{(CH}_2)_{16}\text{-}\overset{\text{CH}_3}{\underset{\text{CH}_3}{\text{N}^+}}\right)_n- \quad \text{Br}^-$$

CH$_2$

(CH$_2$)$_7$ $\qquad$ **6**

COONa

5

Fig. 2a–f. Examples of micellar polymers: **a** block copolymer ("macrosurfactant" [25]; **b** star [35]; **c** graft copolymer [36]; **d** dendrimer [44]; **e** segmented block copolymer [56, 57]; **f** polysoap [61–74]

Despite these similarities, other properties of polysoaps can differ considerably from the ones of standard surfactants, as exemplified by their *intra*molecular aggregation and the usually missing critical micelle concentration "CMC" [24, 46, 51, 52, 65, 75–78].

2.2 Synthetic Strategies to Polysoaps

Although several natural polysoaps have been discovered [79, 80], the vast majority of polysoaps are synthetic polymers. As a basic requirement the polymers must have an appropriate hydrophilic-hydrophobic balance "HLB" to allow for water solubility on one hand (not just dispersion!), but sufficient hydrophobic parts to enable aggregation on the other. Considering the successful and unsuccessful systems in the literature, it seems that the required HLB for polysoaps is slightly more hydrophilic than for low molecular weight surfactants (see also Sect. 4.2).

Many strategies for synthesis have been established applying the full range of preparative polymer chemistry to meet the practical needs (Fig. 3). The pathways chosen have to take into account as well the efficient work up of the final polysoap, as the amphiphilic character can render e.g. the purification from residual reagents and by-products extremely difficult.

Polysoaps with the best defined chemical structure are produced by the polymerization of prefabricated reactive surfactants [51, 61–74, 76, 78, 81–93] (Fig. 3a). This strategy has been widely used, in particular for model studies, as it enables the comparison of analogous monomeric surfactants and polysoaps. Additionally, if the polymerization of the surfactants is performed under micellar conditions, the rate of polymerization and the molecular weights obtained are strongly increased by the aggregation [92, 94–105]. The improved reactivity may even allow the use of poorly polymerizable moieties which would not react, or react sluggishly only by standard polymerization procedures [61–74, 76, 82, 84, 88, 106–117], although the degree of polymerization may still be low. Concerning the influence of micellar polymerization conditions on tacticities, the results reported even for identical systems are contradictory [105, 118]. In any case, the control over the chemical structure must be paid by the often demanding synthesis of the surfactant monomers, and by the difficulties in controlling or even measuring molecular weights of the polysoaps. Examples of polymerizable surfactants are given in Tables 1–5 (Sect. 8).

Alternatively, polysoaps with well defined chemical structures can be prepared by the polyaddition or polycondensation of non-amphiphilic reagents [56, 57, 119, 120] (Fig. 2b). The surfactant fragments are formed in the course of the reaction only. This technique appears particularly attractive for polysoaps of "main chain geometry" (see below and Figs. 1e, 2e and 6d). The problems concerning the molecular weights are the same as for the polymerization of reactive surfactants, and the scope of useful chemical structures seems to be limited.

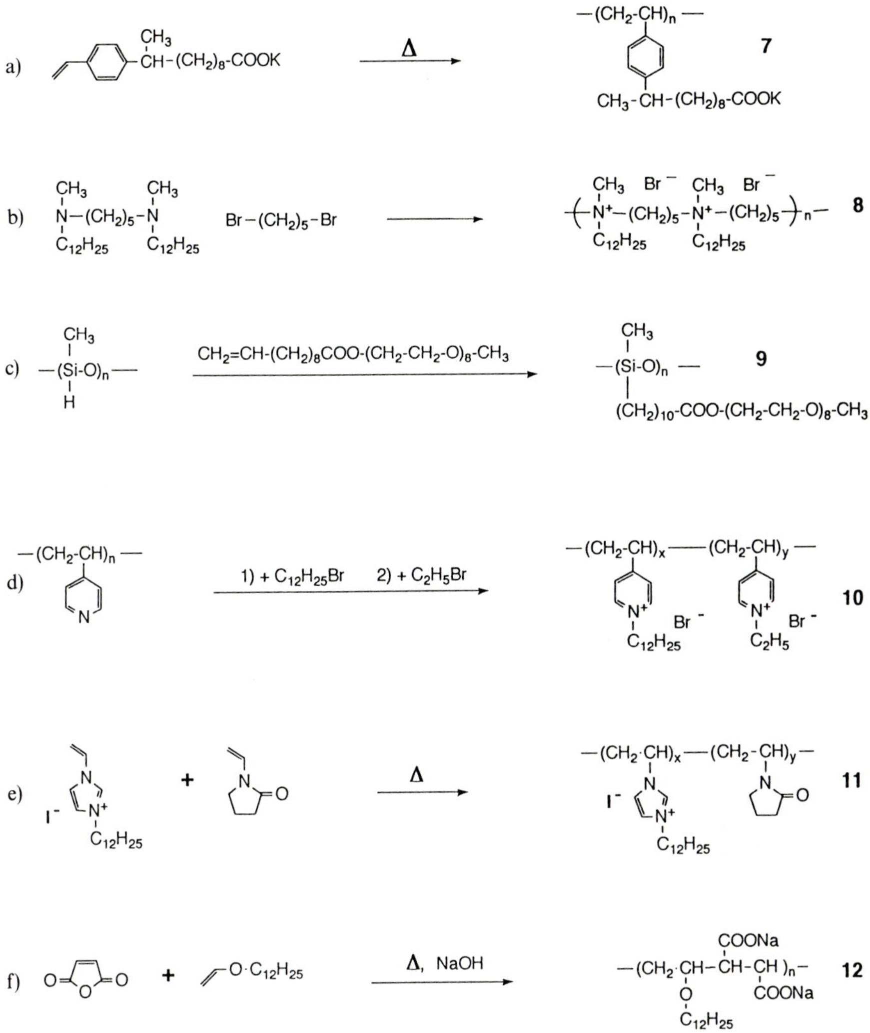

a) $\quad$ CH$_3$
$\quad$ —CH—(CH$_2$)$_8$—COOK $\quad\xrightarrow{\Delta}\quad$ 7
$\quad$ —(CH$_2$-CH)$_n$—
$\quad$ CH$_3$-CH-(CH$_2$)$_8$-COOK

b) $\quad$ CH$_3\quad$ CH$_3$
$\quad$ N—(CH$_2$)$_5$-N$\quad$ Br—(CH$_2$)$_5$-Br $\quad\longrightarrow\quad$
$\quad$ C$_{12}$H$_{25}\quad$ C$_{12}$H$_{25}$
$\quad$ CH$_3$ Br$^-$ CH$_3$ Br$^-$
$\quad$ (N$^+$—(CH$_2$)$_5$-N$^+$—(CH$_2$)$_5$)$_n$— $\quad$ 8
$\quad$ C$_{12}$H$_{25}\quad$ C$_{12}$H$_{25}$

c) $\quad$ CH$_3$
$\quad$ —(Si-O)$_n$—$\quad$ CH$_2$=CH-(CH$_2$)$_8$COO-(CH$_2$-CH$_2$-O)$_8$-CH$_3$ $\quad\longrightarrow\quad$
$\quad$ H
$\quad$ CH$_3$
$\quad$ —(Si-O)$_n$— $\quad$ 9
$\quad$ (CH$_2$)$_{10}$-COO-(CH$_2$-CH$_2$-O)$_8$-CH$_3$

d) $\quad$ —(CH$_2$-CH)$_n$— $\quad$ 1) + C$_{12}$H$_{25}$Br $\quad$ 2) + C$_2$H$_5$Br $\quad\longrightarrow\quad$
$\quad$ —(CH$_2$-CH)$_x$—(CH$_2$-CH)$_y$—
$\quad$ N$^+$ Br$^-\quad$ N$^+$ Br$^-$
$\quad$ C$_{12}$H$_{25}\quad$ C$_2$H$_5$ $\quad$ 10

e) $\quad$ (structures) $\quad$ + $\quad$ (structure) $\quad\xrightarrow{\Delta}\quad$
$\quad$ I$^-$ $\quad$ N$^+$ $\quad$ C$_{12}$H$_{25}$
$\quad$ —(CH$_2$-CH)$_x$—(CH$_2$-CH)$_y$—
$\quad$ I$^-$ $\quad$ N$^+$ $\quad$ C$_{12}$H$_{25}$ $\quad$ 11

f) $\quad$ O (structure) O $\quad$ + $\quad$ O·C$_{12}$H$_{25}$ $\quad\xrightarrow{\Delta,\ NaOH}\quad$
$\quad$ COONa
$\quad$ —(CH$_2$·CH-CH-CH)$_n$— $\quad$ 12
$\quad$ O $\quad$ COONa
$\quad$ C$_{12}$H$_{25}$

Fig. 3a–f. Synthetic strategies to polysoaps: **a** polymerization of reactive surfactants [51]; **b** polycondensation of non-amphiphilic reagents [119, 120]; **c** grafting of surfactant fragments [121, 122]; **d** hydrophobization of preformed polymers [49, 52, 75, 130, 131, 138, 139, 141, 142, 146]; **e** copolymerization of reactive surfactants with polar monomers [156]; **f** copolymerization of hydrophobic and hydrophilic monomers [53]

A different route to polysoaps at least with well defined surfactant fragments is the grafting of prefabricated surfactant fragments onto a reactive prefabricated polymer [50, 87, 121–129] (Fig. 3c). Appropriate choice of the latter enables the preparation of polymers with a controlled degree of polymerization. But to maintain a good control over the chemical structure, very high conversions of the grafting reaction and negligible side reactions are needed, thus severely limiting the scope of this strategy.

A more versatile strategy is the modification of (proto)-hydrophilic pre-formed polymers with hydrophobic reagents, or vice versa (Fig. 3d). This technique is the oldest one, and is still widely in use due to its convenience [46–49, 52, 54, 55, 130–155]. It also allows the control of the degree of polymerization by appropriate choice of the parent polymer. However, the structure of the surfactant fragments, and thus e.g. of the hydrophilic-hydro-phobic balance, may be poorly defined, and purification of such polysoaps may be difficult.

Covenient and versatile as well, the preferred technique at present seems to be copolymerization. Either polymerizable surfactants are copolymerized with small, often polar or hydrophilic, comonomers, creating rather well defined surfactant fragments in the polysoaps [73, 77, 78, 156–168] (Fig. 3e). This strategy is more versatile than exclusive homopolymerization and allows the use of polymerizable surfactants whose homopolymers are not water-soluble, but it still requires considerable synthetic effort. Or, much more facile, simple hydro-phobic monomers are copolymerized with small hydrophilic ones (Fig. 3f), taking full advantage of the choice of commercial or easily accessible monomers [39, 53–55, 153, 169–218].

The chemical structure of such copolymer polysoaps can be very heterogen-eous, being subject to the problems of copolymerization. Also, suitable solvents for the copolymerization reaction may be difficult to find. In fact, the choice of the solvent seems to be crucial, as it strongly influences the copolymerization parameters when using monomers of greatly differing polarity [77, 156, 188, 219–222]. This effect can be exploited for the preparation of random and blocky copolymers from the same pair of monomers depending on the reaction conditions [77, 220], and thus enables an additional structural variation of polysoaps (Fig. 4).

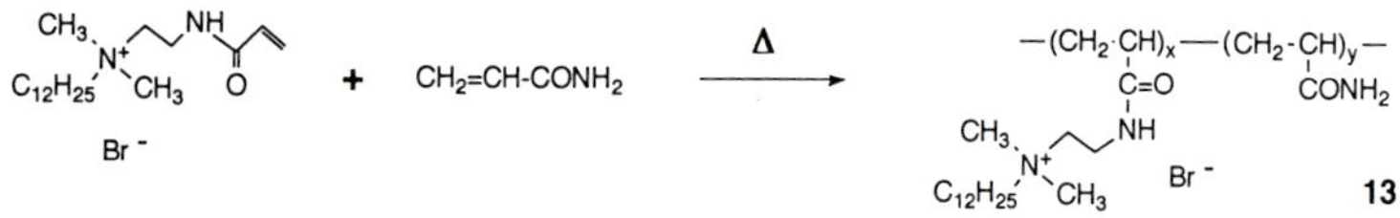

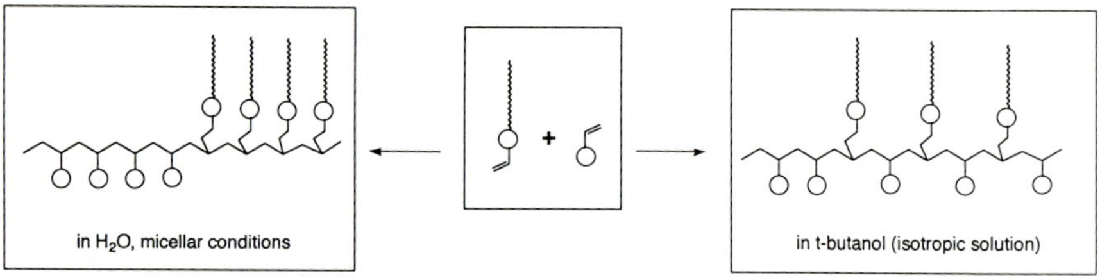

Fig. 4. Preparation of random and blocky copolymers by varying the reaction conditions [77, 220]

2.3 Molecular Architecture of Polysoaps

2.3.1 General

As sketched in Fig. 1, the molecular architecture of polysoaps is characterized by the combination of polymer and surfactant structures. The most obvious structural parts to be varied are the surfactant fragments, with respect to the hydrophilic head and the length and branching of the hydrophobic tail. This corresponds to classical surfactant chemistry, and the majority of systematic investigations of polysoaps have been restricted to such variations. But polysoaps offer additional variations which are characteristic for polymers but do not exist for surfactants. The molecular architecture can be varied e.g. with respect to the polymer geometry [76, 78, 87, 88, 105, 126, 164, 167, 168, 223–228] the nature of the polymer backbone [76, 78, 84, 167, 168, 229, 230] and the incorporation of spacer groups [100–102, 157, 223, 231, 232] controlling the distance of the surfactant fragments from the polymer backbone (Fig. 5). Such variations broaden the synthetic scope considerably, but have been explored and exploited rarely yet.

By attaching the surfactant fragments in different ways to the backbone, various polymer geometries are realized. They include "frontal" attachment at the hydrophilic head group ("head type"), "terminal" attachment at the end of the hydrophobic tail ("tail-end type"), intermediate structures ("mid-tail type") and full incorporation into the backbone ("main chain type") [76, 78, 87, 126] (Fig. 6). By varying the nature of the polymer backbone, the flexibility, the

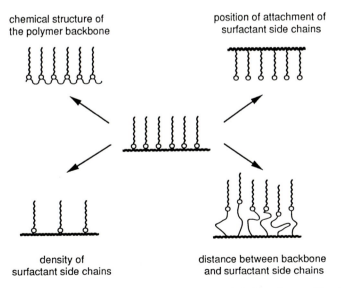

chemical structure of
the polymer backbone

position of attachment of
surfactant side chains

density of
surfactant side chains

distance between backbone
and surfactant side chains

Fig. 5. Structural variables of polysoaps characteristic for polymers. (Reprinted with kind permission from [78]. Copyright 1993 Hüthig & Wepf, Basel)

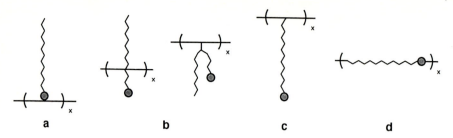

Fig. 6a–d. Geometries of polysoaps: **a** head type; **b** mid-tail type; **c** tail-end type; **d** main chain type

hydrophilicity (and thus the HLB), and the density of the surfactant side chains of the polysoap are modified. The incorporation of flexible side chain spacer groups may improve the aggregation process as shown for a number of self-organized systems such as polymeric amphiphiles [7, 8, 232–235], or side chain liquid crystalline polymers [8, 87, 236, 237].

2.3.2 Surfactant Side Chains

The variations of the surfactant fragments of polysoaps found are plentiful but seldom systematical. In particular many types of head groups have been realized, covering non-ionic, cationic, anionic and zwitterionic ones (Fig. 7, Tables 1–5), although systematic variations are limited [72, 76, 89, 107, 121, 124, 157, 223, 238–251]. They are often confined to the variation of counter-ions [50, 52, 53, 70, 130, 170, 173, 176, 177, 179, 182, 252]. Also, various attempts to functionalize polysoaps are reported, including the incorporation of complexing agents [253–256] chromophores [152, 177, 183, 189, 193, 204, 205, 211, 217, 257–260], mesogens [87, 122–124, 150, 248, 249, 261, 262], redox-active moiet-ies [263–272] and electrically conductive groups [270–274] (see Sect. 2.4). But their effects on the polysoap behaviour are difficult to assess at present. Most systematic investigations of the surfactant fragments in polysoaps are restricted to the influence of the length of the alkyl tails [54, 57, 68, 85–89, 94–97, 100–104, 108, 118–120, 123–125, 133, 135, 143, 144, 152, 153, 155, 161, 162, 169–171, 176, 177, 179–182, 187–196, 199–203, 206, 213, 221, 241–243, 249, 275–293], in analogy to homologous surfactant series. Recently, fluorocarbon hydrophobic tails [216, 218, 222, 294–297] have been used instead of hydrocarbon ones.

Considering hydrocarbon tails, it seems that the minimal length to produce polysoap properties is about C_8 [52, 191, 199, 200]. Polysoap behaviour was reported even for some shorter chains, but then additional hydrophobic units in the backbone must be present (see below). Also, hydrophobic counterions such as tetrabutylammonium [252] or alkyltrimethylammonium ions [186, 204] may induce hydrophobic aggregation for short chain "polysoaps", but in such cases it

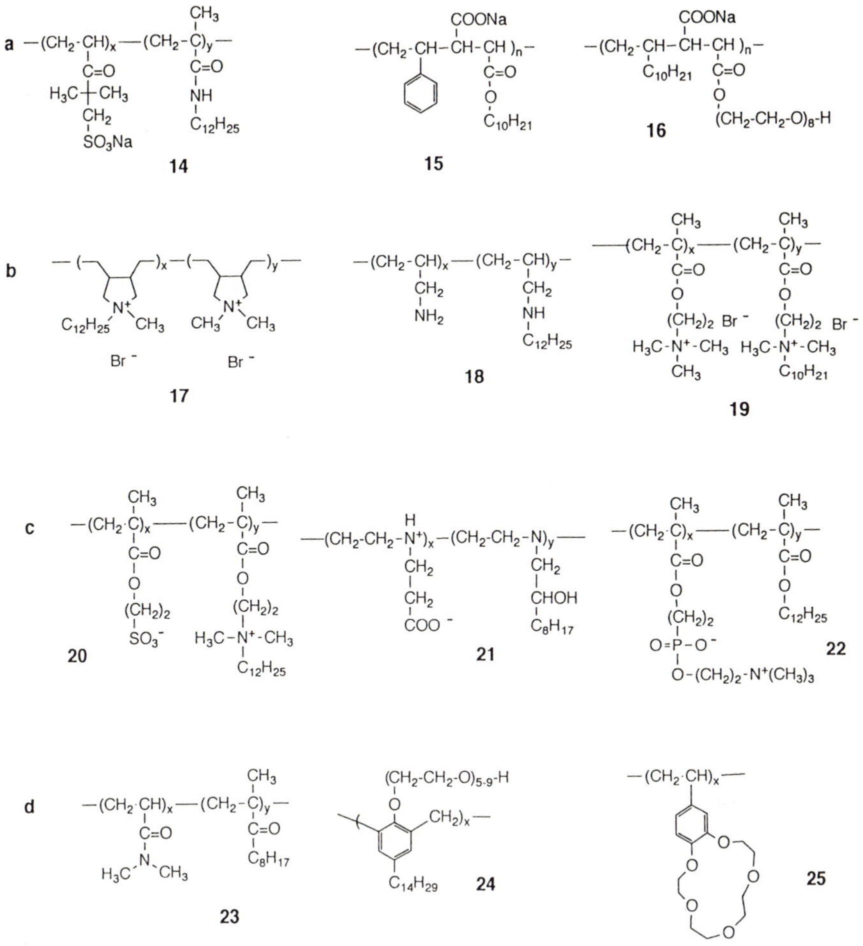

Fig. 7a–d. Examples of polysoaps with different types of head groups: **a** anionic [155, 193, 195, 207, 208]; **b** cationic [152, 161, 167, 199, 200]; **c** zwitterionic [147, 158–160, 209, 210]; **d** non-ionic [194, 241–243, 252, 253]

is disputable whether the amphiphilic ions favour the hydrophobic aggregation of the polymer, or rather the polymer favours the aggregation of the "counterions". Noteably, there are some indications for a maximal useful length of about C_{18} beyond which the hydrophobic chains are "crystallized" and thus the hydrophobic aggregates are no more fluid-like [239, 248, 249, 298]. Within these limits the effects of length variations generally agree well with the known effect of similar variations on low molecular weight surfactants. E.g., increasing the length of the alkyl tails or changing from hydrocarbon to fluorocarbon tails, the hydrophobic association of the polysoaps is promoted.

Specific deviations from the behaviour of low molecular weight surfactants may arise for charged polysoaps, as they represent polyelectrolytes with all implications. For practical aspects this may be of limited importance (although gelling etc. may occur at low concentrations [50]). In the case of fundamental studies however, concentration dependent studies of charged polysoaps become difficult to interpret as the dissociation of the ionic groups and thus the HLB is concentration dependent. Attempts to suppress counterion dissociation by addition of salts often result in precipitation of the polysoaps [50, 299]. Therefore non-ionic and fully zwitterionic polysoaps have been developed to bypass such problems (see Tables 3–5). They have indeed enabled many insights, but often they suffer from some shortcomings of their own. Non-ionic systems are prone to phase separation at elevated temperatures, i.e. they exhibit lower critical solution temperatures [87, 121–124, 126, 231, 251]. Furthermore, most non-ionic head groups, such as oligoethyleneoxides or acylated oligoethylenimines, are very large in comparison to the appropriate hydrophobic moieties yielding an unfavourable ratio of hydrophobic to hydrophilic domains. Alternative non-ionic head groups in the form of sugars or related moieties face solubility problems due to strong hydrogen bonding [246, 247, 250, 300–302]. Similarly, zwitterionic polysoaps often exhibit low solubility in pure water [227]. But by minimizing these problems by proper design, a number of non-ionic and zwitterionic polysoaps suited for systematic investigations have been prepared [78, 167, 168].

Instead of using well defined surfactant fragments, micellar polymers similar to polysoaps have been prepared by combining a large number of small hydrophobic monomers with a small number of charged monomers (e.g. styrene with 2-acrylamido-2-methyl propane sulfonate AMPS [258, 303]). Alternatively, a well balanced ratio of hydrophobic to hydrophilic units is achieved by controlled, limited dissociation e.g. of carboxylic groups, as in poly(methacrylic acid) [186, 304] or in poly(styrene-*alt*-maleic acid) [305, 306] or by controlled, limited quarternization of amino groups, like in poly([thio-1-(diethylamino)methyl] ethylene) [307, 308]. Still, it appears that the hydrophobic domains in such polymers with very short tails are less shielded than in surfactant micelles or in polysoaps with "normal" tail lengths [258, 303].

2.3.3 Steric Requirements of the Polymer Architecture

Systematic variations of the structural variables characteristic of polymers have been addressed only recently. Most of the work was focused on the role of polymer geometry, as the surfactant fragments with their hydrophilic "front part" and their hydrophobic "back side" introduce directionality into the systems [78, 126]. Neglecting polysoaps of the "main chain type", isomeric sets of vinylic surfactant monomers were converted into polysoaps of different geometry (Fig. 6), in which the surfactant fragments are fixed at different positions to the backbone (Fig. 8), but which have identical hydrophilic-

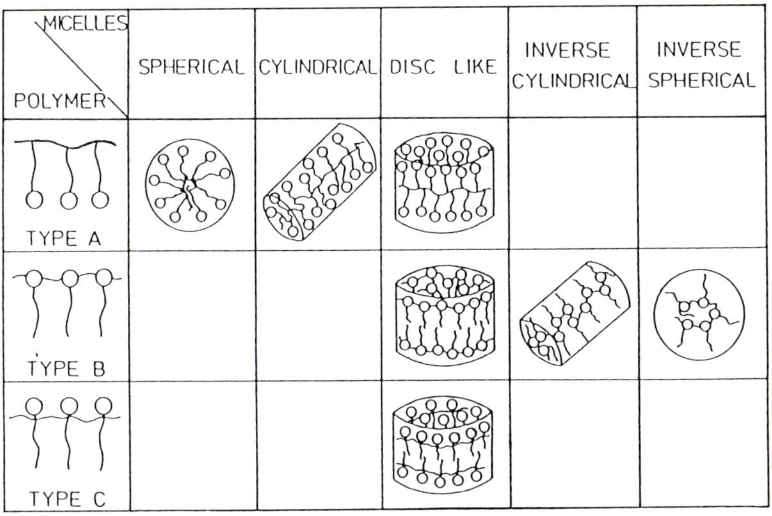

MICELLES / POLYMER	SPHERICAL	CYLINDRICAL	DISC LIKE	INVERSE CYLINDRICAL	INVERSE SPHERICAL
TYPE A					
TYPE B					
TYPE C					

Fig. 8. Scheme of isometric polysoaps of different geometry, and the assumed possible micellar shapes in aqueous solution. (Reprinted with kind permission from [231]. Copyright 1987 Steinkopff Verlag, Darmstadt)

hydrophobic balances. Most surprisingly to the early investigators, whereas all the monomers show comparable surfactant properties, only polymers of the "tail end" geometry (Fig. 5c) are water-soluble and behave like polysoaps. The other isomers do not dissolve in water, but instead are soluble in less polar solvents, in which the water-soluble isomers are insoluble [76, 78, 85, 86, 126, 156, 164, 167, 168, 225, 231, 279, 280]. Hence in case of vinylic polymerized surfactants, no straight forward correlation between solubility and HLB exists: apparently more hydrophilic polymers of the poorly performing "head" geometry are less soluble in water than apparently more hydrophobic ones of the well behaving "tail end" geometry. The opposite is true for organic solvents.

These observations were originally explained by a mismatch between the curvature of spherical or cylindrical micelles and the bending possibilities of the polymer backbone for the different geometries [76, 87], neglecting that many polysoaps give lamellar aggregates in which case this reasoning is not important (Fig. 8). More probably the phenomenon is more general, and independent of the shape of the "polymeric micelles" formed (see Sect. 4.2). It is based on an inherent overcrowding of surfactant fragments at the vinyl polymer backbone, keeping in mind that the C_2-repeat unit of a vinyl backbone has a length of ca. 0.25 nm, and the minimal diameter of a hydrocarbon tail is ca. 0.5 nm [78, 167, 168]: the backbone cannot offer enough space for an "amphiphilic conformation" of the surfactant fragments (Fig. 9a), independent of the shape of the hydrophobic aggregates. Instead, either the hydrophobic or the hydrophilic parts are exposed exclusively at the "shell" of the polymer to yield a "hydrophobic conformation" (Fig. 9b) or a "hydrophilic conformation" (Fig. 9c)

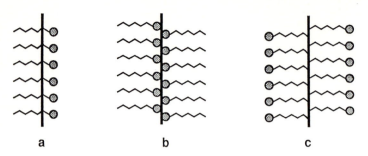

Fig. 9a–c. Possible arrangements of polysoaps in solution (scheme): **a** amphiphilic; **b** hydrophobic **c** hydrophilic. (Reprinted with kind permission from [78]. Copyright 1993 Hüthig & Wepf, Basel)

depending on the polymer geometry. The fragments of different polarity near the backbone are hidden in the polymer "core".

These steric problems with their consequences should affect all vinylic surfactant polymers, independent of the inherent surface curvatures of the different models of "polymeric micelles" (see Sect. 4.2). Thus, vinylic surfactant homopolymers of other than "tail end" geometry should be of very limited use as polysoaps. In fact, very few exceptions to the geometry controlled model of solubility have been reported [82, 106, 128, 225, 289–291]. In these examples, the chemical integrity of the polymers prepared, the attributed structures, or the polymeric nature may be questioned considering the results obtained for very similar compounds [115, 232, 309]. But even if these exceptions are real, this rule will help to design new monomers and polymers.

It should be noticed, however, that metastable aqueous solutions of "head" geometry polymers can be obtained by micellar polymerization of some surfactant monomers with very long alkyl tails [85, 86, 103, 104, 156, 280, 288, 292] (Fig. 10). Apparently, even a poorly realized "amphiphilic arrangement" still represents a local energy minimum, which can be frozen in for long alkyl tails. Such a behaviour parallels the formation of stable insoluble monolayers at the air-water interface of some polymerized surfactants with "head" geometry [168, 245, 232, 310, 311]. This form of self-organization is sterically less demanding, and thus can tolerate a poor "amphiphilic arrangement" (Fig. 11).

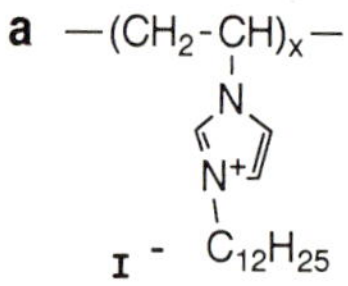

a $-(CH_2-CH)_x-$ **26**

     N

     N$^+$

     I$^-$ C$_{12}$H$_{25}$

b CH_3

     $-(CH_2-C)_y-$

     $C=O$ **27**

     O

     $(CH_2)_2$

     $H_3C-N^+-CH_3$ Br^-

     $C_{16}H_{33}$

c $-(CH_2-CH)_n-$ **28**

     CH_2

     $H_3C-N^+-CH_3$ Cl^-

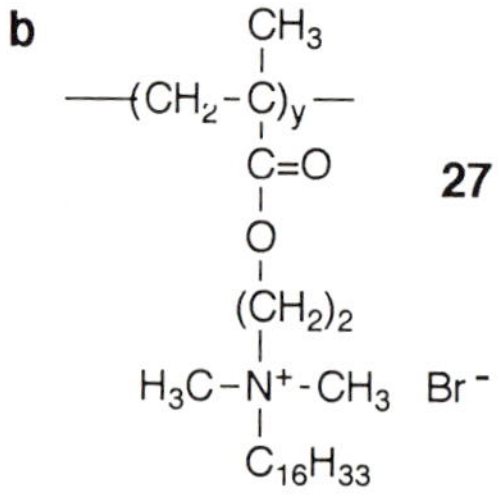

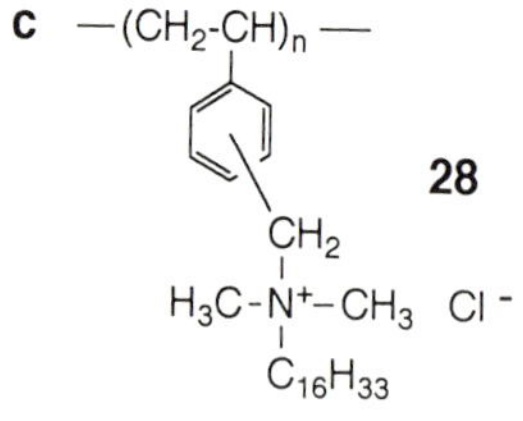

     $C_{16}H_{33}$

Fig. 10a–c. Polysoaps of "head" geometry reported to form metastable aqueous solutions upon polymerization under micellar conditions: **a** [85, 86, 156]; **b** [280]; **c** [103, 104, 288, 292]

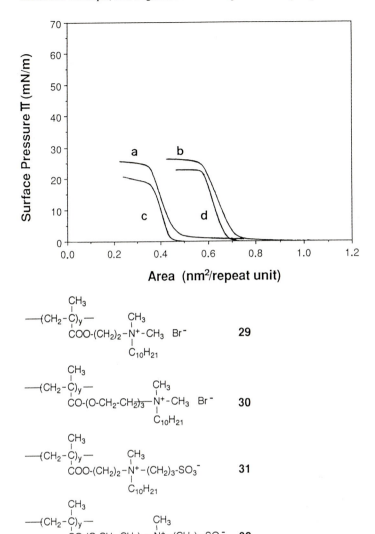

Fig. 11. Stable, insoluble monolayers from polymerized surfactants of "head" geometry. a = **29**, b = **30**, c = **31**, d = **32**; data from [232, 245]

2.3.4 The Spacer Concept

The geometrical constraints of vinyl polymers are the consequence of the excessive density of surfactant side chains. The problem is a general one encountered in functionalized polymers, and can be overcome by reducing the density by the incorporation of spacer groups. In the case of "main chain geometry" the effect of the spacer is straightforward. If the functional groups are in the side chain (e.g. the surfactant fragment for "tail end", "mid-tail" and "head" geometry), side-chain spacers and main chain spacers can be distinguished as illustrated in Fig. 12. Main chain spacers diminish the density of the functional groups at the backbone, whereas side-chain spacers provide additional space by separating the functional groups from the backbone. The spacer concept was particularly successful for some self-organizing polymers such as polymeric liquid crystals [8, 87, 236, 237] and polymeric lipids [7, 8, 13, 56, 232–235]. It has proven useful for polysoaps as well, making any geometry accessible [167, 168].

"Main chain spacers" can be realized in many ways. In fact, they have been used unconsciously from the very early days of the polysoaps in the classical approach to polysoaps, i.e. the incomplete hydrophobization of polar parent polymers (see Fig. 2c), when the non-hydrophobized units of the backbone act as spacer segments [46–49, 52, 139–141]. Another convenient approach is the copolymerization of amphiphilic monomers with small polar ones, or of hydrophobic monomers with small hydrophilic ones [73, 77, 78, 156–168] (Figs. 3e, 3f).

By incomplete hydrophobization of parent polymers and by random copolymerization, the spacer length (or the density of the hydrophobic tails) is easily controlled by the reaction conditions. In agreement with the steric constraints discussed above, the spacers have to be longest for polysoaps of the "head" geometry to guarantee water-solubility. A detailed evaluation of the spacer lengths needed is difficult to derive from the reported examples, as in particular the modified content of hydrophilic "spacer" comonomers shifts the HLB simultaneously. But spacer lengths seem to increase with both the bulkiness of the surfactant fragment and of the main chain spacer. E. g. for hydrophobized poly(sulfobetaine)s of "head" geometry, the minimal spacer length corresponds to a C_6–C_8 repeat unit, whereas for analogs of "mid tail" geometry, the minimal spacer length is roughly a C_4-repeat unit [167]. According to the above discussion, the success of the polysoap copolymers poly(maleate-*alt*-vinylether)

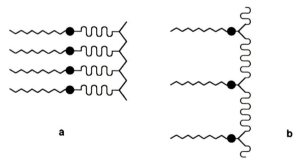

Fig. 12a, b. Scheme of spacer groups in polysoaps: **a** side chain spacer; **b** main chain spacer

and poly(maleate-*alt*-alkene) such as **12, 16, 33** and **34** (Figs. 3f, 7a, 13), can be attributed to their advantageous combination of favourable "mid-tail" geometry, of small head group and of small spacer unit, i.e. a repeat unit of four carbons, in combination with their ease of preparation [53–55, 169–182, 184–187, 189–192, 196, 197, 206–208, 215].

If well defined spacer segments are desired, the synthesis of polysoaps becomes more complicated. Homopolymerization of other than vinylic surfactant monomers, e.g. of epoxides [312], butadienes [168, 313] or vinylcyclopropanes [168] is needed (for examples see Tables 2 and 4). Alternatively, strictly alternating copolymerization enables the preparation of defined repeat units longer than C_2-fragments [54, 82, 168–170, 184–187, 189–192, 195–197, 314], usually of C_4 (Fig. 13). Polyaddition or polycondensation reactions provide defined spacer segments as well, but only few such systems (e.g. hydrophobized ionenes [119, 120] or phenol-formaldehyde resins [241–243, 315]) have been reported (see Figs. 2e, 3b). Presumably this lack is caused by the sensitivity of most of the polycondensations to humidity and protic solvents as encountered in many polysoap systems. Here the use of the thiol/ene addition polymerization [316] may offer a promising new route (Fig. 14).

In addition to considering minimal spacer lengths, the question of maximal lengths arises for which polysoap behavior is still observed. This borderline is very important, as slightly hydrophobized water-soluble polymers are known to aggregate *inter*molecularly thus acting e.g. as thickeners [77, 151, 162, 218, 222, 285, 294–297], whereas even lesser hydrophobized polymers do not aggregate at all. This is in contrast to the *intra*molecular aggregation of polysoaps producing solutions of very low viscosity (see Sects. 3.1 and 4). The maximal spacer length corresponds to a minimal density of hydrophobic groups needed for aggregation which has been called "critical alkyl group content" or "CAC" [52, 75] in analogy with the critical micelle concentration "CMC".

The detailed CAC values depend on the HLB of the polysoap. For a given head group, CAC values decrease with increasing alkyl lengths; typically they are in the range of 20 mol% for octyl tails and of 10 mol% for dodecyl tails [52, 75, 133, 145, 152, 167, 276, 317]. However 3% of octadecyl tails are not sufficient [151]. Analogously, when the length and the content of the hydrophobic tails are kept constant, CAC values increase with increasing hydrophilicity of the polysoaps, and the hydrophilicity of the head group may determine whether the polymer behaves as a polysoap, or not. Thus the polysoap character of poly(maleic acid-*alt*-alkylvinylether)s of medium chain length depends on the extent of deprotonation, as the anionic $-COO^-$ groups are much more hydrophilic than $-COOH$ groups [318] (compare Sect. 3.1). Whereas polymers with decyl tails behave like polysoaps at any degree of dissociation, polymers with hexyl tails act as polysoaps only when less than 40%, polymers with pentyl tails when less than 30% and polymers with butyl tails when less than 20% of the $-COOH$ groups are deprotonated [54, 176, 177, 181]. As for the latter examples, much hydrophobicity derives from the polymer backbone in the partially dissociated state, and these polymers may be considered as intermediate cases

copolymer	(surfactant) monomer	comonomer
33	$CH_3-(CH_2)_9-O-CH=CH_2$	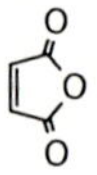 maleic anhydride
34	$CH_3-(CH_2)_9-CH=CH_2$	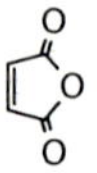 maleic anhydride
35	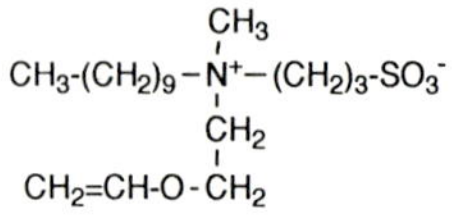 $\begin{array}{c} CH_3 \\ CH_3-(CH_2)_9-\overset{+}{N}-(CH_2)_3-SO_3^- \\ CH_2 \\ CH_2=CH-O-CH_2 \end{array}$	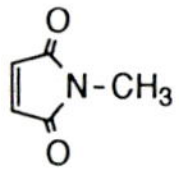 N-methylmaleimide
36	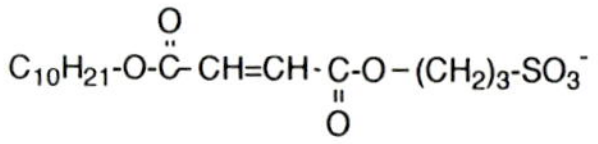 $C_{10}H_{21}-O-\overset{O}{\overset{\|}{C}}-CH=CH-\overset{\|}{C}-O-(CH_2)_3-SO_3^-$	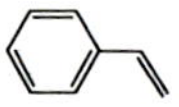 styrene
37	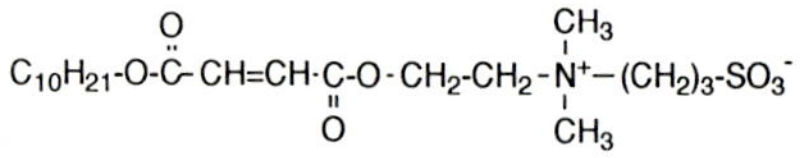 $C_{10}H_{21}-O-\overset{O}{\overset{\|}{C}}-CH=CH-\overset{O}{\overset{\|}{C}}-O-CH_2-CH_2-\overset{CH_3}{\overset{\|}{\underset{CH_3}{N^+}}}-(CH_2)_3-SO_3^-$	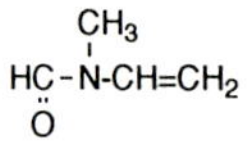 $\begin{array}{c} CH_3 \\ H\overset{\|}{C}-N-CH=CH_2 \\ O \end{array}$
38	$\begin{array}{c} O \quad CH_3 \\ CH_3-(CH_2)_9-N-\overset{\|}{C}-CH_2-\overset{+}{N}-(CH_2)_3-SO_3^- \\ CH_2 \quad CH_3 \\ CH_2=CH \end{array}$	$O=S=O$
39	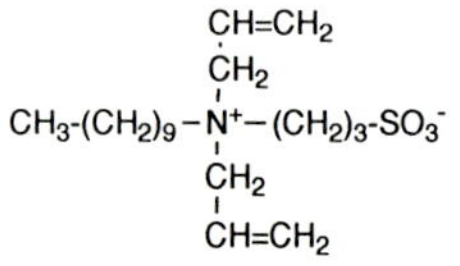 $\begin{array}{c} CH=CH_2 \\ CH_2 \\ CH_3-(CH_2)_9-\overset{+}{N}-(CH_2)_3-SO_3^- \\ CH_2 \\ CH=CH_2 \end{array}$	$O=S=O$

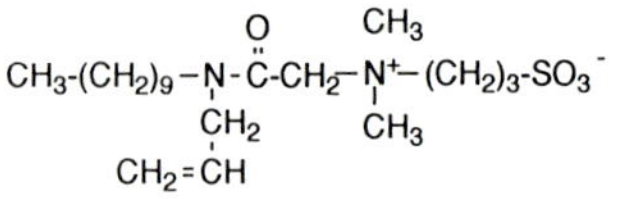

Fig. 13. Examples of reactive surfactants yielding polysoaps with elongated repeat units by alternating copolymerization. **33**: [54, 169, 170, 182, 186, 190, 196]; **34**: [191, 192]; **35, 37, 38**: [168]; **36**: [314]

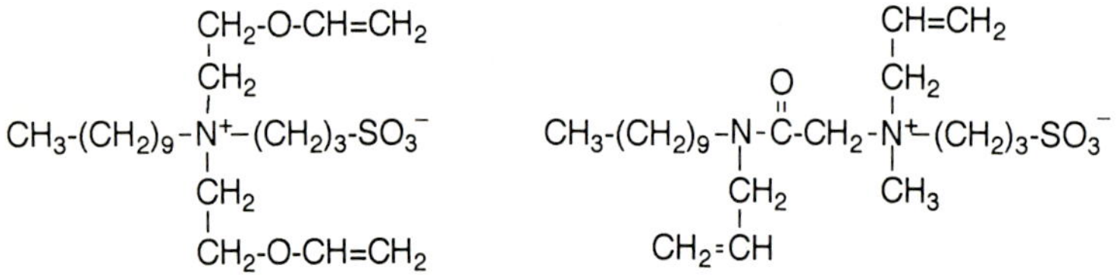

$$CH_3-(CH_2)_9-\overset{\overset{\displaystyle CH_2-O-CH=CH_2}{\displaystyle CH_2}}{\underset{\underset{\displaystyle CH_2-O-CH=CH_2}{\displaystyle CH_2}}{\overset{+}{N}}}-(CH_2)_3-SO_3^-$$

$$CH_3-(CH_2)_9-\overset{\overset{\displaystyle O}{\|}}{N}-\overset{\|}{C}-CH_2-\overset{\overset{\displaystyle CH=CH_2}{\displaystyle CH_2}}{\underset{\underset{\displaystyle CH_3}{}}{N^+}}-(CH_2)_3-SO_3^-$$

Fig. 14. Diolefinic reactive surfactants for polyaddition reactions with dithiols [316]

between the longitudinal knotting of surfactant units and the lateral knotting found in classical polysoaps (see Figs. 1e and 1f).

The alternative to "main chain spacers" is the separation of backbone and surfactant fragments by "side chain spacers" (Fig. 12) which should allow the use of vinyl homopolymers for less advantageous geometries. However in contrast to other self-organized systems such as polymeric liquid crystals, "side chain spacers" work poorly for micellar polymers, probably because the steric requirements in polysoaps are much more stringent. The backbone must neither interfere with an efficient packing nor with the correct orientation of the side chains. Thus, extensively long spacers are needed to achieve water-solubility [100, 102, 157], much longer than the spacer group equivalent of six to twelve carbons which are sufficient for polymeric liquid crystals or for polymeric lipids [7, 8, 87, 232–237]. The "head type" polymers having only standard tri- to octaethyleneoxide spacers (Table 5) are only able to swell, but do not dissolve in water [126, 231, 232, 251] (see also Fig. 37).

2.3.5 Flexibility of the Polymer Backbone

The problem of backbone flexibility has not yet been addressed for polysoaps. The vast majority of polymers reported have very flexible backbones. Clearly there must be gradual differences in the flexibility of the polysoaps prepared, depending on the main chain spacer and the reactive moieties employed. E.g., polysoaps based on polystyrene [51, 83, 104, 130, 238, 277, 278, 288, 292, 319–323] or poly(N,N-dialkyl,N,N-diallyl-ammonium salt)s [161, 168, 199, 200] should differ from polyvinyl-esters [224, 225, 232], polybutadienes [168, 313] or aliphatic polysulfones [164, 168].

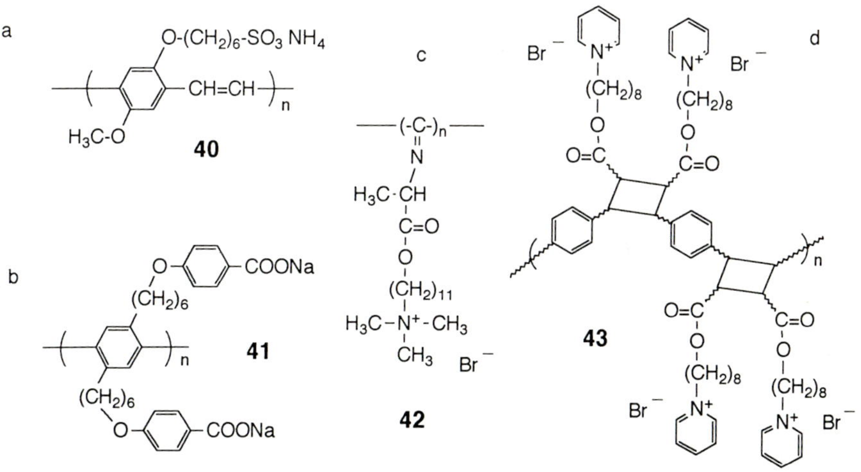

Fig. 15. Examples of amphiphilic polymers with a rigid polymer backbone (polysoap behaviour not reported). a: [273]; b: [274]; c: [329]; d: [330]

Still the differences should be minor compared to the ones to be expected from the use of rigid-rod polymers. But very few attempts to synthesize polymers with semi-rigid or rigid backbones are described, whose chemical structure could be considered as polysoaps. In case of "main chain geometry" (Fig. 6d), this may be due to the inherent poor solubility of rigid-rod polymers. The preparation of such water-soluble rigid-rod polymers gains increasing interest [324–328] but it still at its infancy. The preparation of stiff polysoaps linking surfactant fragments laterally (Figs. 6a–c) seems more promising, as the side chains are known to improve the solubility of rigid-rod polymers. But there are but singular notes on such compounds [273, 274, 329, 330] (Fig. 15).

2.4 Functional Polysoaps

A number of reports deal with functionalized polysoaps, including the incorp-oration of chromophores, mesogens, redox-active moieties and electrically conductive groups (Fig. 16). Three basic intentions can be identified: i) the functional unit serves as probe to monitor certain properties of the parent polysoap; ii) the polysoap provides a suitable matrix for the functional unit, e.g. by compartmentalization, clustering, orientation etc.; iii) the combination of functional unit and polysoap creates new collective features in the system, e.g. improving or modifying the self-organization.

Incorporations of probe molecules and chromophores are the most wide-spread functionalizations with various goals aimed at. As mostly limited degrees of functionalization are wanted, such polysoaps are conveniently prepared either by copolymerization [331] or by grafting reactions onto polysoaps or polysoap precursors.

The majority of studies aim at labelling of polysoaps, to study their behaviour as discussed in Sects. 3 and 4. UV/visible-probes, fluorescence probes and ESR probes have been attached. For example azo dyes were used as labels taking advantage of their trans-cis photoisomerization [203, 217, 260]. More frequently fluorescent labels have been fixed such as dansyl [174, 175, 177, 287], naphthyl [211, 259], anthracenyl [183, 189] or pyrenyl groups [204, 259]. For ESR studies, e.g. nitroxy labelled polysoaps based on poly(ethyleneimine) are described [145].

Cases where the polysoap serves as matrix for the functional unit attached include the incorporation of catalytic sites [256, 332, 333], of naphthyl groups used as photochemical antennas [211], and of pyrenyl and naphthyl groups for energy transfer processes [193, 258, 259]. Similarly, viologen moieties were fixed to polysoaps for electron transfer processes, and to study solar energy conver-sion [257, 264, 265]. The use of polysoaps as matrix is reflected as well by functionalized polysoaps prepared for pharmaceutical applications [334], e.g. in form of copolymers with anticytostatica [335], or as polysoaps carrying tar-geting groups for drug delivery [336]. Other functional polysoaps are made for specific binding by bearing crown ethers [253–255] and saccharides. The latter

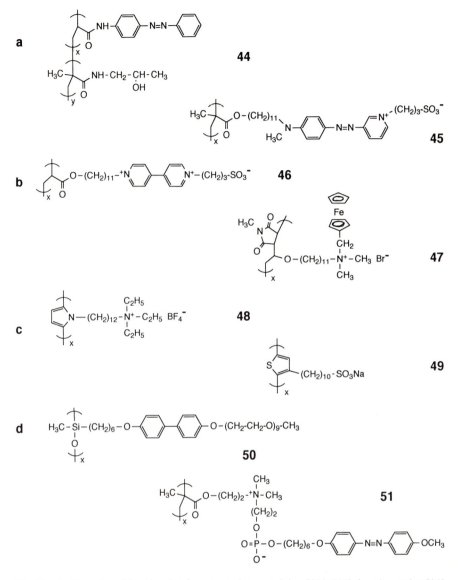

Fig. 16a–d. Examples of functional polysoaps: **a** dye containing [205, 331]; **b** redox-active [268, 269]; **c** electrically conductive [271, 272]; **d** containing mesogens [123, 262]

have been considered for molecular recognition [246, 247, 250], immune reactions and problems linked to biocompatibility [230, 337], but the majority of polysoaps bearing sugar moieties were only prepared for their use as nonionic hydrophilic units (see Table 3 and Fig. 39) rather than as functional ones up to now.

The combination of functional units and polysoap to create new collective properties represent the least undertaken functionalizations yet. The synthesis of such functional polysoaps often requires the preparation of particular homo-polymers to obtain a high content of functional groups. By analogy with other self-organization processes, incorporation of mesogenic units takes a prominent place [87, 122–124, 150, 248, 249, 261, 262]. Mesogenic azo dyes were also linked to polysoaps with the idea of photochemical switching between different forms of aggregation [205]; the reported effects however are small. More marked changes of the aggregation are expected by the incorporation of redox-active moieties, in particular if the systems incorporated enable the variation of the number of charged groups present, thus modifying the HLB. This was studied by synthesizing polysoaps bearing viologen, ferrocene, or nicotinamide moieties in high concentrations [266–269]. An alternative use of functional polymerizable surfactants and their polymers bearing redox-active moieties is oriented towards conductive polymers derived from polypyrrol or polythio-phene [270–274]. However, most of the latter examples do not yield water-soluble polymers.

3 Properties of Polysoaps in Aqueous Solution

The section reviews some prominent properties of polysoaps such as viscosity in solution, surface activity, solubilization and dynamic properties. As the majority of the studies are singular, the reports do not always allow a clear opinion as to whether the amphiphilic polymers studied are indeed polysoaps or not, as an unequivocal experimental proof may be missing. In these cases, the "polysoap" character was tentatively attributed due to the strong similarity of the chemical structures of the compounds to well known polysoaps. Or the systems are discussed because the authors claim the term "polysoap" for the systems studied (even if convincing evidence is missing). On the other hand, a number of doubtful cases where the systems studied could possibly represent polysoaps have been omitted to minimize the speculative character of the discussion. Hence the selection made is somewhat arbitrary. Furthermore, many investiga-tions are confined exclusively to one analytical technique and/or one property of the system. Therefore a full evaluation of the data and comparisons are difficult. The reader is strongly recommended to consult the original literature if inter-ested in getting a full picture.

3.1 Viscosity

The viscosity of aqueous polysoap solutions is characteristically low up to high polymer contents (see Figs. 17–19). In fact this property initiated the study of

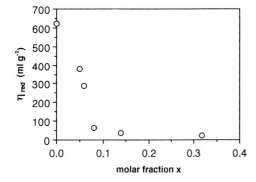

Fig. 17. Intrinsic viscosities [η] of poly(4-vinylpyridinium bromides) quaternized with ethyl and dodecyl groups in 0.0223N aqueous KBr, as function of the molar fraction of dodecyl groups x (fraction of ethyl groups y $\approx$ 0.9-x, fraction of not quaternized pyridines z $\approx$ 0.1). data taken from [52].

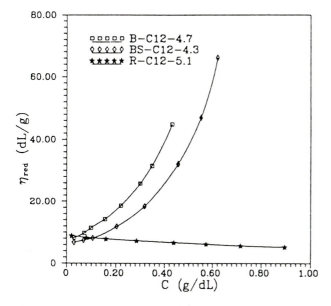

Fig. 18. Concentration dependence of the reduced viscosities of the cationic copolymer soaps **13**; fraction of surfactant monomer in the copolymer ca. 0.05. from top to bottom: $\square$ = blocky copolymer, $\diamond$ = microblocky copolymer, * = random copolymer. (Reprinted with kind permission from [77]. Copyright 1993 American Chemical Society)

polysoaps [46]. It is considered to be a key feature distinguishing ionic polysoaps from ordinary polyelectrolytes [50, 52, 75, 133]. The exceptionally low viscosities are explained by the *intra*molecular aggregation of the hydrophobic side chains, keeping the hydrodynamic radius small.

This behaviour contrasts with the one of slightly hydrophobized water-soluble polymers which act as thickeners [162, 220, 285, 312]. For them, the substantial increases in viscosity are attributed to *inter*molecular aggregation,

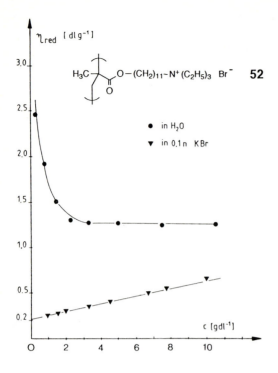

Fig. 19. Concentration dependence of the reduced viscosities of the polysoap 52 in water (●) and in 0.1 N KBr (▼). (Reprinted with kind permission from [126]. Copyright 1989 American Chemical Society)

presumably because the number and hydrophobicity of the scattered substituents is insufficient to create intramolecular aggregates.

Interestingly, the viscosity of polysoap solutions is frequently subject to important ageing effects [75, 99, 130, 163, 284] and temperature effects [75, 130], dropping asymptotically towards its final value. The decrease can amount to 90% of the initial value and can endure for up to one month [99, 163]. These effects are not well understood but are putatively attributed to conformational changes. The phenomenon has to be kept in mind when evaluating viscosity data, and should always be verified – or excluded – in viscosity studies of polysoaps.

As discussed before, the borderline between polysoaps and polyelectrolytes or thickeners is determined by the hydrophilic–hydrophobic balance HLB, as well as by the length and the density of the hydrophobic tails chosen. The longer they are, the lower is the content of hydrophobic tails – the "Critical Alkyl Group Content" CAC [52, 75] – needed in order to produce polysoap behaviour (cf. Sect. 2.3.2). For poly(2-vinylpyridine) and poly(4-vinylpyridine), about 20% of derivatization with octyl tails and about 10% of derivatization with dodecyl tails are needed as a minimum to obtain the characteristic low viscosities [49, 52, 75, 133, 141, 317] (Fig. 17). Comparable CAC values are obtained for derivatized poly(vinyl-imidazol)s [140] and for poly(allylamine)s [152]. Similarly in anionic copolymers of poly(sodium 2-acrylamido-2-

methylpropanesulfonate), low viscosities characteristic of polysoaps are achiev-ed at about 60 mol% of N-phenylacrylamide content, 15 mol% of N-dodecyl-methacrylamide content and 10 mol% of N-pyrenylacrylamide content [193].

Concerning the importance of the HLB, e.g. for poly(vinylether-alt-maleate)s the critical length of the hydrophobic tails yielding the typical low viscosities depends on the degree of dissociation of the carboxyl groups [54, 171]. The tails have to be enlarged to counterbalance the increasing charge in the polymer with increasing neutralization. Only if the tail lengths exceed octyl polysoap behaviour is observed independently of the degree of neutralization [186, 190, 192, 196]. An analogous effect may be responsible for the markedly reduced viscosities seen for polysoaps with low, just sufficient hydrophobe content when a certain amount of hydrophobic solubilizate is added [47, 49, 130]: the incorporated hydrophobic compounds mimic a decrease of the HLB.

Viscosity studies on perfluoroctyl substituted hydroxyethyl cellulose suggest that the CAC values are considerably lower for fluorocarbons than for hydro-carbons [297]. The lower CAC values can be explained by the increased hydrophobicity of fluorocarbon tails.

An additional parameter defining the borderline between polyelectrolytes or thickeners on one side, and polysoaps on the other side, is the distribution of hydrophilic and hydrophobic segments in the macromolecules. Studying the viscosities obtained for copolymers of surfactant monomers with acrylamide, random copolymers (i.e. with random distribution of hydrophobic tails) behave as polysoaps: they show low viscosities pointing to intramolecular aggregation. In contrast, blocky copolymers of the same average chemical composition are prone to intermolecular aggregation thus acting as viscosifiers, although the HLB of the two types of copolymers is identical [77] (Fig. 18). Analogous effects were observed for copolymers of acrylamide and ethylphenylacrylamide [220].

Viscosity studies of ionic polysoaps in pure aqueous solution usually suffer from the polyelectrolyte behaviour experienced at low concentrations [46, 49, 52, 75, 78, 98, 99, 126, 130, 193, 219, 229, 338]. Indeed, as the dissociation of the ionic groups varies with the concentration, meaningful concentration dependent studies become difficult. The problem was frequently overcome by addition of salt (Fig. 19). But ternary systems are created, making the systems even more complex, and many ionic polysoaps tend to precipitate in brine [50, 54, 299]. These problems can be avoided by zwitterionic polysoaps, facilitating the interpretation of concentration dependence [78].

Much confusion had arisen from the polyelectrolyte-like viscosity of ionic polysoaps at low concentrations. Based on the old polyelectrolyte model of coil-to-rod transition [339–342], it was suggested that the strong increase of reduced viscosities at low concentrations would indicate a transition of intramolecularly aggregated coils to extended conformations of the polysoaps which are no more aggregated. This would mean that the polysoap characteristics would be suspended at low concentrations [126, 229, 292].

This interpretation clearly disagrees with the presence of hydrophobic aggregates in solutions of charged polysoaps down to high dilutions, as evid-

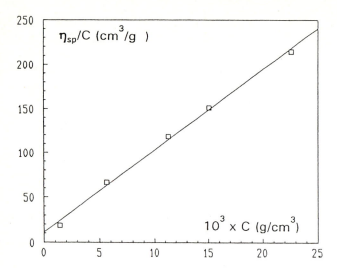

Fig. 20. Concentration dependence of the reduced viscosities of the metastable solutions of poly-soap **28** in water. (Reprinted with kind permission from [292]. Copyright 1993 American Chemical Society)

enced by many other analytical techniques, in particular by qualitative and quantitative solubilization studies [72, 78, 343] (cf. Sect. 3.3). Considering recent theories of polyelectrolytes which assume the formation of superstructures to explain the marked maximum of the reduced viscosity at low concentrations without postulating a coil-to-rod transition [344–346], the apparent contradiction is easily resolved: polyelectrolyte behaviour does not imply the absence of micellar structures.

Still, there is the most interesting phenomenon that the cationic polymer poly(N-hexadecyl-N,N-dimethy-N-vinylbenzyl ammonium chloride) **28** exhibits very low reduced viscosities but does *not* show polyelectrolyte behaviour in aqueous solution [103, 292]: the plot of reduced viscosity vs concentration is strictly linear, and is insensitive to added salt (Fig. 20). Importantly, this "head" type vinyl polymer without main chain spacer is not water-soluble and thus not a "true" polysoap, but forms only metastable aqueous solutions (see Sect 2.2.4). Similar results were reported for alkylated poly(vinylimidazoles) such as **26** [347]. It may be speculated that such solutions represent rigid "molecular latexes" rather than flexible "polymeric micelles", and further studies on such systems would be most interesting.

3.2 Surface Activity

The surface activity of polysoaps has rarely been studied. This is surprising because surface activity is one of the key features in soap performance, as reflected in the more general term "surfactant" used for natural and synthetic soaps and for similar compounds. The few reports do not provide a clear, coherent view at present, but nevertheless they illustrate substantial differences to low molecular weight surfactants. Unfortunately, experimental work is hampered as the measurements are very sensitive to trace impurities [348]. Therefore an accurate evaluation of the experimental data may be difficult, or may even be fooled by artefacts. Also, extensive equilibration times may be necessary [349, 350]. The latter may be attributed to slow diffusion [144, 351], or to conformational changes of the polysoaps [144].

The surface activity of low molecular weight surfactants is easily character-ized by following the aqueous surface tension as a function of the logarithm of concentration as derived from the transformation of Gibbs' equation [28]. In the very dilute concentration regime, molecularly dissolved surfactants are in equilibrium with a soluble monolayer at the interface. Increasing enrichment of the surfactant at the interface causes a continuous decrease of surface tension with increasing concentration. But passing the critical micelle concentration CMC, all newly added surfactants will be incorporated into the micelles. As a consequence the surface activity of the molecular dissolved surfactants is kept constant above CMC (in a first approximation), and the equilibrium concentra-tion at the interface, too. In consequence, the surface tension stagnates above CMC, producing the classical shape of the semilogarithmic surface tension vs concentration curves (Fig. 21).

In the existing studies on the surface activity of polysoaps, three cases can be distinguished:

i) the polysoaps decrease only marginally the surface tension, if at all [50, 72, 82, 113, 131, 228, 229, 245, 300, 317] (Fig. 21);

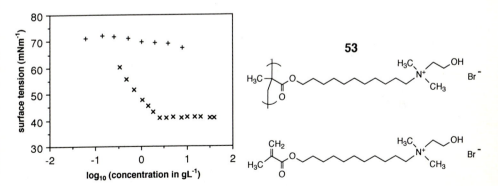

Fig. 21. Surface activity of surfactants and polysoaps in water at 25 °C, exemplified by a cationic methacrylate: + = polysoap **53**, X = corresponding surfactant monomer. (Data taken from [245])

ii) the polysoaps induce a continuous, notable decrease of surface tension with increasing concentration, but no CMC or a break point resembling a CMC is observed [55, 78, 131, 164, 167, 168, 193, 228, 245, 350, 352]; (Figs. 22, 23).

iii) the polysoaps induce a decrease of surface tension with increasing concentration exhibiting a break point resembling a CMC [143, 144, 193, 213, 241, 242, 251, 353].

There are some additional reports stating considerable surface activity of polysoaps, but without specifying the concentration dependence [152, 194, 301].

The first two cases lacking any indication for the presence of "CMC"s are in agreement with the idea of intramolecular hydrophobic aggregation. The third case points to an intermolecular aggregation of the polysoaps, similar to low molecular weight surfactants. However it is difficult to decide by the present data whether this rare case is indeed real or artificial, and whether the break points in the semilogarithmic plots of surface tension vs concentration indeed reflect the onset of aggregation. It should be noted that many of such "CMC" values reported are virtually independent of the length of the alkyl tails [143, 144, 241, 242] which is difficult to rationalize, or the surface tension starts to decrease at much higher concentrations than those where hydrophobic aggregation can be detected by the solubilization experiments [213]. Also, reported break points from surface tension studies interpreted as "CMC" or "pre-CMC" are in contradiction to solubilization studies where no "CMC" is observed (compare [66, 67, 71] with [65]).

The appearance of surface activity and an apparent "CMC" could well be caused by low molecular weight contaminants, originating e.g. from insufficient removal of educts or as by-products of the synthesis, or from partial decomposition [143, 144, 213]. The strong effects of hydrophobic counterions bound to polyelectrolytes on the surface tension are well known [350] and traces many suffice to provoke the effect. Alternatively, apparent "CMC"s of polysoaps have been reported when low oligomers are involved [241, 242, 251, 353]. Thus dimers or trimers etc. might be responsible for the effect, exhibiting intermolecular micellization (see Sect. 6).

The appearance of "CMC"s for polysoaps bearing particularly short hydrophobic tails [193] is rather a semantic problem as these examples do not match the original definition of polysoaps anymore. If the hydrophobe tails are too short, no intramolecular aggregation can take place as evident from viscosity measurements [24, 133, 193], and intermolecular aggregation is needed to reduce hydrophobic interactions.

Although originally it was assumed that intramolecular aggregation would result in a complete lack of surface activity, the first two cases where no "CMC" is observed seem to represent extremes of the same general behaviour. In fact, gradual transitions from hardly to strongly surface-active polysoaps can be found if the chemical structure is gradually modified or if certain additives are gradually fed in. Molecular parameters which influence surface activity are the polysoap geometry, the hydrophilicity of the head group, and size and content of the hydrophobic groups. The rules do not always agree with the ones observed in low molecular weight surfactants.

Concerning the role of the polysoap geometry, the reported data suggest that the surface activity increases with increasing proximity of the backbone and the surfactant head group [78, 167, 228]. The surface activities reported for polysoaps of the "tail end" geometry are generally lower than those reported for analogues of the "head" geometry (Fig. 22). This holds true even if the "head" type polysoap has a considerably higher HLB, counting the various molecular fragments on the polymer: e.g. the "head" copolymers **19** with choline methacrylate reduce the surface tension much more strongly than the "isomeric tail end" homopolymer **53**.

This behaviour is opposite to that of low molecular weight surfactants for which a higher HLB at a given concentration will cause higher surface tensions [251, 281, 354]. But it parallels the problem of the different polysoap geometries in realizing an optimal amphiphilic arrangement in solution: "tail end" geometry enables a more efficient "shielding" of the hydrophobic aggregates, thus reducing the tendency to adsorb at interfaces. This picture may also explain why "tail end" polysoaps bearing the same surfactant moiety but having different polymeric backbones exhibit virtually identical reductions of aqueous surface tension (Fig. 23), although the monomers behave quite differently. In agreement with the above discussion of shielding efficiency, random copolymers show higher activity than graft copolymers of identical composition [228]. These observations are instructive examples for how the polymer architecture can modify – or even override – the properties expected from simple cumulation of the fragment's properties.

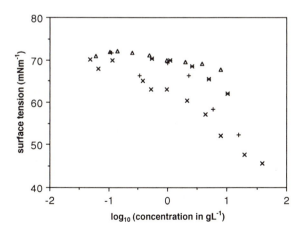

Fig. 22. Surface activity of closely related cationic polysoaps in dependence on their geometry and on their main chain spacer in water at 25 °C. (Data from [167, 245]). "Tail end" geometry: $\triangle$ = polysoap **53**. "Head" geometry: X = copolymer soap **19a** (hydrophobe content y = 0.4), + = copolymer soap **19b** (y = 0.2), * = copolymer soap **19c** (y = 0.1)

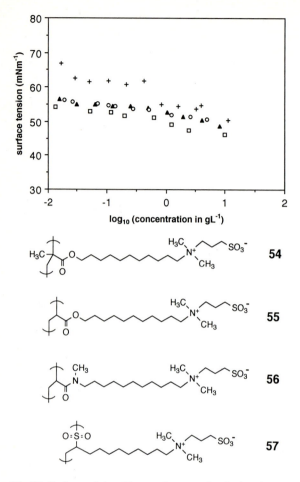

Fig. 23. Surface activity of four analogous zwitterionic polysoaps of tail-end geometry with different polymer backbones. (Data from [164, 167, 245]). ▲ = polymethacrylate **54**, □ = polyacrylate **55**, ○ = polyacrylamide **56**, + = polysulfone **57**

Looking however at polysoaps of one geometry only, the available data suggest that decreasing HLB causes increasing surface activity. E.g. for "head" type copolymers made from hydrophobic monomers (or from surfactant monomers respectively) with hydrophilic comonomers, the surface activity increases with increasing hydrophobe content [152, 167, 193, 228, 350, 355] (see copolymers **19**, Fig. 22). This observation also fits into the general scheme discussed above: the shielding of the hydrophobic groups in "head" type polysoaps becomes more difficult with less strongly hydrophilic groups available. The simultaneously reduced length of the main chain spacer group however seems to be of minor importance for the effect. If the polar backbone is made from uncharged acrylamides instead of charged units and the hydrophobic tails are incorporated in the form of complete surfactant fragments (not only of pure hydrophobic tails), the HLB is kept approximately constant, and the increasing density of hydrophobic tails in the polysoaps results only in marginal changes of the surface activity [167].

There are very few studies on the influence of the length of the hydrophobic tails in polysoaps. The reported increase of surface activity for some ionic polysoaps with increasing length of the hydrophobic tails corresponds to the behaviour of standard surfactants [152, 155, 193]. However, decreasing surface activity with increasing length was reported for substituted acrylamides [194]. Another strange result was reported from some nonionic liposaccharide polysoaps for which surface activity increases monotonously with the length of the hydrophobic tails, but the values measured for monomers and polymers are identical [301].

In analogy with the above discussion of the influence of the hydrophobe content, increasing hydrophilicity of the head groups coincides with decreasing surface activity. This trend goes along with the one of low molecular weight surfactants, but the effects are much more pronounced. E.g. zwitterionic polysoaps of "tail end" geometry exhibit moderate surface activity, whereas cationic analogues with the higher HLB are only slightly surface active [245] (Fig. 24).

The observed effects of added salts [131, 208, 299] are much stronger than for low molecular weight surfactants. They can be rationalized in the same way. Added salt enhances the surface activity of cationic polysoaps [131, 299] by depressing the dissociation of the ionic head groups and reducing their hydrophilicity. Similarly, the surface activity of some anionic polysoaps based on hydrophobized maleic acid copolymers is increased by lowering the pH, which reduces the dissociation of the pendent carboxylates and thus the HLB [208]. Similar (though much weaker) effects are found even for polymethacrylic acid [304]. In agreement with this discussion, the surface activity of zwitterionic polysoaps is decreased when salt is added [299] due to their anti-polyelectrolyte character [356], contrasting well with ionic polysoaps.

The rule concerning the influence of the hydrophilicity of the head groups may serve as a useful guideline, but some exceptions should be noted. In contrast to their ammonium analogue **53**, the viologen polysoaps **58** exhibit notable surface activity which is similar to one of their sulfobetain analogue **54** (Fig. 24), though **58** has the highest HLB adding the various molecular fragments and in agreement with the increasing CMCs of the respective monomers. Perhaps the tendency of the polymer bound viologene moieties to aggregate, – as evidenced from electrochemical studies [268] – is intervening.

For certain hydrophobic substances added to polysoap solutions, a notable reduction of the surface tension is observed [131]. As more detailed studies were missing until now, the reasons are not clear. But the action of organic compounds as cosurfactants seems not unreasonable.

The influence of the molecular weight of polysoaps on the surface activity has been touched occasionally [131, 229, 357]. It seems that for high molecular

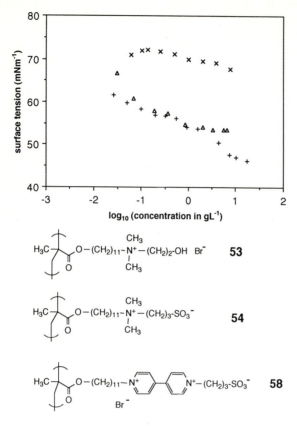

Fig. 24. Surface activity of poly(methacrylate) polysoaps of tail-end geometry with head groups of increasing hydrophilicity. (Data from [245, 267]). + = **54**, X = **53**, △ = **58**

weight samples the influence is minor if existing at all (Fig. 25). However for oligomers below the critical size for intramolecular aggregation, differences in the molecular weight are important, as will be discussed in Sect. 6.

3.3 Solubilization

Solubilization [20, 358–360] was defined by Elworthy, Florence and Macfarlane as "the preparation of thermodynamically stable isotropic solutions of substances normally insoluble or slightly soluble in a given solvent by the introduction of an additional amphiphilic component or components" [358]. Like surface activity, the ability to solubilize water-insoluble compounds represents a key property in the performance of surfactants [21]. Therefore, solubilization by polysoaps has raised interest from the very beginning [46–51].

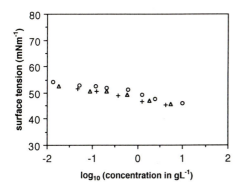

Fig. 25. Surface activity of zwitterionic poly(acrylate) **55** of different molecular weights [357]. $\bigcirc$: $M_w = 27 \times 10^3$; $\triangle$: $M_w = 42 \times 10^3$; $+$: $M_w = 110 \times 10^3$. (Apparent molecular weights by GPC, standard poly(sodium methacylate))

Efficient solubilization is bound to the presence of hydrophobic aggregates, in particular of micelles. This implies for surfactants subject to intermolecular aggregation – such as low molecular weight surfactants – that concentrations above the CMC must be employed [358–360]. This may cause problems if dilute solutions are needed (e.g. for reasons of stability of biological solutions, of toxicology, or of waste water treatment). In addition, as dilution of solutions of intermolecularly aggregating surfactants below the CMC will disrupt the micelles, dilution will also lead to the deposition of the solubilized material, a problem often encountered in rinsing processes. Here, polysoaps may be the perfect match to standard surfactants for special applications (see Sect. 7), as polysoaps are generally capable of solubilization at any concentration [25, 50, 51, 53, 65, 76–78, 106, 112, 113, 186, 196, 197, 245, 361] (Fig. 26).

Basically, two types of solubilization studies can be distinguished. In the first type, solubilization is only a tool to incorporate probe molecules into the polysoaps, to learn of the nature of hydrophobic domains available, e.g. their size, form and polarity etc. ("qualitative solubilization"). The majority of such studies employs dye probes whose spectral properties are sensitive to the environment of the chromophore [20, 362–366]. The second, less frequent type

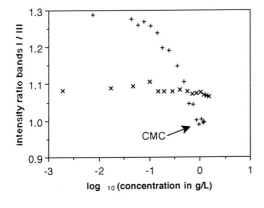

Fig. 26. Solubilization of pyrene by the zwitterionic polysoap ($\times$) and its monomer ($+$). Concentration dependence of the intensity ratio of fluorescence bands I (372 nm) and III (383 nm). Lower ratios I/III indicate a more hydrophobic environment of the dye. (Data from [245])

of studies is focussed on the solubilization capacity of the polysoaps ("quantitative solubilization").

3.3.1 Solubilization of Probe Molecules

Various spectroscopic techniques and probes have been used to investigate solubilization of probe molecules, mostly using UV/visible spectroscopy, fluorescence spectroscopy, ESR spectroscopy [64, 74, 217, 287] and NMR-spectroscopy [367–369]. Fluorescence spectroscopy is particularly versatile [370], as various static and dynamic aspects can be covered by studying excitation and emission spectra, excimer or exciplex formation, quantum yields, quenching, fluorescence life-times, fluorescence depolarization, energy transfer etc.

"Qualitative solubilization" studies are by no means trivial [362]. Like all studies based on probe molecules, they suffer from inherent uncertainties. On one hand the probe content has to be kept as small as possible to minimize the perturbation of the system (which can hardly be excluded rigorously). On the other hand it is generally not known where the actual solubilization site of the probe is, and whether the small number of probe molecules will indeed occupy representative sites in the system or – perhaps – will accumulate in exceptional locations thus monitoring non-representative – though real – properties and events. This is particularly important if very large and bulky probes such as ET_{30} are employed [212] which are about of the same order of size as the aggregate. Furthermore, assumptions often have to be made to enable a treatment of the acquired data which cannot be verified [196, 370]. Therefore results derived from such studies may be ambiguous and the conclusions even misleading, and several techniques should preferably be combined. But despite all these many shortcomings, there are virtually no other techniques available at present which would allow comparable studies.

Pyrene derivatives are the widest used probes for qualitative solubilization [365] by virtue of the solvatochromic shifts of the absorption bands [255], the excimer formation [145, 186], the polarity dependent quantum yields [197] and fluorescence life-times [185–187, 196, 197, 202, 215, 292], and the pyrene fluorescence fine structure [65, 74, 78, 103, 112, 167, 224, 363, 371]: the intensity ratio of the fluorescence bands I at 372 nm and III at 383 nm is a convenient measure for the polarity of the environment of the pyrene label ("py"-scale; "I/III" values increase with polarity, cf. Fig. 27). As, however, the fluorescence of pyrene is very sensitive to the experimental set-up [372], absolute I/III values reported by different groups are difficult to compare.

The following will focus on studies exploring the nature of the solubilization sites. Studies concerning dynamics in polymeric "micelles", the overall shape and aggregation numbers are included in Sects. 3.5 and 4.2.

The simplest information obtained by "qualitative" solubilization is the existence (or the absence) of hydrophobic microdomains, using probes which are sensitive to the polarity of the environment. Such solubilization experiments are

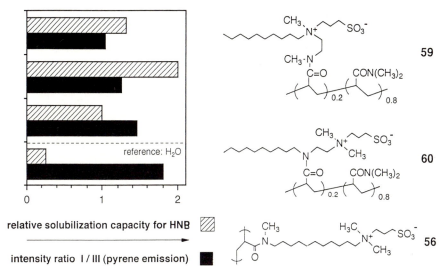

relative solubilization capacity for HNB

intensity ratio I / III (pyrene emission)

Fig. 27. Concentration dependent intensity ratio of fluorescence band I (372 nm) and band III (383 nm) of pyrene solubilized by analogous polysoaps of different geometry, compared to their relative solubilization capacity for 4-hexyloxy-nitrobenzene. From top to bottom: **59** ("head" geometry), **60** ("mid tail" geometry), **56** ("tail end"); data from [167, 343]

easy to perform, and many techniques and probes will be useful. Solvatochromic shifts of absorbance and emission spectra are particularly simple to detect. It should be noted however that extent and magnitude of the solvatochromic shifts are often difficult to rationalize [343, 364, 366]. Furthermore, shifts due to solubilization by hydrophobic aggregates may be hard to distinguish from effects due to specific (e.g. electrostatic) binding of the probes [142, 154, 215, 238] which is not linked to the presence of micelles. Also the technique does not differentiate at this simple level between the formation of intramolecular hydrophobic aggregates and intermolecular ones.

If the probes are slightly water-soluble, an additional problem arises from partitioning of the probes between the aqueous and the "micellar phase". Thus sometimes the assumed "CMC"s, because of changing I/III values of pyrene fluorescence [213, 353], may be apparent only because of the unfavourable partition of the solubilizate between the aqueous and the "micellar phase" at low soap concentrations [154, 292].

In more elaborate studies with pyrene it was shown that the hydrophobicity of the solubilization sites increases with the length of the alkyl tails incorporated [103, 128, 185, 186, 196, 292]. Similar results were obtained with various dansyl derivatives [152] and with methyl orange [200]. Such behaviour agrees well with that of low molecular weight surfactants [185]. Additionally, the polarity of the sites seems to depend on the rigidity of the hydrophobic tails. In copolymer soaps of "head geometry" based on polyacrylamide, the polarity around the pyrene probes increases with cyclododecyl < dodecyl < adamantyl tails, which sequence was attributed to the efficiency of steric shielding of the solubilizate [202]. An analogous explanation was given for the decreased polarity around dansyl based probes with increasing degree of hydrophobic substitution of poly(allylamine) [152]. The decreasing polarity of the solubilization sites with increasing degree of hydrophobic substitution is corrobated by studies of pyrene solubilized by hydrophobized poly(ethylvinylether-*alt*-maleic acid) [154] and methyl orange solubilized by hydrophobized poly(diallylammonium bromide) [200].

The nature of the head group seems to be of lesser importance, as the degree of dissociation in poly(vinylether-*alt*-maleic acid)s and poly(alkene-*alt*-maleic acids) does not change the I/III values [185, 186, 197]. This is somewhat surprising as, according to ^{1}H-NMR studies, aromatic probes reside at the "micellar interface" [373]. Also, this result contrasts markedly with the increasing polarity in polysoaps based on poly(allylamine) with increasing protonation as sensed by dansyl probes [287]. In the case of poly(undecenoic acid), the I/III values increases strongly around a pH of 8. The enhanced polarity was attributed to a transition between tightly and loosely packed micelles with increasing pH, i.e. with increasing deprotonation of the –COOH groups. The difference to the constant I/III values found for the simultaneously studied poly(octadecene-*alt*-maleic acid) was attributed to a reduced hydrophobicity of the undecenoate residues in comparison with the C_{16} tails [197]. Still, the presumably comparably hydrophobic poly(decyl-vinylether-*alt*-maleic acid) does not show such a transition either [196]. This makes one wonder whether geometrical effects may intervene as polysoaps of "tail-end" and of "mid-tail" type are compared.

Indeed, concerning the polymer geometry, structural variations of vinylic polysoap isomers suggest that solubilization sites in "polymeric micelles" are more polar for polysoaps of "tail end" geometry than for ones of "head" geometry [167] (Fig. 27). These observations can be rationalized by the hydrophobicity profile of the respective polysoap structures. As the polymer backbone generally bears some polar groups, it has to be considered as a fragment of intermediate hydrophobicity, partially interfering with the polarity profile of the parent surfactant side chain (cf. Fig. 29). The exact quality of the solubilization sites depends on the chemistry of the backbone chosen.

Alternatively, the geometrical effects on solubilization sites can be explained by a simple "excluded" volume effect, forcing the probe closer to the "micellar surface" in "tail end" geometry. This explanation is supported by the findings that rather high polarities were observed for a number of vinylic polysoaps of the "tail-end" type in comparison to their monomers [64, 65, 74, 164, 245]. In contrast, polysoaps of the "head type" may show similar or even less polar environments than analogous monomers [152, 167, 187].

3.3.2. Solubilization Capacity

"Quantitative" solubilization by polysoaps was studied in the very early days [46–51]. However, correlations between the molecular structure of polysoaps and their absolute solubilization capacities are problematic because – as for low molecular weight surfactants [358, 359] – the capacities depend in a complex way on both the surfactant and the solubilizate chosen, and may be modified by additional factors such as pH, added salts or ionic strength. Also, the uptake of considerable amounts of solubilizate may induce changes of the aggregate structure [49], thus rendering the interpretation of the studies even more difficult. As even solubilization by standard low molecular weight surfactants is still poorly understood, it is not surprising the solubilization by polysoaps is a rather empirical topic at present.

Most notably, the solubilization capacity of polysoaps is not correlated with their surface activity [78, 343]. Thus any combination of these two properties can be realized in polysoaps, even unusual ones such as low surface activity with high solubilization capacity. The phenomenon is not understood, but might be related to different conformations taken at the gas-water interface and in solution (cf. Sect. 6.1), or to cosurfactant effects of the solubilizates.

As mentioned above, there seems to be no lower limit for solubilization by high molecular weight polysoaps. Thus starting in the origin, the solubilization capacity increases linearly with the concentration for most polysoaps [46–48, 51–53, 78, 112, 113, 153, 196, 253] (Fig. 28). Exceptional cases show a break point with increasing capacity at intermediate concentrations [112, 113, 281] which were attributed to the transition from purely intramolecular aggregation to additionally superposed intermolecular aggregation [281].

In contrast, oligomeric polysoaps behave similarly to low molecular weight surfactants, i.e. they require a minimal soap concentration for solubilization [67, 71, 73] (cf. Sect. 6). Analogously, in poly(alkylvinylether-*alt*-maleic acid) of short tail length in which aggregation is favoured with decreasing dissociation of the carboxyl groups, solubilization capacities improve with decreasing pH values [54, 215].

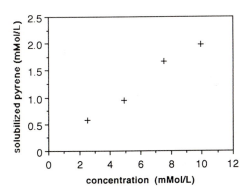

Fig. 28. Concentration dependence of the solubilization of pyrene by polysoap poly(sodium maleate-*alt*-decylvinylether) **33** at 25 °, degree of neutralization $\alpha = 0$; data from [196]

The performance of polysoaps in comparison to low molecular weight analogs depends on the system polysoap/solubilizate chosen. Notably, longer equilibration times than for low molecular weight surfactants may be needed to reach maximal solubilization. Molar ratios of solubilizate per hydrophobic chain of 0.01–0.05 are frequently reported [53, 54, 72, 215, 361], but the ratio can amount to 0.15 or higher [343]. Keeping in mind that the hitherto used solubilizates were mainly chosen due the ease of quantitive analysis, these values do not represent upper limits. In comparison to their monomers or to low molecular weight analogs, all possible cases – i.e. comparable [51, 72, 78, 343, 361], improved [50, 53, 78, 186, 343, 374] or inferior [78, 343] solubilization capacities – have been reported. The molar mass seems to be of little importance for high polymers [50, 213]. Major deficiencies of polysoap performance can be avoided by optimizing certain molecular parameters, but otherwise few predictions can be made at present.

Concerning the structure of the surfactant side chains, the solubilization capacity increases with increasing length of the alkyl tails in the few systems studied [54, 135, 281], a relationship which is widespread for standard surfactants [358, 359]. These findings parallel the decrease of polarity of the solubilization sites discussed in the previous section. However, a general correlation between polarity of the solubilization sites and solubilization capacities is not valid [78, 343] (Fig. 27). As for the head groups, their charge and their counterions seem to have little influence on the solubilization capacities in general if the hydrophilic–hydrophobic balance HLB and steric requirements are not markedly changed [53, 72, 78, 343]. However, specific interactions between the head groups and the solubilizate may play a crucial role [255, 343]. This is most evident for electrostatic effects, i.e. cationic soaps favour anionic solubilizates, but reject cationic ones, and vica versa [53, 78, 185, 343, 361]. Other supporting interactions have been noticed as well, e.g. between aromatic compounds [375].

The role of the hydrophobe content seems rather straightforward: as the content of hydrophobic chains increases, the solubilization capacity increases as well. In some systems the solubilization capacity increases faster than the hydrophobe content [46, 213]. But often the dependence can be approximated by a linear relationship, i.e. the solubilization capacity per hydrophobic chain is approximately constant [78, 343]. Only in cases of extreme steric crowding a maximum of the capacity as function of the hydrophobe content is found [343]. Recalling that main chain spacers are needed for polysoaps of "mid-tail" and "head" geometry (see Sect. 2.3.4), these findings imply the existence of an optimal spacer length for such polysoaps: it must be sufficiently long to provide water-solubility, but as short as possible to allow for high performance.

The effects of the polymer geometry and of the nature of the backbone are interdepending. Considering Fig. 29, only a hydrophobic backbone will provide a hydrophobic interior of polymeric "micelles" for polysoaps of the "tail end" geometry, whereas only a hydrophilic backbone will provide a favourable polarity profile for polysoaps of the "head type". Accordingly, "optimized"

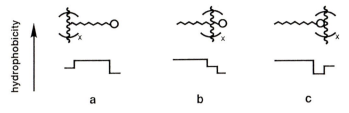

Fig. 29a–c. Schematic polarity profile in the surfactant side chains of polysoaps: **a** "tail end" geometry; **b** "mid tail" geometry; **c** "head" geometry. (Reprinted with kind permission from [78]. Copyright 1993 Hüthig & Wepf, Basel)

polysoaps of different geometry should be employed for good results. Indeed, the capacity of analogous "tail end" polysoaps increases strongly in the series polysulfone ≤ poly(acrylamide) < poly(acrylate) < poly(methacrylate), i.e. with increasing hydrophobicity of the polymer backbone [78, 343].

When optimized polysoaps bearing the analogous surfactant structure were used, only gradual differences in solubilization capacity were found. More polar solubilizates which are assumed to reside close to the "micellar surface" are somewhat more efficiently solubilized by polysoaps of "tail end" geometry. In contrast, solubilisates of amphiphilic structure are somewhat more efficiently solubilized by polysoaps of "mid tail" geometry. Polysoaps of "head" geometry fall shorter in both cases [78, 343]. Similar comparative studies for pure aromatic compounds and hydrocarbons are not available. The differences observed may be due to the respective positions of the polymer backbones, occupying space which is needed to accomodate the solubilizate. Notably, the results imply that " the optimal polysoap" structure does not exist, but the systems of choise will depend on the problem adressed.

In the above discussion, polysoaps of "Head" geometry emerge as least efficient solubilizing agents. Nevertheless, the capacities observed are still fair, and therefore such polysoaps are useful solubilizing agents too. "head" type polysoaps may, for practical purposes, even prove to be the agent of choice by virtue of their straight forward synthesis (cf. Sect. 2.2).

3.4 Emulsifying and Dispersing Properties

The problem of emulsification and stabilization of dispersions is still to be addressed. At present, only mostly qualitative statements are available. E.g., good emulsifying properties are implied for the reported use of polysoaps in emulsion polymerization [376] and for the stabilization of latexes [50, 214]. Good emulsifying properties are reported as well for natural polysoaps [79, 80]. Oligomeric polysoaps are efficient emulsifiers for liquid hydrocarbons [82]. High dispersing efficiency for alumina particles is claimed for some polymerized surfactants [377, 378]. But structure-property relationships are still to be established. In the case of copolymers of acrylic acid and butyl and dodecyl-acrylate respectively, dispersing properties for carbon black improve with increasing hydrophobe content and increasing length of the hydrophobic tails. In contrast, the dispersing properties for TiO_2 improve with decreasing hydrophobe content and decreasing length of the hydrophobic tails [153]. Considering the discussion in the previous chapter, emulsifying/dispersing and solubilizing abilities may thus be opposed. Such an opposing behaviour was also observed for the dispersing and solubilizing power of surfactant copolymers based on acrylamides [167]. Whereas the solubilization capacity of 4-(4'-butyl-phenylazo)-N,N-diethylaniline increases with the hydrophobe content [343], the dispersing capability of small solid dye particles decreases [357]. The parent polymers without hydrophobe content however are poor dispersants or emulsifiers.

3.5 Dynamic Properties

Micelles of low molecular weight surfactants are known to be very dynamic structures, although the various fragments of the molecules within a micelle are subject to some restrictions of mobility in comparison to molecular solution [23, 379–381]. As the hydrophilic head groups are "anchored" at the micellar surface", NMR-studies show that the mobility and the order parameter decreases along the alkyl tail from its end towards the head group [379–385].

Analogously, the mobility of the various fragments of polysoaps is somewhat restricted due to hydrophobic clustering [68, 69, 145, 275]. But the dynamic profile seen for standard micelles is superposed by the dynamic restrictions imposed by the polymer backbone as the backbone tends to immobilize the side chain fragments close by [276, 386, 387]. Whereas the general profile stays the same in polysoaps for "head" geometry as in standard surfactants, i.e., the head groups are the least mobile segments [388, 389], polysoaps of "tail end" geometry show a dynamic profile of the surfactant fragments opposite to the one of the monomeric analogues. Here the end of the hydrophobic tails are the least mobile segments in the system (Fig. 30) [68, 69]. The implications of this difference for the performance of polysoaps are not clear at present. They might influence e.g. the type of solubilization sites available [64, 65, 74, 78, 164, 245] and thus the catalytic performance of polysoaps [361], but interpretations are speculative at present.

The restricted mobilities of the hydrophobic segments and the dynamic profile are also reflected in the shape of NMR-spectra of vinylic polysoaps in aqueous solution. The signals of protons in the proximity of the polymer backbone are strongly broadened [193, 258, 303, 355] or virtually invisible [39, 227] (Fig. 31). This effect decreases with decreasing density of the hydrophobic tails [193, 303, 355, 357] and with decreasing molecular weight.

The problem of restricted mobility not only concerns the various fragments of the polysoap itself, but the mobility of solubilized material is affected as well ("microviscosity"). Studying the motion of the ESR-spin probe, polysoaps

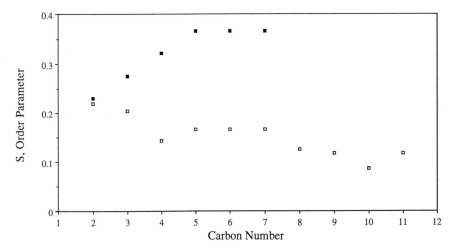

Fig. 30. Order parameter for micellar monomeric (□) and oligomeric (■) sodium undecenoate **6** as a function of the position in the hydrophobic tail ($C_1 = -COONA$). (Reprinted with kind permission from [69]. Copyright 1990 Academic Press, Orlando)

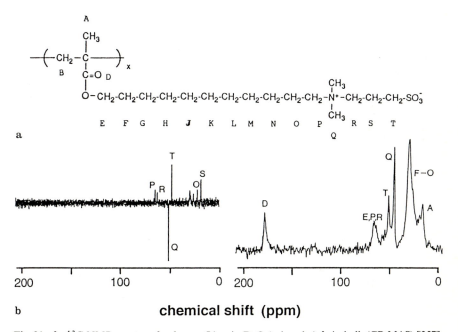

Fig. 31a, b. ^{13}C-NMR spectra of polysoap **54: a** in D_2O (spin-echo); **b** in bulk (CP-MAS) [357]

reduce the mobility of solubilized materials more strongly than micelles of low molecular weight surfactants do [64, 74, 152, 287], thus pointing to a higher microviscosity. Notably, the results vary with the probe chosen and the restriction of mobility is not limited to intramolecular aggregation but is seen for intermolecular aggregates as well [298]. The general behaviour is corroborated by measurements of fluorescence life-times of solubilized dyes [152, 217, 287]. Additional evidence for more restricted mobility in polysoaps in comparison to micelles of low molecular weight surfactants comes from fluorescence quenching and fluorescence depolarisation studies [185]. There seem to be steric effects of the polymer backbone, too, as the more densely packed dodecylated poly(allylamine) gives rise to stronger mobility restrictions than dodecylated poly(ethyleneamine) [145, 287].

Increasing the hydrophobe content and the length of the hydrophobic tails, the mobility of solubilized probes becomes more and more reduced according to both ESR- and fluorescence studies [152, 187, 189, 287]. However, for a series of anionic copolymer soaps of poly(sodium 2-acrylamido-2-methylpropane sulfonate), the mobility of the ESR-probes was found to be approximately the same for various hydrophobic tails such as n-dodecyl, adamantyl and naphthyl residues [217]. But within this series of compounds, the results obtained with various techniques are not consistent. In contrast to the ESR-studies, the rates of fluorescence quenching of pyrene suggest increasing rigidity in the sequence n-dodecyl < cyclododecyl < adamantyl (which intuitively may appear the most reasonable) [202], whereas the rate and extent of trans- to cis-isomerization of azo dyes suggest a rigidity scale adamantyl < cyclo-dodecyl < n-dodecyl [203, 260]. These apparently contradicting results raise some doubts about the unambiguity of the information obtained by such techniques because they depend on the probe used in a way which cannot be anticipated.

In addition to the motion of the surfactant fragments within a "micelle", dynamic exchange of surfactants between micelle and isotropic solution, and dynamic formation and breakdown of micelles occur in micelles of low molecular weight surfactants [20, 23, 390]. The two processes give rise of two relaxation times. None of these topics has yet been studied thoroughly for polysoaps. Thus it is unknown whether all or only a fraction of the hydrophobic tails aggregate [186, 196], and, if the latter is true, whether exchange of "unimer units" and dynamic micelle formation take place or not. These questions are clearly linked to the nature of "polysoap micelles" (see Sect. 4.2). Investigations of energy transfer in polysoaps point to partial aggregation with fast exchange rates of clustered and not clustered tails [258, 259], but the scarcity or data at present precludes general statements.

4 Aggregation in Aqueous Solution

4.1 Micelles

Micelles of low molecular weight surfactants are highly dynamic aggregates which are formed spontaneously above a critical concentration, the so-called

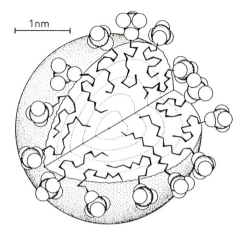

1nm

Fig. 32. Model of the standard ionic micelle (sodium dodecyl sulfate SDS). (Reprinted with kind permission from [381]. Copyright 1985 Steinkopff Verlag, Darmstadt)

critical micelle concentration CMC. They are in equilibrium with molecularly dissolved surfactants ("unimers") and a soluble surfactant monolayer at the interface. Though some problems are still to be resolved, major progress has been made in the past years to understand micelle formation, dynamics, shapes and aggregation numbers, as documented in a number of recent reviews [20, 23, 360, 381].

In the simplest cases, micelles can, on average, be envisaged as fluctuating aggregates of spherical shape, with aggregation numbers ranging from 10 to 200 (Fig. 32). The detailed numbers depend on the hydrophilic–hydrophobic balance HLB and the size of the surfactant; in special cases, cylindrical and disc-shaped micelles of extreme size can also be formed. Characteristically, the surfactant molecules in micelles have a mobility comparable to the liquid state, and undergo rapid exchange with the non-aggregated unimers in the solution. Furthermore, individual micelles have a limited, short lifetime, being continuously broken down and reformed.

4.2 Polymeric Micelles

4.2.1 Models

Basically, the aggregation of the hydrophobic tails in polysoaps is assumed to take place *intra*molecularly. In particular the solubilization of probe molecules has been employed to verify the intramolecular aggregation, demonstrating the lack of an equivalent of the critical micelle concentration CMC up to extreme dilutions [50, 51, 53, 65, 72, 76–78, 106, 112, 113, 126, 145, 186, 196, 197, 245, 343, 361]. Even more convincing proof was obtained by using covalently fixed probe molecules [152], because the problems of the probe's partition between the aqueous and the "micellar phase" (with the non-solubilized probe molecules dominating the signals at low soap concentration [292]) are avoided. A most elegant proof of intramolecular aggregation has been given by the lack of energy transfer in a mixed solution of two differently labelled polysoaps [217, 259].

A critical concentration for the onset of solubilization (pointing to intermolecular aggregation), or CMC-like breaks in the surface tension curves have been reported only for oligomeric polysoaps [73, 74, 109, 110, 213, 251, 353] or for polymers whose hydrophobe content is below the CAC [212, 214, 298, 391].

At high concentrations, the primary intramolecular aggregation of polysoaps may be superposed by a secondary intermolecular one [128, 217, 208, 392]. The latter process seems to depend on the molecular weight [392] and on the length of the hydrophobic tails [192], but only few data are available yet. Superposed intermolecular aggregation seems to be favoured as well by high ionic strength [208].

Whereas the debate about the structure of standard micelles has now been mostly settled [23], the structure of "polymeric micelles" at the molecular level is still a matter of discussion. The few experimental studies available provide diverse results without a unified picture. This is partially due to the complexity of the phenomena involved, and partially due to the lack of straightforward analytical techniques. Still, three major models [78] have been proposed or implicitly used, which will be referred to as "local micelle", "regional micelle" and "molecular micelle" in the following (Fig. 33).

Emerging from the pioneering work of Strauss, the model of the "local micelle" (Fig. 33a) assumes the intramolecular aggregation of a limited number of neighbouring surfactant side chains in polysoaps [54, 55, 132, 180, 181, 215]. Being independent of the degree of polymerization P_n, the model allows for a gradual transition from more to less aggregated macromolecules [54, 55, 132, 133]. Also an exchange between aggregated surfactant fragments and "free unimers" is not excluded, similar to micelles of low molecular weight surfactants. The model imposes high steric demands on the polysoaps, and high flexibility of the polymer backbone would be favourable for an efficient aggregation of the surfactant fragments. But due to unavoidable steric problems, polysoaps should exhibit more hydrophobic contacts than analogous low molecular weight surfactants. Therefore, the HLB of polysoaps should be more to the hydrophilic side according to this model.

The model of the "molecular micelle" (Fig. 33c) originates in the work of Elias, and has been treated theoretically in recent years [195, 207]. It assumes the intramolecular aggregation of virtually all surfactant side chains of a given macromolecule into one aggregate [84, 87, 126, 185, 197, 322]. Therefore there should be no dynamic process corresponding to the exchange between micelle-bound surfactant and free unimers. In fact the model was applied in the attempts to "fix" micelles of low molecular weight surfactants for imaging by electron microscopy (which were unsuccesful due to the fast exchange kinetics) [84, 393]. As the aggregation number would be identical to the degree of polymerization P_n, the shape and properties of the "molecular micelles" should be controlled by

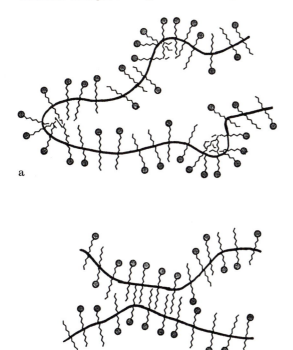

a

b

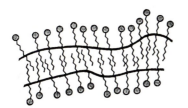

c

Fig. 33a–c. Models of polymeric micelles formed by polypolysoaps: **a** "local micelle" **b** "regional micelle"; **c** "molecular micelle". (Reprinted with kind permission from [78]. Copyright 1993 Hüthig & Wepf, Basel)

P_n rather than by the HLB and the size of the surfactant fragments. E.g. above a critical value of P_n, spherical aggregates could no longer be realized [126] (cf. Fig. 8). Compared to the "local micelle", the "molecular micelle" requires less flexibility of the polymer backbones, and thus the hydrophobic tails should be better shielded from contacts with water molecules. Therefore the HLB of polysoaps should be closer to that of analogous low molecular weight surfactants.

An intermediate situation between these extremes is found in the model of the "regional micelle" (Fig. 33b) [78, 203, 217]. Here the aggregation of a few neighbouring surfactant side chains is superposed by the aggregation of individual segments of the polymer. The resulting properties would be similar to the ones of the "local micelle", except that the severe geometric restrictions of the latter are avoided due to the more efficient regional aggregation. Similar to the model of the "molecular micelle", the HLB of polysoaps could be close to that of analogous low molecular weight surfactants. Also this model would easily account for a gradual transition from intra- to intermolecular aggregation.

4.2.2 Experimental Data and Aggregation Numbers

Up to now, none of the models for "polysoap micelles" can be definitively confirmed or rejected. Perhaps there exists more than a single structure depending on the molecular architecture of the polysoaps employed. More experiments will be needed to resolve this problem.

Fluorescence quenching studies are interpreted to support the model of the "local micelle", as aggregation numbers well below the degrees of polymerization studied were measured [103, 104, 186, 196, 215, 224, 225, 292]. The presence of "regional micelles" would be consistent with the data, too. The combination of viscosimetric and solubilization experiments [78], potentiometric titrations [180, 181] and neutron scattering studies [190–191] also support the "local micelle" or the "regional micelle" models.

From time-resolved fluorescence quenching (TRFQ) studies, small aggregation numbers of the hydrophobic side chains in "polymeric micelles" were proposed [103, 184, 186, 196, 292]. The numbers are virtually independent of the concentration and degree of polymerization, but surprisingly increase with temperature [196]. They also increase with the length of the hydrophobic chains, ranging from 15 to 60 [103, 196, 292]. In analogy, increasing the hydrophilicity of the head groups by deprotonation, and thus increasing the HLB, was found to reduce the aggregation numbers for some polysoaps (Fig. 34) [196], whereas in other cases no changes were observed [184]. Even for the identical systems studied the aggregation numbers determined may differ substantially [184, 215]. In any case it should be noted that TRFQ studies suffer from several assumptions made which are difficult to verify. E.g., the ratio of aggregated and non-aggregated hydrophobic tails has to be known, but is hardly available [196]. Also notable are reports that TRFQ experiments indicate the presence of hydrophobic pockets with high aggregation numbers in some polyelectrolytes, although hydrophobic side chains are absent, and other polysoap characteristics are missing [198].

Potentiometric titrations give comparable aggregation numbers as obtained by TRFQ. The significance of the numbers obtained however may be questioned as larger aggregation numbers have been found for shorter hydrophobic tails [180, 181].

Small angle neutron scattering experiments yield aggregation numbers of polymeric micelles greater than the degree of polymerization [190–192]. This suggests the overlap of intra- and intermolecular aggregation as observed by light scattering investigations of polysoaps [392, 394] or by energy transfer studies [217].

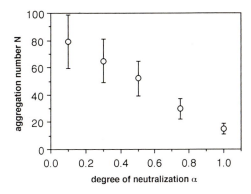

Fig. 34. Aggregation number N in polymeric micelles of polysoap poly(sodium maleate-*alt*-decylvinylether) **33** as function of the degree of neutralization α by NaOH at 25 °C, derived from fluorescence quenching. (Data from [196])

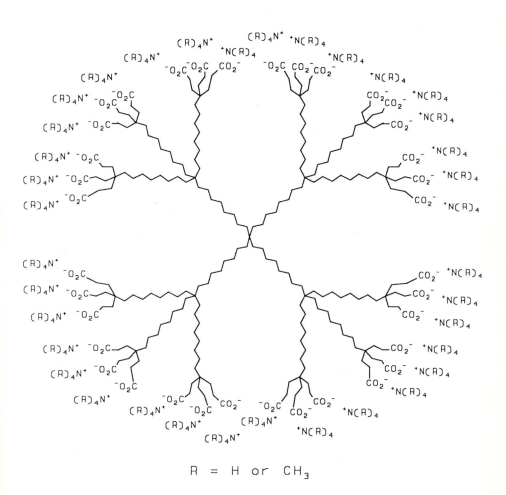

$$R = H \text{ or } CH_3$$

Fig. 35. Cascade polymer **61** forming unimolecular micelles. (Reprinted with kind permission from [43]. Copyright 1991 VCH Verlagsgesellschaft, Weinheim)

In contrast to the above discussion, theoretical treatments favour the model of the "molecular micelle" [195, 207]. At least an analogue to the latter has been realized recently in some dendrimers, arborols (Fig. 35) and hypercrosslinked polymers, replacing the self-organization of hydrophobic moieties by covalent bonding [41–45].

Chromatographic studies were interpreted in terms of molecular micelles [38]. Also, some TRFQ studies support the idea of "molecular micelles" as the experiments yielded aggregation numbers for the hydrophobic tails in "polymeric micelles" equal to the degrees of polymerization employed [106, 185, 197]. But at least in some cases [106, 185], the discrepancy with the above mentioned TRFQ studies may be apparent only because of the low degrees of polymerization used: they are of the same range as the aggregation numbers, so that "local micelles" and "molecular micelles" become indistinguishable. For very low molecular weights, even "molecular aggregation" of the hydrophobic tails is not sufficient, and intermolecular aggregates are encountered [74, 110, 117].

Recent attempts to visualize polysoaps by optical and electron microscopy do not clarify the situation. Generally the micrographs show large clusters [84, 96, 161, 193, 361]. Intermolecular aggregation in addition to intramolecular aggregation is presumably observed, and preparational artefacts must be taken into account. Exceptionally, cryo-transmission electron micrographs [395–397] visualize globular and threadlike structures of the polysoaps similar to giant micelles [288]. But the resolution of the micrographs does not allow an unambiguous distinction between "strings of beads" created from "local micelles" or "regional micelles" and extended cylindrical structures of real "molecular micelles".

Fig. 36. Texture of the lyotropic mesophase of zwitterionic polysoap p-XX between crossed polarizers [357]

4.3 Lyotropic Liquid Crystals

For low molecular weight surfactants the capability to form lyotropic liquid crystals (LC) in water is an essential property [318]. Although the appropriate temperature and concentration ranges of the liquid-crystalline mesophases can be small, mesophases are always found when carefully searched for.

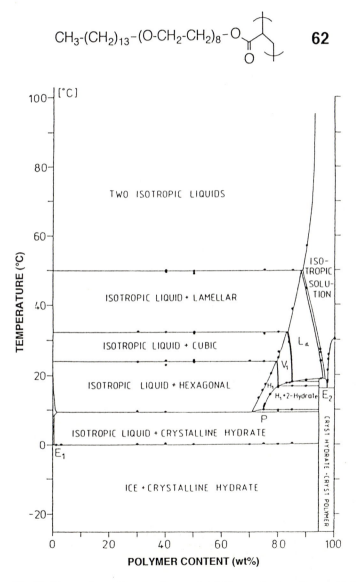

Fig. 37. Binary phase diagram of water-insoluble non-ionic polysoap **62**. (Reprinted with kind permission from [231]. Copyright 1987 Steinkopff Verlag, Darmstadt)

In this light, it is not surprising that there are many reports on polymeric lyotropic LCs formed by polysoaps at high concentrations [62, 63, 70, 87, 121–124, 126, 172, 173, 178, 230, 231, 240, 244, 249, 261, 300, 331, 398–404], particularly when polymerized surfactants are involved (Figs. 36, 37).

As for low molecular weight surfactants, the superstructures are assumed to be formed by "micellar" aggregates [126]. But it seems that the formation of lyotropic liquid crystals is supported by the additional presence of thermotropic mesogens [87, 122–124, 126]. Lamellar, hexagonal, cubic and even nematic and cholesteric mesophases were reported for binary systems, the latter being exceptional. Lyotropic mesophases were also observed in non-aqueous solvents [240, 400, 401, 405]. If polymerizable surfactants are studied, not only the phase diagram but also the types of mesophases observed for the monomer and the polymer may be different.

However, for the majority of the polysoaps studied so far, lyotropic mesophases have not yet been observed. Notably, lyotropic mesophases have rarely been reported for polysoaps of other geometry than "tail end" type. It may thus be possible that the capability to self-organize into lyotropic liquid crystals is not a general feature of polysoaps, contrasting with standard surfactants.

There is a second particularity to be noted: many virtually water-insoluble polymeric amphiphiles can be swollen to yield polymeric lyotropic mesophases, even if the miscibility gap is broad (Fig. 37). Such behaviour seems to be widespread for vinylic polymerized surfactants with side-chain spacers [126, 231, 331]. I.e., neither polysoap behaviour implies the capability to form lyotropic mesophases, nor the presence of lyotropic mesophases the classification as polysoap.

5 Aggregation in the Solid State

In addition to the interest in lyotropic mesophases, some studies have looked into the self-organization and aggregation of polysoaps in the dry, i.e. generally in the solid state. In the case of ionic polysoaps, various lamellar super-structures are observed, the structure of which can be very complicated [331, 406–408] (Fig. 38). Although a number of surfactant monomers exhibit thermo-tropic mesophases [168, 331, 406], mesophases do not normally occur for the polymers and have only been occasionally reported [59, 127, 150, 226, 408]. In the examples described, the incorporation of true mesogenic moieties is neither necessary nor particularly advantageous for mesophase formation. Still, due to the hygroscopy of the polymers, the distinction between dried samples with small residual water content and truly anhydrous samples is difficult. Thus, even the occasional reports on thermotropic mesophases of ionic polysoaps [203] may be erroneous, as rigorous drying of some nearly identical ionic polysoaps

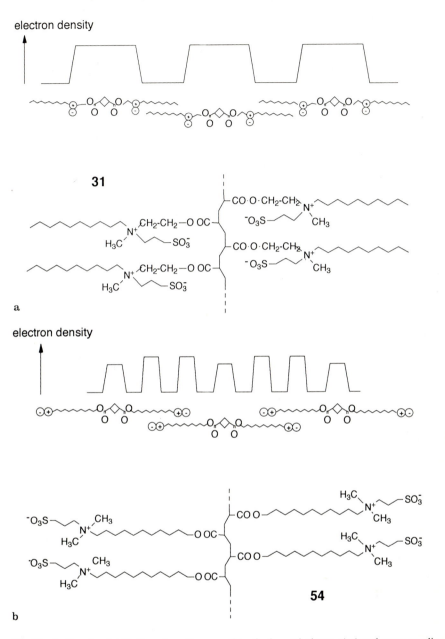

Fig. 38a, b. Density distribution along the normal to the layered planes $\rho(x)$ and corresponding models of the packing of two zwitterionic polysoaps in the anhydrous solid state (model adapted from [407]): **a 31** ("head" type); **b 54** ("tail-end" type)

resulted in the disappearance of the mesophases seen [172, 173, 249, 331, 399] which accordingly are lyotropic ones.

For non-ionic polysoaps the situation is quite different. Whereas in the case of oligoethyleneoxide head groups thermotropic mesophases seem to be absent [87, 121–124, 126, 231, 403, 409], polysoaps with liposaccharide [230, 240, 300] surfactant fragments or with lipopeptide [244, 400–402] ones frequently show lamellar mesophases (e.g. of smectic A type) and even nematic ones (Fig. 39). It should be emphasized that the thermotropic mesophases here are not the result of mesogenic groups being present, but are the consequence of the amphiphilic character of the polysoaps and the resulting microphase separation.

Speculating on the reasons for the different behaviours of the different classes of polysoaps, ionic interactions may be too strong to allow for a mesophase without plasticizing water. On the other hand, interactions in oligoethylene-oxides may be too weak to ensure self-organization above the melting point. Only the presence of interactions of intermediate strength, such as H-bonding or strong dipole–dipole interactions, installs the right balance of interactions needed.

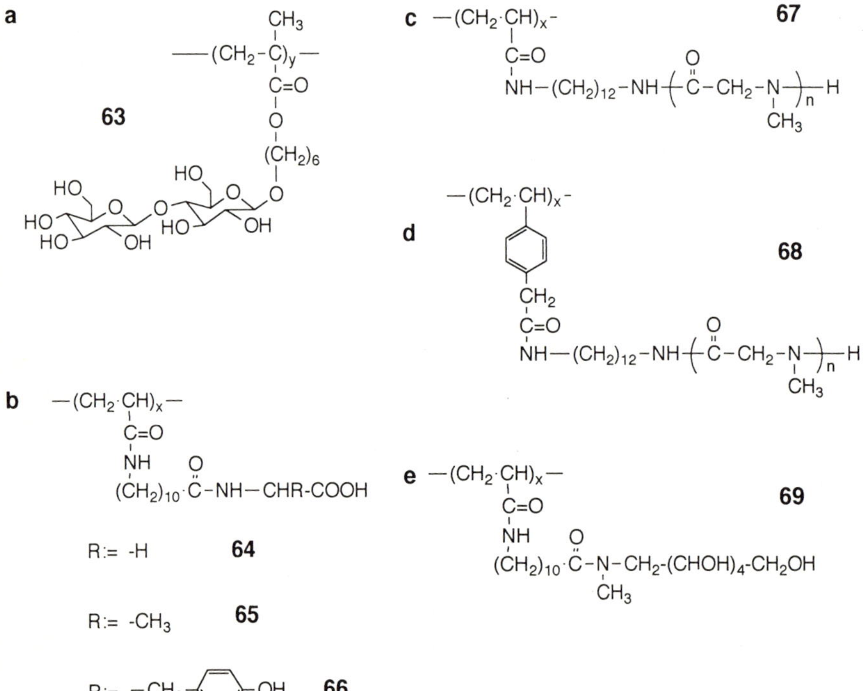

Fig. 39a–e. Non-ionic polysoaps capable of forming thermotropic liquid crystals: **a** [230]; **b** [401]; **c** [244]; **d** [402]; **e** [240]

6 Molecular Weight Effects and the Behaviour of Oligomers

In the previous sections some characteristic differences between low molecular weight surfactants and polysoaps were presented, which were attributed to covalent linkage of the surfactant fragments (instead of reversible linkage by secondary valence forces), modified dynamics, proximity effects and intra-molecular aggregation. As pointed out for high molecular weight polysoaps, there are no indications of major differences between samples of different molecular weights concerning properties such as surface activity or solubilization. But of course the question arises of how the various properties of surfactants evolve with increasing degree of polymerization in the intermediate, i.e. the oligomeric regime; e.g. whether there is a gradual evolution of properties or a discrete one etc. This problem has been addressed in recent years from the aspect of speciality surfactants, without the intention to investigate polysoaps. Although the studies often appear incomplete as studies of polysoaps, and sometimes they are difficult to interpret, the results may be useful for the comprehension of the latter.

6.1 Defined Oligomeric Surfactants

Defined "oligomeric" surfactant structures obviously have stimulated the im-agination of many colloid chemists, as a number of imaginative names were suggested for them: bola-surfactants [410–412], gemini-surfactants [413–415], tentacle molecules [416–417], octopus molecules [418], hexapus [419], multi-armed amphiphiles [420–422]. . . and so on (Figs. 40, 41). The semantic confu-sion is aggravated by the use of different terms for identical structures (e.g. tentacle, octopus molecules, and multi-armed) and of identical terms for differ-ent structures (e.g. gemini). Also, some of the names are misleading, e.g. if "octopus" is applied to tetrameric structures [418, 423].

As for the polysoaps, surfactant oligomers can be distinguished by their geometry. In fact two type of surfactant "dimers" can be identified in the literature, the α,ω-dipolar surfactants (often referred to as "bola-surfactants") representing dimers of the "head type" geometry, and twinned surfactants (often referred to as "gemini-surfactants") (Fig. 40) representing dimers of the "tail end type" geometry. Both classes of speciality surfactants have generated much interest in the past decade, and a number of systematic studies exist now, covering even some polymerizable derivatives (cf. Table 2).

Generally, both types of dimeric surfactants still show well-defined, clear CMCs, though in selected cases the CMC is less obvious to standard methods, in particular to conductometric methods [415, 424, 425]. The values of CMCs determined by different methods agree well, and the counter-ion effects are the usual ones [426, 427]. But, strikingly, CMC values of both bola-surfactants and of gemini-surfactants are lowered by one order of magnitude or more in

a
b
c
d
e
f
g

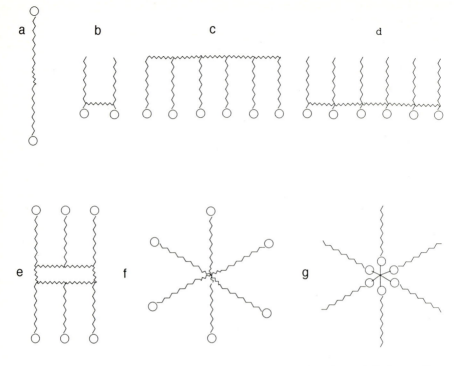

Fig. 40a–g. Various types of "oligomeric" surfactants: **a** "bola" dimer; **b** "gemini" dimer; **c** linear hexamer ("tail end" type) **d** linear hexamer ("head" type); **e** cyclic hexamer; **f** star-shaped hexamer ("tail end" type); **g** star-shaped hexamer ("head" type)

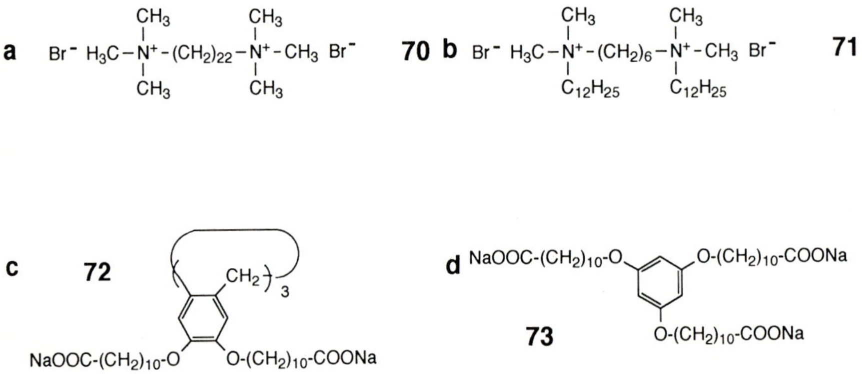

a
Br^- $H_3C-N^+(CH_3)_2-(CH_2)_{22}-N^+(CH_3)_2-CH_3$ Br^-
70
b
Br^- $H_3C-N^+(CH_3)(C_{12}H_{25})-(CH_2)_6-N^+(CH_3)(C_{12}H_{25})-CH_3$ Br^-
71
c
72
CH_2
3
$NaOOC-(CH_2)_{10}-O$
$O-(CH_2)_{10}-COONa$
d
$NaOOC-(CH_2)_{10}-O$
$O-(CH_2)_{10}-COONa$
73
$O-(CH_2)_{10}-COONa$

Fig. 41a–d. Examples of "oligomeric" surfactants: **a** "bola" dimer [426–428]; **b** "gemini" dimer [435, 445, 448]; **c** cyclic hexamer of "tail end" type [419]; **d** star-shaped trimer of "tail end" type [420]

comparison to the corresponding "monomers" [239, 424, 428–434]. The reduction of the surface tension appears slightly lower than for the monomers, but the surface activities are still substantial [330, 405, 413, 424, 427, 429–435]. Interestingly, above the CMC a number of dimeric surfactants do not show the standard plateau of surface tension but exhibit a small but continuous decrease with increasing concentration.

From surface tension studies of bola-surfactants, it is concluded that they exhibit wicket-like conformations at the gas-water interface [239, 428, 430, 431]. In micelles and liquid crystals however, a stretched conformation is preferred [436, 437]; this implies that surface tension data and interfacial tension data do no more describe the "micellar" interface with all the implications for solubilization (compare Sect. 3.4). In fact, some reports stress the extremely low solubilization capacity of bola-surfactants [431, 432, 437, 438], although others obtain capacities comparable to the ones of the "monomers" [430]. Also noteworthy, solubilized fluorescence probes indicate a more polar environment for the solubilizates than in micelles of the "monomers" [430–432], but micellar aggregation numbers of the bola-surfactants are comparable or only slightly lower [429, 432, 438, 439]. In exceptional cases, very high aggregation numbers and the existence of an additional pre-CMC are observed [440].

NMR-studies indicate a reduced segmental mobility and higher order parameter for the central hydrocarbon fragments of bola-surfactants [437] compared to the tail ends of the corresponding monomers. The differences however are less pronounced as for oligomeric polysoaps [68, 69].

The majority of reports on gemini-surfactants [441–445] indicate increasing CMC values with increasing length of the hydrocarbon tails, as expected from HLB. Less evident, the surface tensions at the CMC are sometimes notably higher for higher homologues, in particular for long alkyl tails [442, 443]. Recently some surprising results were reported for gemini-surfactants with rigid spacer groups, such as increasing CMC with increasing length of the hydrophobic tails [413, 415, 446]. "Premicellar" aggregation was suggested as an explanation, based on solubilization results indicating efficient solubilization much below the CMC values which are deduced from surface tension data. But the exact positioning of some of the CMCs may also be disputable, revising the surface tension data reported. More studies are awaited to clarify the unusual findings (see also below).

The properties of gemini-surfactants are markedly influenced by the length of the linking unit, which represents the equivalent to the main chain spacer unit in "head type" polysoaps: the CMC does not decrease monotonously with increasing spacer length, as would be expected from the decreasing HLB, but shows a maximum value at intermediate spacer length (C_6–C_8) [433, 435, 445, 447, 448]. The surface tensions obtained at the CMC show a minimum in the same range of spacer length [435], and the solubilization capacity a maximum [448]. The coincidence with the optimal spacer length of "head type" polysoaps discussed in Sect. 2.3.4 is striking. Other effects on the CMC which cannot be explained by the HLB changes have also been observed with more complex spacer groups [434].

Virtually all examples reported for higher oligomeric surfactants than "dimer" belong to the "tail end type", preventing an analysis of the evolution of geometric effects. This is the more true, as the majority of the "tail end" structures represents "stars" [416, 417, 420, 422, 425, 449] rather than linear or cyclic oligomers [418, 419, 423]. Most attempts to produce "head-type" trimers suffer from the poor water solubility of the products [450], but recently some water-soluble examples have been reported [427].

The trimeric surfactants and higher oligomers investigated show pronounced peculiarities compared to "monomeric" surfactants. CMC-like transitions become less clear, in particular of studied by surface tension and conductivity measurements [416–422, 425]. The reduction of the surface tension by such oligomers is often quite low [416–420, 422, 423, 425], and the shape of the plots of surface tension vs the logarithm of the concentration can be bizarre [417, 425] though not necessarily for all systems studied [423]. For "hexamers" the surface activity becomes very weak [419]. Thus solubilization studies appear to be more appropriate for CMC determinations but have been seldom performed. Interestingly, the CMC values assigned from conductivity and surface tension data level approximately at the level of the "dimers" [422, 425] (Fig. 42), whereas the CMC value derived from solubilization data for a "hexamer" surfactant is lowered by at least a factor of 100 compared to the "monomer" [419]. The reason for such discrepancies is obscure. There are indications that solubilization and surface tension or conductimetric studies respectively do not give consistent CMC values for such uncoventional surfactants [425] which may be due to the formation of very small aggregates at low concentrations [417]. The surprising behaviour of trimeric surfactants of the "head" type was

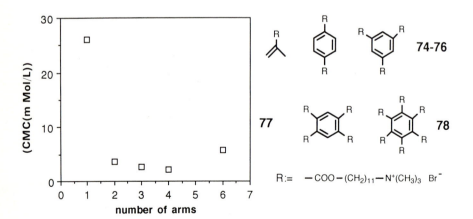

Fig. 42. Dependence of critical micelles concentration CMC of cationic oligomeric surfactants on the number of surfactant arms ("degree of oligomerization"), as determined by surface tension measurements and conductimetry. (Data from [422])

explained analogously [427]. They exhibit increasing CMCs with increasing length of the hydrophobic tails deduced from surface tension data, but show efficient solubilization much below the apparent CMC values, in analogy with the reports on some gemini-surfactants mentioned above [413, 415].

Although no full picture of the above referred studies can be obtained, some general trends are visible: oligomerization of surfactants tends to decrease surface activity and CMC, spacer effects in contrast to simple HLB considerations are evident, and the mobilities of the various molecular fragments change considerably. All these general trends parallel the differences seen between standard surfactants and polysoaps in Sects. 3 to 5.

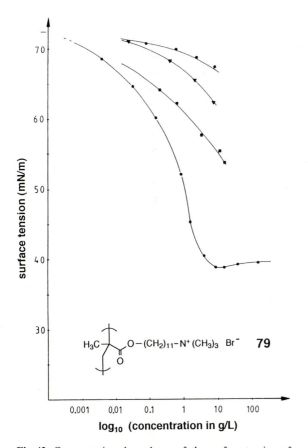

Fig. 43. Concentration dependence of the surface tension of aqueous solutions of polymerized methacrylic surfactant **79** as function of the amount of initiator used. (Adapted with kind permission from [229]). *Top to bottom*: polymer (0.5 mol % AIBN used), polymer (1 mol % AIBN used), polymer (5 mol % AIBN used), monomer

6.2 Oligomeric Mixtures

Oligomeric mixtures are usually obtained by oligomerization of surfactant monomers, most frequently of such bearing allylic moieties as polymerizable group (e.g. 6, Fig. 2f). In contrast to the above mentioned "multi-armed" surfactants, such mixtures are composed of linear oligomers. The average degree of polymerization is often difficult to determine exactly, but generally lies in the range of 2–10 surfactant units [61, 65, 67, 74, 82, 84, 109, 251, 315], sometimes up to 20 units [68, 69, 71, 110, 117].

Molecular weight distributions have to be assumed to be broad. Better defined oligomers are obtained by polymer analogous reactions of reactive surfactants onto well-characterized polymer backbones [126, 451] (see Fig. 44).

Typically for such oligomers, CMCs are often found [74, 109, 110, 117, 213, 251, 452], i.e. intermolecular aggregation takes place. This is not only visible by surface tension data, but also by solubilization studies [67, 71, 74, 110, 117, 452]. Due to distribution of molecular weights, it cannot be decided whether CMCs are still present for the higher oligomers or whether only the low oligomeric parts are responsible for the phenomenon. But as pointed out in Sect. 4.1, it is reasonable to expect intermolecular micellization for oligomers if their degree of polymerization is lower than the aggregation numbers expected in "polymeric micelles", assuming that they are in the same range as those of standard surfactants.

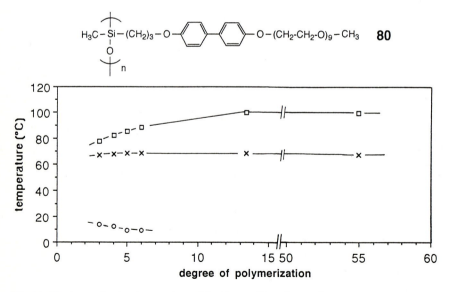

Fig. 44. Maximum clearing temperature of liquid crystalline phases of a non-ionic polysoap **80** as function of the degree of polymerization: $\bigcirc$ cubic I_1 phase, $\times$ hexagonal H_1 phase, $\square$ lamellar L_α phase. (Data from [451]; see also [126])

This influence of the degree of polymerization of the surface activity of oligomers has been addressed by Wagner (Fig. 43). With increasing degree of polymerization, the surface activity finally vanishes [229]. Unfortunately the data do not allow a quantitative estimation of the degrees of polymerization involved, but the trend is clear.

A quantitative evaluation of the effects of the degree of polymerization on the properties of polysoaps was undertaken for the phase diagrams of polymeric lyotropic liquid crystals. Here the property shifts level off within a degree of 10–15 (Fig. 44) [126, 451]. This value may serve as a first approximation, but, as additional studies, e.g. on viscosity, surface tension or solubilization are missing, more studies are needed.

7 Applications of Polysoaps

Theoretically, polysoaps may be of interest for all the potential applications mentioned for micellar polymers in the introduction. In practice, the attempted uses of polysoaps cover a much narrower range. Besides a number of diverse suggestions, proposed uses are basically classical colloidal applications, medical or pharmaceutical applications, and catalytic systems.

Concerning colloidal applications, the uses as dispersants or emulsifiers are the most frequent ones. But although there exist several studies on the use of polymerizable surfactants [82, 91, 93, 219, 251, 314, 333, 337, 352, 452–477] – i.e. of the precursors for polysoaps – the use of polysoaps in emulsion polymerization is still the exception [376]. In both cases improved latex stabilities and improved resistance to moisture of film-formed latexes were aspired to, but at least in the former case the results were often disappointing [93, 251, 470–472]. Polysoaps are used to stabilize lattices [214, 230]. The opposite use for flocculation has also been tested, [284]. Polysoaps have also been studied as viscoelastic fluids [108].

Potential medical or pharmaceutical applications have attracted more attention. Studies cover improved efficiency and delivery [334–336, 478–480], improved stability [334], and lower toxicity [50, 315, 334, 374, 481] of polysoap bound drugs. The use of polysoaps to fix or stabilize enzymes, antibodies etc. is also discussed [214, 230]. Cationic polysoaps were also studied for antimicrobial activities [277, 278].

The majority of the work is focused on the use of polysoaps in micellar catalysis [482–484], in particular as models for esterases [57, 71, 136–140, 332, 361]. In addition, the catalytic activities in diverse other model reactions were investigated [141, 200, 211, 308, 317, 333, 361, 463, 464, 485–487]. Various photochemical systems for energy harvesting [183, 217, 259], or for charge separation [217, 257–259, 263] can be subsumed by applications of polysoaps in catalysis as well. Frequently, enhanced catalytic activity is observed compared to micelles of low molecular weight surfactants [113, 200, 258, 361, 485, 487].

Table 1. Anionic polymerizable surfactants

STYRENE DERIVATIVES AND RELATED MONOMERS

Structure	References
$CH_2=CH$—⟨aryl⟩—$CH-(CH_2)_8-COOK$, CH_3	[51, 83]
$CH_3(CH_2)_n-OOC-CH=CH$—⟨aryl⟩—SO_3M n:= 10,12 M:= Na, K, NH_4	[91]
$CH_2=CH$—⟨aryl⟩—$(CH_2)_{10}-COONa$	[261, 398, 399]
$CH_3(CH_2)_{11}-OOC-C=CH$—⟨aryl⟩—SO_3K, CH_3	[459]
⟨aryl⟩—$O-(CH_2)_{12}-SO_3Na$, $CH=CH_2$	[461, 462]

ALLYLICS

Structure	References
$CH_2=CH-(CH_2)_8-COOM$ M:= Na, K	[61–74, 109, 117, 467, 475]
$CH_2=CH-(CH_2)_9-OOC-CH_2$, $CH_2=CH-(CH_2)_9-OOC-CH-SO_3Na$	[114, 476]
$CH_2=CH-(CH_2)_9-COONa$	[68]
$CH_2=CH-(CH_2)_9-OOC-CH_2$ + $CH_3-(CH_2)_{10}-OOC-CH-SO_3Na$; $R_1-CH-SO_3Na$, R_2-CH	[114, 476]
$CH_2=CH-(CH_2)_{10}-COONa$	[68]
$CH_3-(CH_2)_x-CH-SO_3Na$, $COO-CH_2-CH=CH_2$ x:= 13,15	[82, 309, 455]
$CH_2=CH-(CH_2)_{11}-COONa$	[110]
$CH_3-(CH_2)_{11}-OOC-CH_2-CH-SO_3Na$, $COO-CH_2-CH=CH_2$	[470–472]
$CH_2=CH-CH_2-CH-CH=CH-CH_2-COONa$, CH_3	[109]
$CH_2=CH-(CH_2)_9-SO_3Na$	[72, 361]
$CH_3(CH_2)_8$—⟨aryl⟩—$O-CH_2-CH-(O-CH_2CH_2)_{10}-O-SO_3NH_4$, $CH_2-O-CH_2-CH=CH_2$	[116]
$CH_2=CH-(CH_2)_9-O-SO_3Na$	[72, 111, 352, 361, 453]
$CH_2=CH-(CH_2)_8-COO-CH_2-CH_2-SO_3Na$	[458]

ACRYLATES

$CH_2=CH-COO-(CH_2)_{15}-COOK$ [223]

$CH_3-(CH_2)_5-CH-(CH_2)_{10}-COOK$
$\quad\quad\quad\quad\quad |$
$\quad\quad\quad\quad CH_2=CH-COO$ [223]

$CH_2=CH-COO-\langle aryl \rangle-CH-(CH_2)_8-COOK$
$\quad\quad\quad\quad\quad\quad\quad\quad\quad |$
$\quad\quad\quad\quad\quad\quad\quad\quad\quad CH_3$ [51, 83]

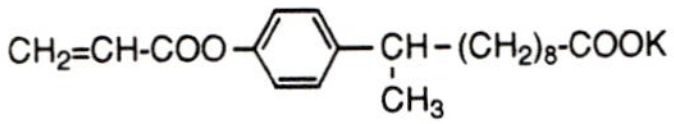

$CH_2=CH-COO-(CH_2)_{11}-SO_3Na$ [486]

METHACRYLATES

$CH_3-(CH_2)_9-CH=CH-CH-CH_2-COOK$
$\quad\quad\quad\quad\quad\quad\quad\quad\quad\quad\quad CH_3$
$\quad\quad\quad\quad\quad\quad\quad\quad\quad\quad\quad |$
$\quad\quad\quad COO-CH_2-CH_2-O-C-CH=CH_2$
$\quad\quad\quad\quad\quad\quad\quad\quad\quad\quad\quad\quad ||$
$\quad\quad\quad\quad\quad\quad\quad\quad\quad\quad\quad\quad O$ [223]

(steroid structure)   [248]

$CH_2=C-COO-\langle aryl \rangle-(CH_2)_n-COONa$
$\quad\quad |$
$\quad\quad CH_3$ n:= 6,12   [95]

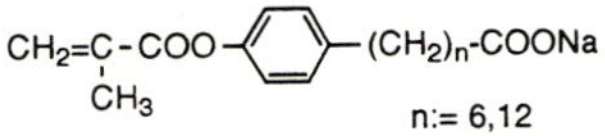

$CH_3-(CH_2)_n-O-\underset{||\,\,O}{\overset{O\,\,||}{P}}-O-CH_2-CHOH-CH_2-O-\underset{}{\overset{O\,||}{C}}-C=CH_2$
$\quad\quad\quad\quad\quad\quad\quad |$
$\quad\quad\quad\quad\quad\quad\quad ONa \quad\quad\quad\quad\quad\quad\quad\quad\quad\quad CH_3$
$\quad\quad\quad\quad\quad\quad\quad\quad n:= 7, 11, 13, 15$ [289, 291, 492]

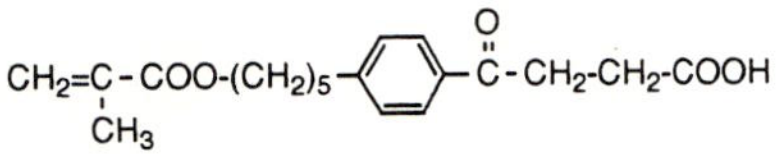

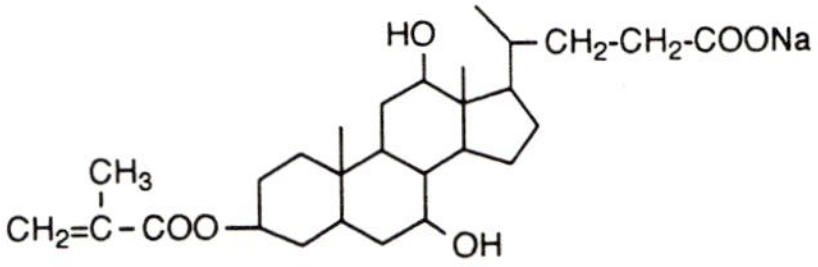

$CH_2=C-COO-(CH_2)_5$
$\quad\quad |$
$\quad\quad CH_3$
$\quad\quad\quad\quad\quad\quad\langle aryl \rangle-\overset{O\,||}{C}-CH_2-CH_2-COOH$ [491]

Continued

ACRYLAMIDES

$CH_2=CH-CONH-(CH_2)_{10}-COOM$ [83, 98, 163, 283, 384, 456, 466, 494]
M:= Na, ·,

$CH_2=CH-CON-(CH_2)_{10}-COONa$ [99, 284, 493, 494]
    |
    R   R:= -CH_3, C_2H_5

$CH_2=CH-CON-CH(CH_2)_x-COONa$ x+y:= 8 [456]
    |
    (CH_2)_y-CH_3

$CH_3-(CH_2)_x-CH-(CH_2)_y-COOM$ [454, 457, 467, 477, 495, 496]
    |
$CH_2=CH-CONH$ M:= Na, K   x+y:= 15

$CH_2=CH-CONH-(CH_2)_{15}-CONH-C-COONa$ [402]
                       |
                       R
R:= -H, -CH_3, -CH_2⟨⟩-OH

$CH_3(CH_2)_n-CH-CH_2-SO_3Na$ n:= 9, 13 [314, 162, 497]
         |
$CH_2=CH-CONH$

$CH_3-(CH_2)_9-N-CH_2-CH_2-SO_3Na$ [91]
           |
           $CH_2=CH-C=O$

METHACRYLAMIDES

$CH_2=C-CONH-(CH_2)_{10}-COOH$ [498]
    |
    CH_3

$CH_2=C-CON-(CH_2)_{10}-COONa$ R:= CH_3, C_2H_5, C_3H_7 [84]
    | |
    CH_3 R

$CH_2=C-CONH-(CH_2)_{10}-CONH-⟨⟩-SO_3Na$ [455]
    |
    CH_3

$CH_2=C-CONH-(CH_2)_{10}-CONH-(CH_2)_n-SO_3Na$ n:= 2, 4 [309, 455]
    |
    CH_3

$CH_3-(CH_2)_9-N-CH_2-CH_2-SO_3Na$ [91]
           |
      $CH_2=C-C=O$
           |
           CH_3

OTHERS

$CH_3\text{-}(CH_2)_n\text{-}OOC\text{-}CH_2\text{-}\underset{\underset{CH_2}{\parallel}}{C}\text{-}COOM$ M:= Na, K [499–501]
n:= 7,11,13,15,17

$CH_3\text{-}(CH_2)_n\text{-}OOC\text{-}CH{=}CH\text{-}COO\text{-}CH_2\text{-}CH_2\text{-}SO_3Na$ [314, 502]
n:= 11,13,15,17

—$(CH_2)_n\text{-}SO_3Na$ n:= 6,10 [272]

$CH_3\text{-}(CH_2)_9$ SO_3M [91]
$CH_2{=}CH\text{-}C{=}O$

$CH_3\text{-}(CH_2)_n$ $N\text{-}(CH_2)_3\text{-}SO_3K$ [270]
n:= 12,18

Table 2. Cationic polymerizable surfactants

STYRENE DERIVATIVES

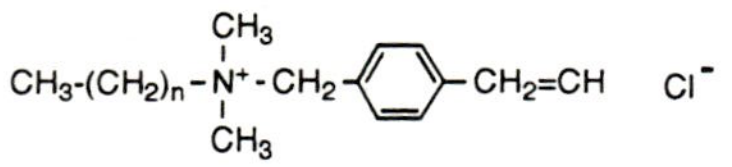

$CH_3-(CH_2)_n-N^+-CH_2$—⬡—$CH_2=CH$ Cl^- n:= 7,11,15

[103, 104, 277, 278, 292, 320, 323, 333, 375, 463, 480]

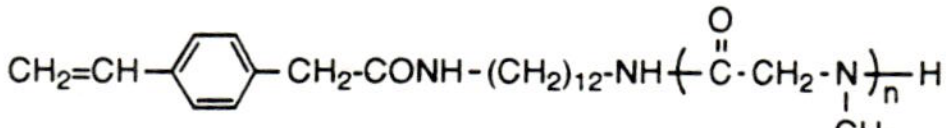

$CH_2=CH$—⬡—$CH_2-CONH-(CH_2)_{12}-NH(-C-CH_2-N)_n-H$ [244]

n:= 6-18

VINYLPYRIDINES AND VINYLIMIDAZOLS

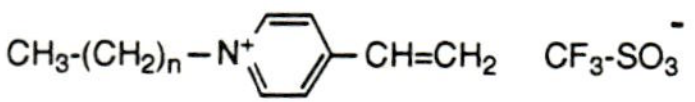

$CH_3-(CH_2)_n-N^+$ $CH=CH_2$ $CF_3-SO_3^-$ [221, 503, 504]

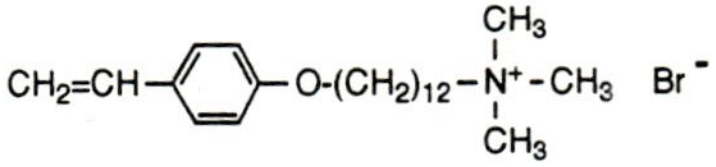

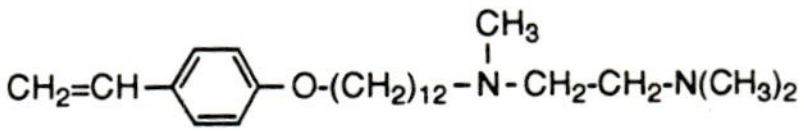

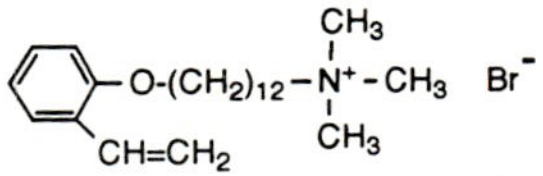

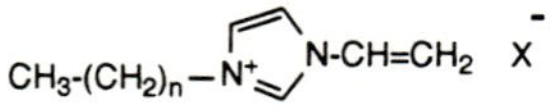

n:= 7,11

$CH_3-(CH_2)_n-N^+$ $N-CH=CH_2$ X^- n:= 6,11,15   [85, 86, 115, 156, 275, 347, 505]

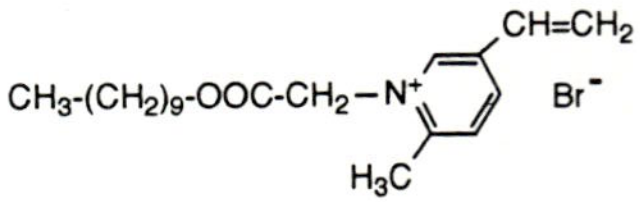

x:= I^-, Br^-

$CH_2=CH$—⬡—$O-(CH_2)_{12}-N^+-CH_3$ Br^- [333, 463]

$CH_2=CH$—⬡—$O-(CH_2)_{12}-N-CH_2-CH_2-N(CH_3)_2$ [256]

⬡—$O-(CH_2)_{12}-N^+-CH_3$ Br^- [333, 464]

$CH_3-(CH_2)_9-OOC-CH_2-N^+$ Br^- [91]

ALLYLICS

$CH_2=CH-(CH_2)_9-N^+(CH_3)_3$ Br^- [72, 361, 453]

$CH_3-(CH_2)_{15}-OOC-CH_2-N^+-CH_2-CH=CH_2$ Br^- [96, 107]
(with CH_3 above and CH_3 below on N)

$CH_2=CH-(CH_2)_9-OOC-(CH_2)_{10}-N^+(CH_3)_3$ Br^- [239]

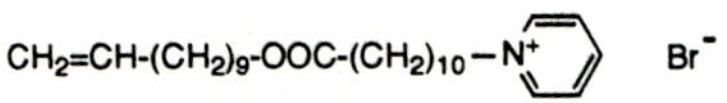

$CH_3-(CH_2)_n-N^+-CH_3$ Cl^- n.= 9, 15   [161]
(with $CH_2-CH=CH_2$ above and $CH_2-CH=CH_2$ below)

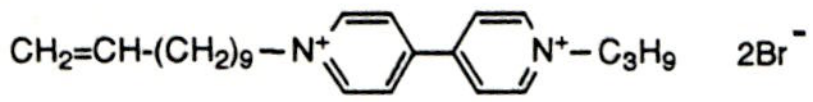

$CH_2=CH-(CH_2)_9-OOC-(CH_2)_{10}-N^+$(pyridinium) Br^- [239]

$CH_3-(CH_2)_{15}-OOC-CH_2-N^+-(CH_3)_3$ Br^- [96, 107]
(with $CH_2-CH=CH_2$ above and $CH_2-CH=CH_2$ below)

$CH_2=CH-(CH_2)_9-N^+$(biphenyl pyridinium)$-C_4H_9$ $2Br^-$ [343]

$CH_3-(CH_2)_{15}-OOC-CH_2-N^+-CH_2-CH=CH_2$ Br^- [96, 107]
(with $CH_2-CH=CH_2$ above and $CH_2-CH=CH_2$ below)

$CH_3-(CH_2)_n-N^+-CH_2-CH=CH_2$ Br^- n.= 11,13,15,17   [106, 108, 115]
(with CH_3 above)

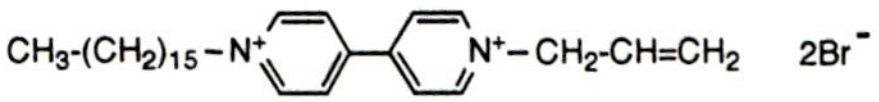

$CH_2=CH-(CH_2)_9-N^+-CH_2$(ferrocene)$-CH_3$ Br^- [268]
(with CH_3 above)

$CH_3-(CH_2)_{15}-N^+$(biphenyl pyridinium)$-CH_2-CH=CH_2$ $2Br^-$ [264]

ACRYLATES

$CH_2=CH-COO-(CH_2)_{11}-N^+(CH_3)_3$ Br^- [76, 88, 89, 223-225, 229, 488, 499]

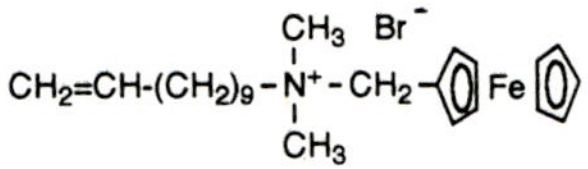

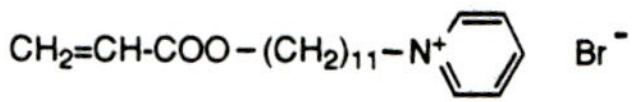

$CH_2=CH-COO-(CH_2)_{11}-N^+$(pyridinium) Br^- [76, 89, 223]

$CH_3-(CH_2)_{15}-N^+-CH_2-CH_2-OOC-CH=CH_2$ Br^- [290]
(with CH_3 above)

$CH_2=CH-COO-(CH_2)_{11}-N^+$(biphenyl pyridinium)$-N^+-(CH_2)_2-CH-CH_3$ $2Br^-$ [269, 343]
(with CH_2OH)

$CH_2=CH-COO-(CH_2)_{11}-N^+$(biphenyl pyridinium)$-C_3H_9$ $2Br^-$ [269, 343]

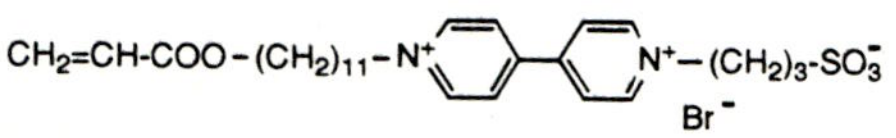

$CH_2=CH-COO-(CH_2)_{11}-N^+$(biphenyl pyridinium)$-N^+-(CH_2)_3-SO_3$ Br^- [267-269, 343]

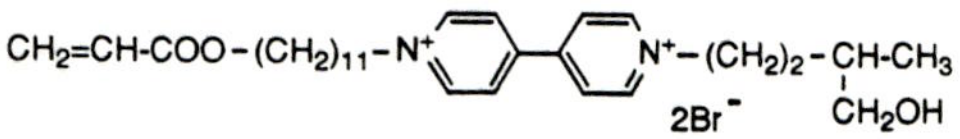

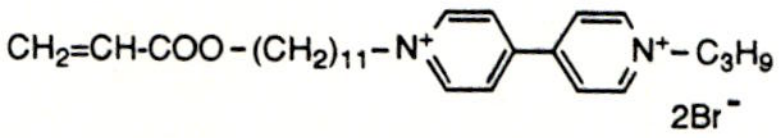

Continued

METHACRYLATES

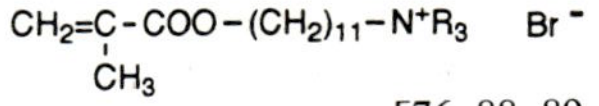

$CH_2=C-COO-(CH_2)_{11}-N^+R_3$ Br^- $R:=CH_3, C_2H_5$ [76, 89, 223]
      |
      CH_3

[76, 88, 89, 105, 126, 223, 226, 229, 491, 499]

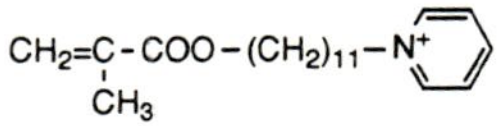

$CH_2=C-COO-(CH_2)_{11}-N^+$ Br^- [76, 89, 223] 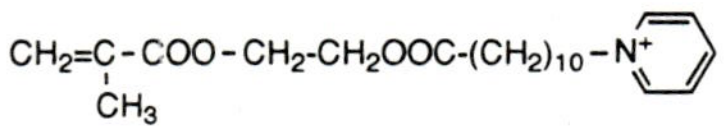
      |
      CH_3

$CH_2=C-COO-CH_2CH_2OOC-(CH_2)_{10}-N^+$ Br^- [76, 89, 223]
      |
      CH_3

$CH_2=C-COO-CH_2CH_2OOC-(CH_2)_{10}-N^+-(CH_3)_3$ Br^- [76, 89, 223] 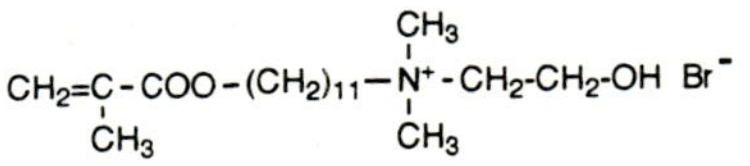
      |
      CH_3

$CH_2=C-COO-(CH_2)_{11}-N^+-CH_2-CH_2-OH$ Br^- [78, 227, 245, 267–269, 343]
      | |
      CH_3 CH_3
                                     CH_3

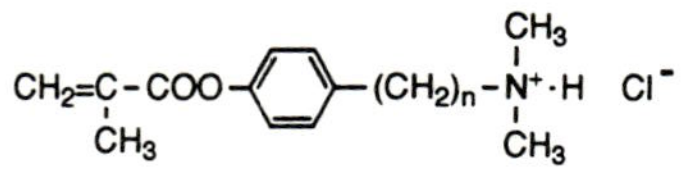

$CH_2=C-COO-(CH_2)_{11}-N^+$ N^+-CH_3 $2Br^-$ [264]
      |
      CH_3

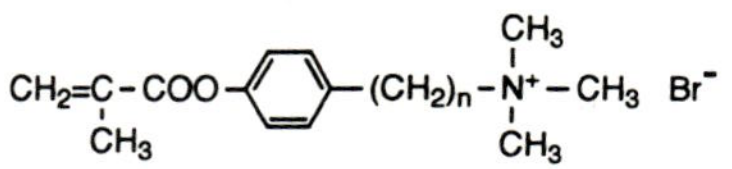

$CH_2=C-COO-(CH_2)_{11}-N^+$ $N^+-(CH_2)_3-SO_3^-$ [267–269]
      | Br^-
      CH_3

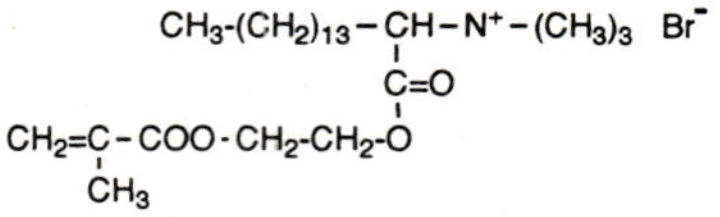

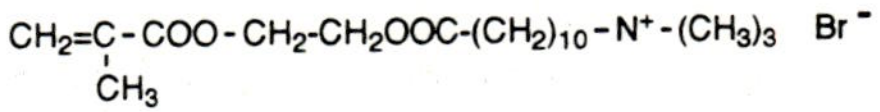

$CH_3-(CH_2)_n-N^+-CH_2-CH_2-O-C-C=CH_2$ Br^- n:= 7,8,9,11,13,15,17
                  | || |
                  CH_3 O CH_3
                  CH_3

[76, 88, 105,118, 167, 219, 226, 227, 232, 245, 279, 280, 290, 333, 343, 406, 463, 469]

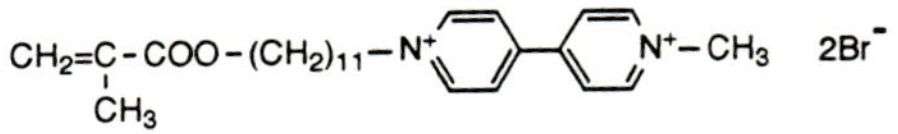

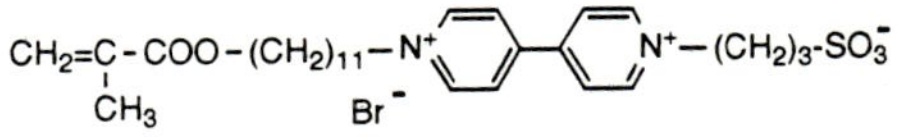

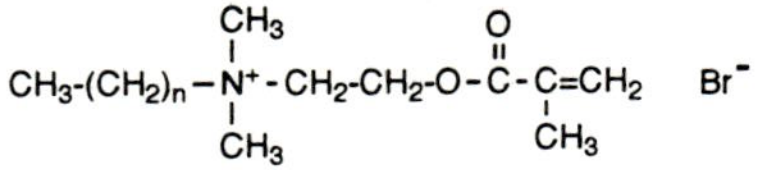

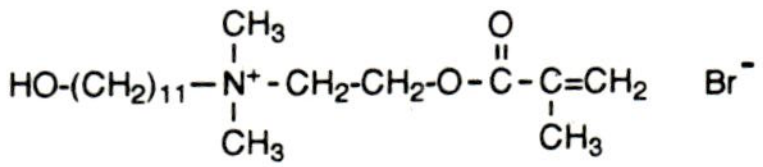

$HO-(CH_2)_{11}-N^+-CH_2-CH_2-O-C-C=CH_2$ Br^- [76, 89]
                  | || |
                  CH_3 O CH_3
                  CH_3

$CH_3-(CH_2)_n-OOC-CH_2-N^+-CH_2CH_2-O-C-C=CH_2$ X^- [96, 97, 107, 286, 505]
                          | || |
                          CH_3 O CH_3
                          CH_3

n:= 7, 9,11,15   X:= Cl, Br

METHACRYLATES

$CH_2=C-COO$—⟨benzene⟩—$(CH_2)_n-N^+-H$ Cl^- [95]
      | |
      CH_3 CH_3
                                CH_3

n:= 3, 7, 13

$CH_2=C-COO$—⟨benzene⟩—$(CH_2)_n-N^+-CH_3$ Br^- [95]
      | |
      CH_3 CH_3
                                CH_3

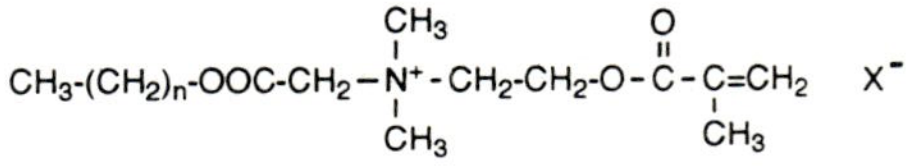

n:= 3, 7, 13

$CH_3-(CH_2)_{13}-CH-N^+-(CH_3)_3$ Br^- [223]
                  |
                  $C=O$
                  |
                  O
                  |
$CH_2=C-COO-CH_2CH_2O$
      |
      CH_3

$CH_3\text{-}(CH_2)_{13}\text{-}CH\text{-}N^+$ Br^-
$\overset{|}{C=O}$
$CH_2=C\text{-}COO\cdot CH_2\cdot CH_2\cdot O$
$\overset{|}{CH_3}$ [223]

$C_{10}H_{21}\text{-}OOC\text{-}(CH_2)_{10}\text{-}N^+\text{-}CH_2\text{-}CH_2\text{-}O\text{-}C\text{-}C=CH_2$ Br^-
with CH_3 groups on N and O / CH_3 [239]

$\text{OOC}\text{-}R\text{-}N^+\text{-}CH_2\text{-}CH_2\text{-}O\text{-}C\text{-}C=CH_2$ Br^-
CH_3 / CH_3 / O / CH_3

$R:=\text{-}CH_2\text{-},\,\text{-}(CH_2)_{10}\text{-},\,\text{-}O\text{-}CH_2\text{-}CH_2\text{-}$ [249]

ACRYLAMIDES AND METHACRYLAMIDES

$CH_3\text{-}(CH_2)_{11}\text{-}N^+\text{-}CH_2\text{-}CH_2\text{-}NH\text{-}C\text{-}CH=CH_2$ Br^-
CH_3 / CH_3 / O [77]

$CH_2=C\text{-}CONH\text{-}(CH_2)_n\text{-}NH\left(\text{C-CH}_2\text{-N}\right)_x\text{-}H$
R / O / CH_3
$R:= H, CH_3$ $x:= 6\text{-}45$ [401]

$CH_3\text{-}(CH_2)_n\text{-}N^+\text{-}CH_2\text{-}CH_2\text{-}CH_2\text{-}NH\text{-}C\text{-}CH=CH_2$ Br^-
CH_3 / CH_3 / O
$n:= 11, 13, 15, 17$ [108, 162, 290]

$CH_2=C\text{-}CONH\text{-}(CH_2)_n\text{-}N^+$
CH_3 with pyridinium-biphenyl-pyridinium $N^+\text{-}CH_3$ $2Br^-$
$n:= 6, 10$ [266]

$CH_3\text{-}(CH_2)_{15}\text{-}N^+\text{-}CH_2\text{-}CH_2\text{-}CH_2\text{-}NH\text{-}C\text{-}CH=CH_2$ Br^-
CH_3 / CH_3 / O [290]

OTHERS

$CH_3\text{-}(CH_2)_{11}\text{-}N^+\text{-}CH_2\text{-}COO\text{-}CH=CH_2$ Br^-
CH_3 / CH_3 [224, 225]

$CH_3\text{-}CH=CH\text{-}CH=CH\text{-}COO\text{-}CH_2\text{-}CH_2$ CH_3 Br^-
$CH_3\text{-}CH=CH\text{-}CH=CH\text{-}COO\cdot CH_2\cdot CH_2$ N^+ CH_3 [313]

Continued

OTHERS

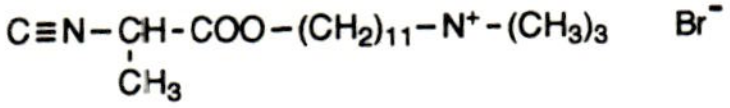

C≡N-CH-COO-(CH$_2$)$_{11}$-N$^+$-(CH$_3$)$_3$ Br$^-$ [329]
     |
     CH$_3$

N-(CH$_2$)$_{12}$-N$^+$-(C$_2$H$_5$)$_3$ BF$_4^-$ [271]

CH$_2$=CH-O-(CH$_2$)$_{11}$-N$^+$-CH$_2$-⬡Fe⬡   [269]
                            |
                            CH$_3$ Br$^-$
                            |
                            CH$_3$

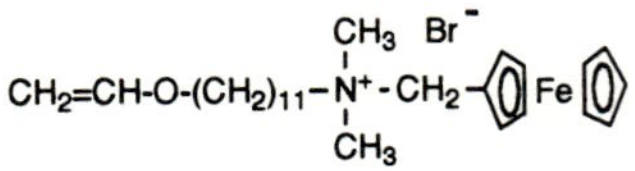

MULTIPOLAR SURFACTANTS ("BOLA"-FORM)

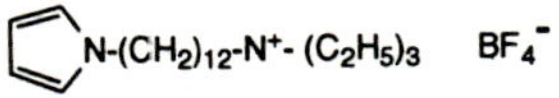

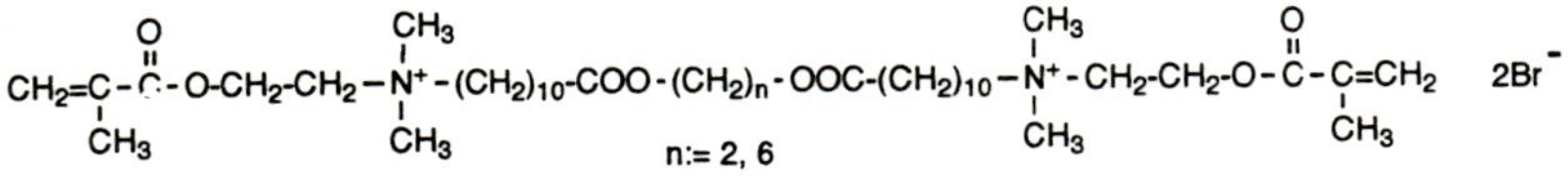

CH$_2$=C-C-O-CH$_2$-CH$_2$-N$^+$-(CH$_2$)$_{10}$-COO-(CH$_2$)$_n$-OOC-(CH$_2$)$_{10}$-N$^+$-CH$_2$-CH$_2$-O-C-C=CH$_2$ 2Br$^-$ [239]
     | ‖ | | ‖ |
     CH$_3$ O   CH$_3$ n:= 2, 6   CH$_3$ O   CH$_3$

CH$_2$=C-C-O-CH$_2$-CH$_2$-N$^+$-(CH$_2$)$_{10}$-COO-CH$_2$ CH-OOC-(CH$_2$)$_{10}$-N$^+$-CH$_2$-CH$_2$-O-C-C=CH$_2$ 2Br$^-$ [239]
     | ‖ | | | ‖ |
     CH$_3$ O   CH$_3$ (CH$_2$)$_9$-CH$_3$ CH$_3$ O   CH$_3$

                         CH$_3$ CH$_3$
                          | |
H$_3$C-N$^+$-(CH$_2$)$_{11}$-OOC-CH=CH-CH=CH-COO-(CH$_2$)$_{11}$-N$^+$-CH$_3$ 2Br$^-$ [499]
      | |
      CH$_3$ CH$_3$

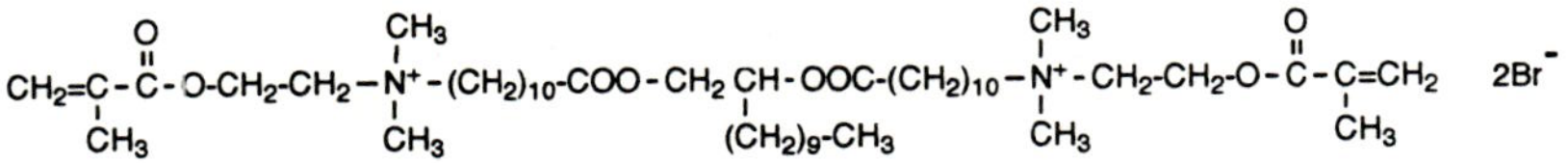

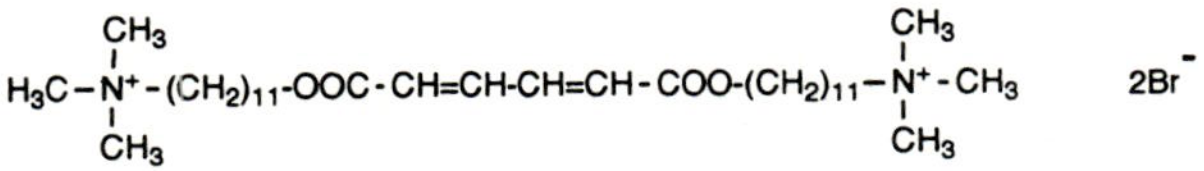

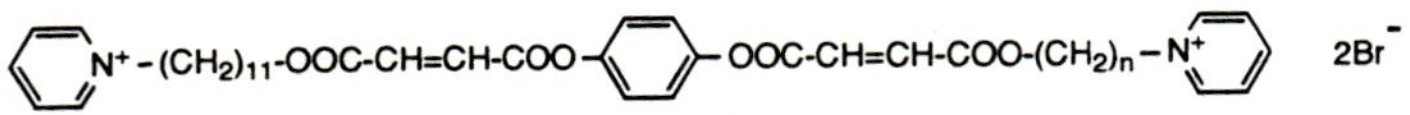

N$^+$-(CH$_2$)$_{11}$-OOC-CH=CH-CH=CH-COO-⬡-OOC-CH$_2$-CH=CH-COO   2Br$^-$ [506]

N$^+$-(CH$_2$)$_n$-OOC-CH=CH-⬡-CH=CH-COO-(CH$_2$)$_n$-N$^+$ 2X$^-$ n:= 6,8, 11   X:= Cl, Br   [330, 425, 506]

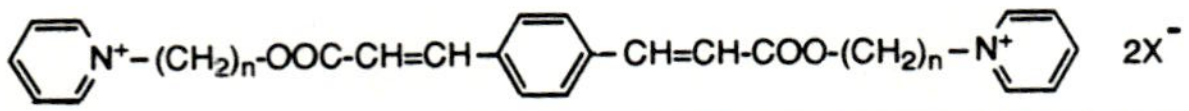

Table 3. Nonionic polymerizable surfactants

STYRENE DERIVATIVES

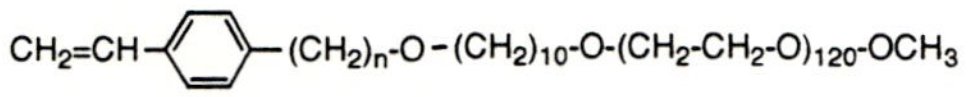

$CH_2=CH-\langle phenyl \rangle-(CH_2)_n-O-(CH_2)_{10}-O-(CH_2-CH_2-O)_{120}-OCH_3$ [101]

n:= 1,2,7

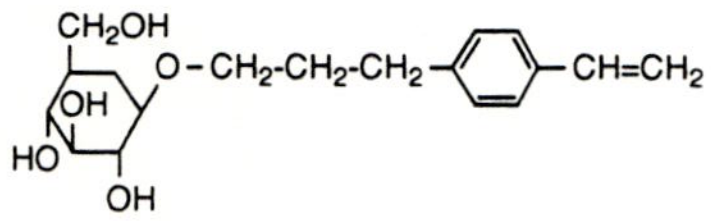

[319, 321, 322]

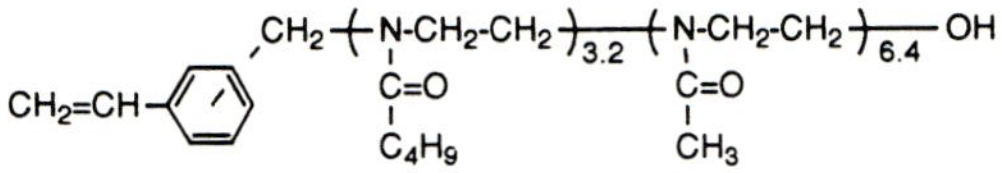

$CH_2=CH-\langle phenyl \rangle \cdots CH_2 \left(N-CH_2-CH_2 \right)_{3.2} \left(N-CH_2-CH_2 \right)_{6.4} OH$
with C=O / C4H9 and C=O / CH3 groups   [474]

$CH_2=CH-\langle phenyl \rangle \cdots CH_2 \left(N-CH_2-CH_2 \right)_{5.8} \left(N-CH_2-CH_2 \right)_{3.2} OH$
with C=O / CH3 and C=O / C4H9 groups   [474]

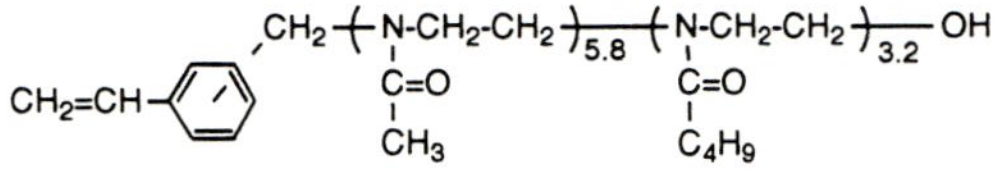

ALLYLICS

$CH_2=CH-(CH_2)_8-COO-(CH_2-CH_2-O)_n-CH_3$ [121, 122]

n:= 4, 6, 8

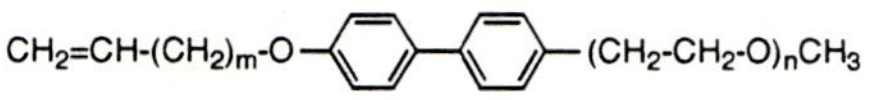

$CH_2=CH-(CH_2)_m-O-\langle biphenyl \rangle-O-(CH_2-CH_2-O)_n-CH_3$ [122–124, 451]

m:= 1   n:= 2,4,6,8,9,11
m:= 2,3,4   n:=9

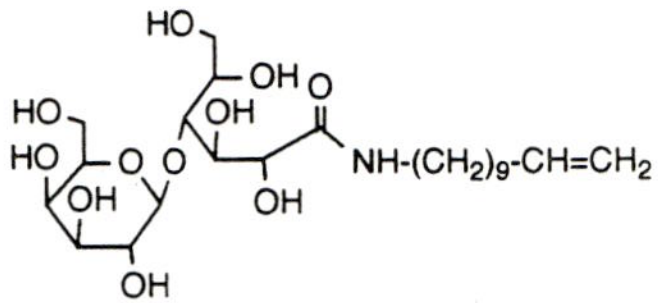

$NH-(CH_2)_9-CH=CH_2$ [111, 112]

ACRYLATES

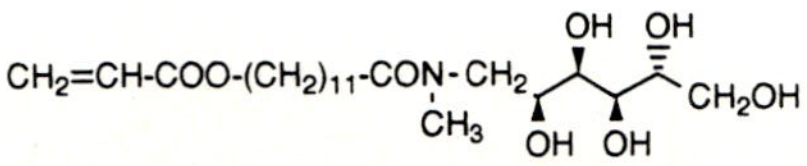

$CH_2=CH-COO-(CH_2)_{11}-CON-CH_2 \cdots CH_2OH$ with CH_3 and OH groups   [300]

$CF_3-(CF_2)_7-S(=O)(=O)-N(C_2H_5)(CH_2-CH_2-O)_{11-14}-OOC-CH=CH_2$ [489]

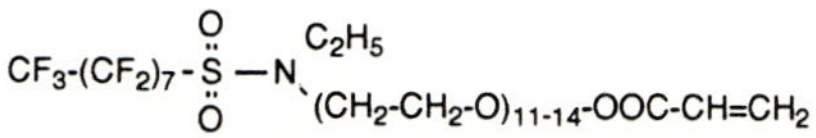

Continued

METHACRYLATES

$CH_2=C-COO-(CH_2)_{11}-O-(CH_2-CH_2-O)_8-CH_3$
    CH_3 [392, 403, 507]

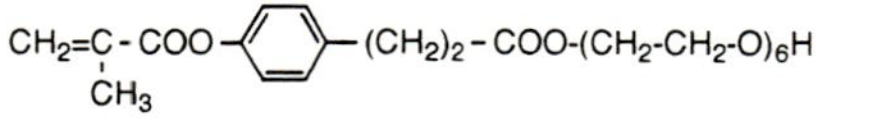

$CH_2=C-COO-(CH_2)_2-COO-(CH_2-CH_2-O)_6H$
    CH_3 [95]

$CH_2=C-COO-(CH_2)_2-COO-(CH_2-CH_2-O)_8-CH_3$
    CH_3 [95]

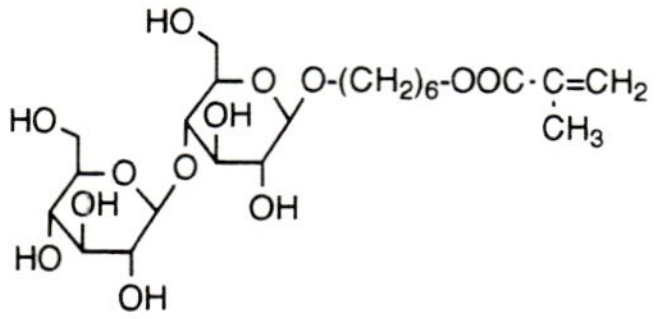

[230, 337]

$O-(CH_2)_6-OOC-C=CH_2$
            CH_3

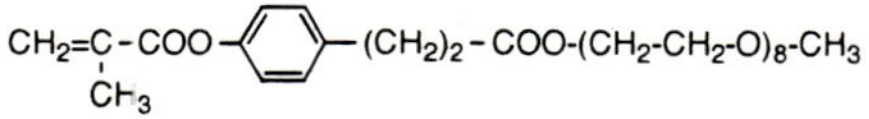

ACRYLAMIDES

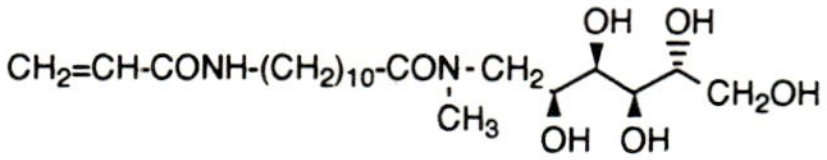

$CH_2=CH-CONH-(CH_2)_{10}-CON-CH_2$
        CH_3 [240]

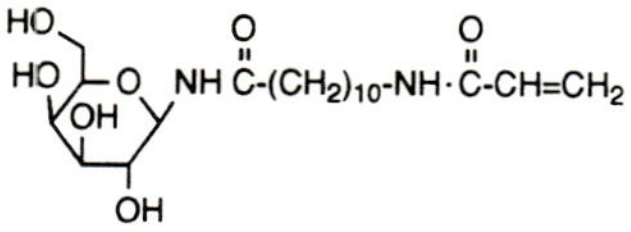

$NH-C-(CH_2)_{10}-NH-C-CH=CH_2$
[240]

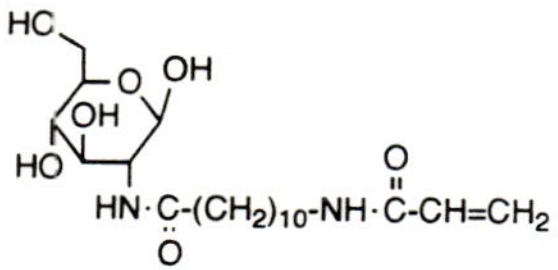

$HN-C-(CH_2)_{10}-NH-C-CH=CH_2$
[240]

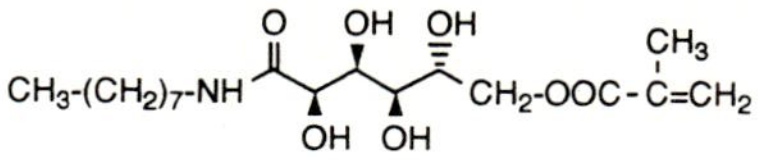

[302]

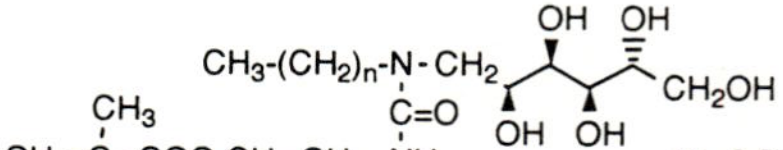

[301]

n:= 3,5,7,9

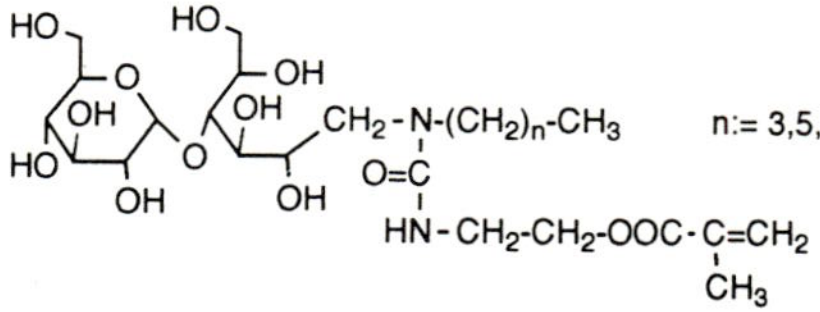

n:= 3,5,7,9   [301]

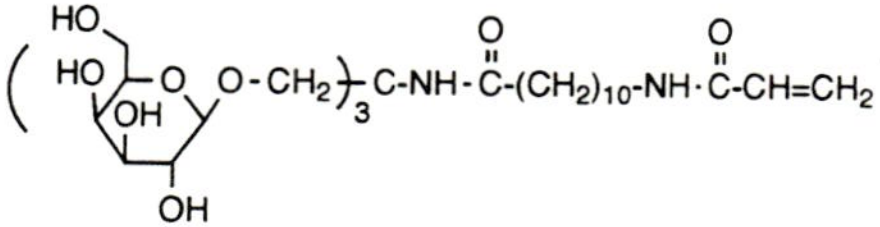

[481]

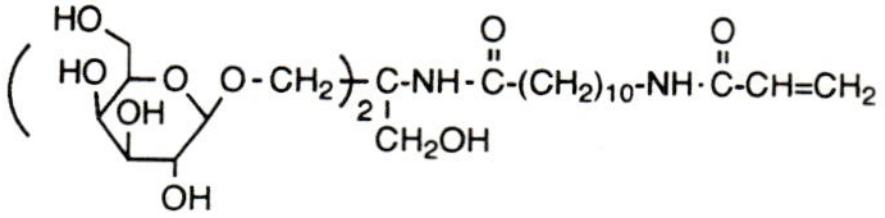

[481]

METHACRYLAMIDES

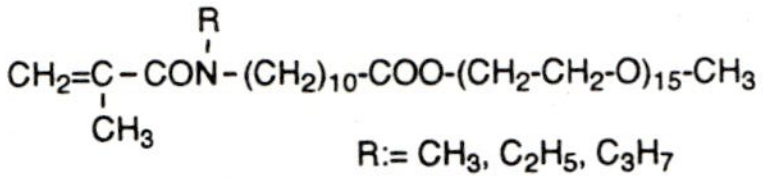
CH2=C-CON-(CH2)10-COO-(CH2-CH2-O)15-CH3
R
CH3
R:= CH3, C2H5, C3H7

[84, 393]

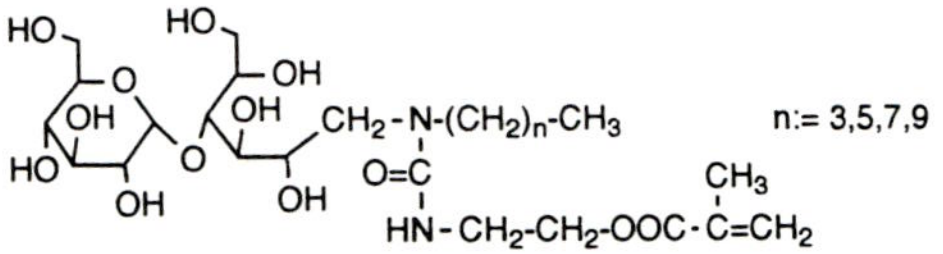
HO
OH
CH2-N-(CH2)n-CH3
O=C
HN-CH2-CH2-OOC-C=CH2
CH3
n:= 3,5,7,9

[301]

OTHERS

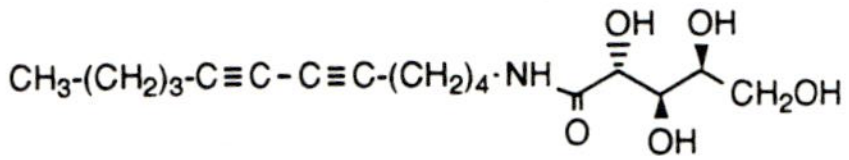
CH3-(CH2)3-C≡C-C≡C-(CH2)4·NH
OH OH
CH2OH
O OH

[250]

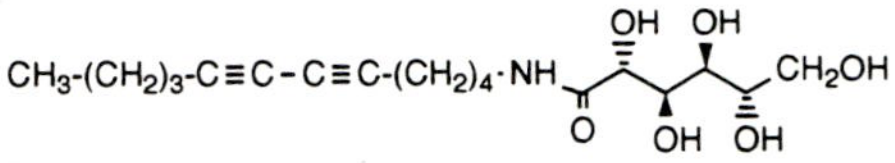
CH3-(CH2)3-C≡C-C≡C-(CH2)4·NH
OH OH
CH2OH
O OH OH

[246]

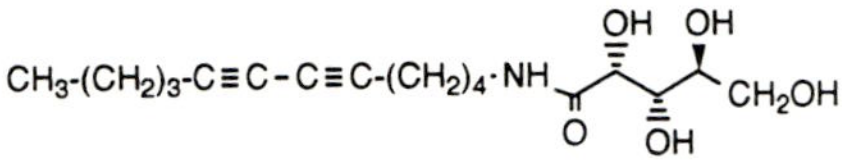
CH3-(CH2)3-C≡C-C≡C-(CH2)4·NH
OH OH
CH2OH
O OH

250]

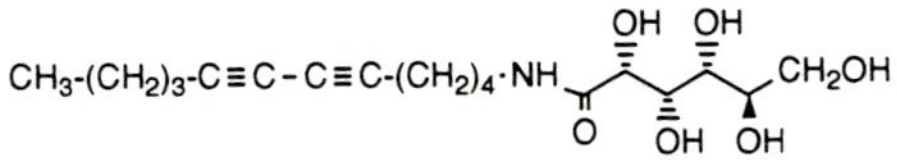
CH3-(CH2)3-C≡C-C≡C-(CH2)4·NH
OH OH
CH2OH
O OH OH

[246]

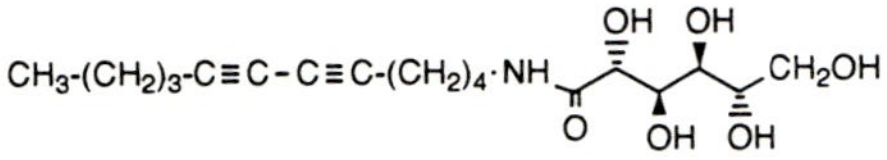
CH3-(CH2)3-C≡C-C≡C-(CH2)4·NH
OH OH
CH2OH
O OH OH

[246, 250]

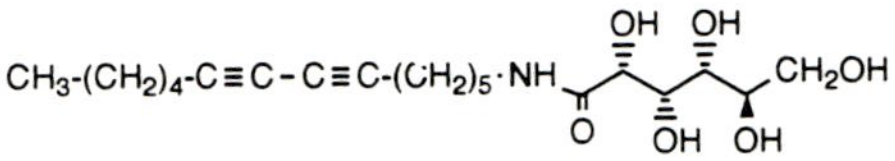
CH3-(CH2)4-C≡C-C≡C-(CH2)5·NH
OH OH
CH2OH
O OH OH

[247]

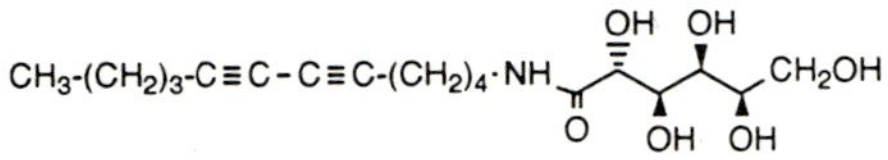
CH3-(CH2)3-C≡C-C≡C-(CH2)4·NH
OH OH
CH2OH
O OH OH

[246, 250]

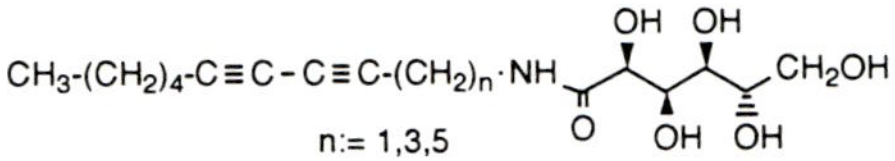
CH3-(CH2)4-C≡C-C≡C-(CH2)n·NH
OH OH
CH2OH
O OH OH
n:= 1,3,5

[247]

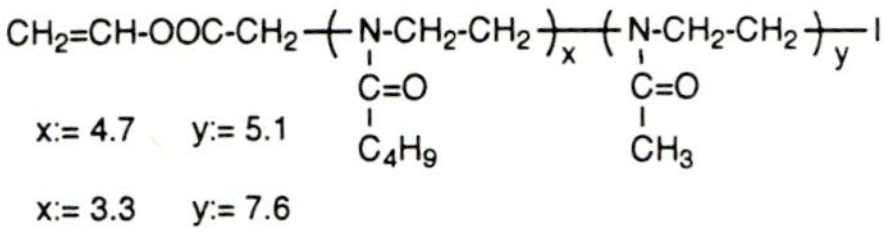
CH2=CH-OOC-CH2-(N-CH2-CH2)x-(N-CH2-CH2)y-I
C=O
C4H9
C=O
CH3
x:= 4.7 y:= 5.1
x:= 3.3 y:= 7.6

[473]

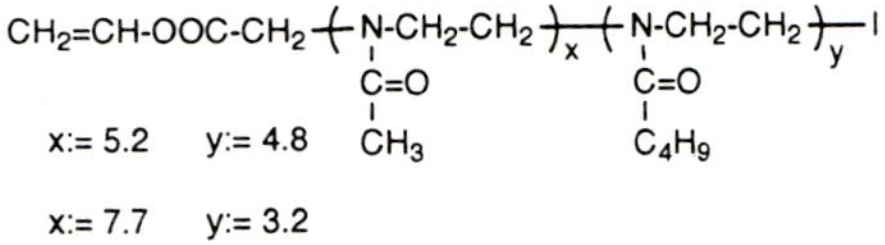
CH2=CH-OOC-CH2-(N-CH2-CH2)x-(N-CH2-CH2)y-I
C=O
CH3
C=O
C4H9
x:= 5.2 y:= 4.8
x:= 7.7 y:= 3.2

[473]

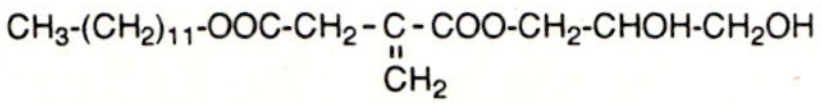
CH3-(CH2)11-OOC-CH2-C-COO-CH2-CHOH-CH2OH
CH2

[404]

Table 4. Zwitterionic polymerizable surfactants

STYRENE DERIVATIVES

CH₂=CH—⟨C₆H₄⟩—(CH₂)₅-O-P(=O)(O⁻)-O-CH₂-CH₂-N⁺(CH₃)₃   [465]

CH₂=CH—⟨C₆H₄⟩—(CH₂)₄-COO-CH₂-CH-CH₂-O-P(=O)(O⁻)-O-CH₂-CH₂-N⁺(CH₃)₃
  |
  OH   [468]

CH₃-(CH₂)₁₄-COO-CH—
CH₂=CH—⟨C₆H₄⟩—(CH₂)₄-COO-CH-CH₂-O-P(=O)(O⁻)-O-CH₂-CH₂-N⁺(CH₃)₃   [465, 468]

(CH₂)₄-COO-CH₂
CH₂=CH—⟨C₆H₄⟩—(CH₂)₄-COO-CH-CH₂-O-P(=O)(O⁻)-O-CH₂-CH₂-N⁺(CH₃)₃   [465, 468]

CH₃
|
CH₃-(CH₂)₉-N⁺—(CH₂)₃-SO₃⁻
|
CH₂
|
⟨C₆H₄⟩-CH=CH₂   [331]

ALLYLICS

CH₃
|
CH₂=CH-(CH₂)₉-N⁺—(CH₂)₃-SO₃⁻
|
CH₃   [164, 343]

CH₂CH=CH₂
|
CH₃-(CH₂)ₙ-N⁺-(CH₂)₃-SO₃⁻
|
CH₂CH=CH₂   [168, 331]

O   CH₃
‖ |
CH₃(CH₂)₉-N-C-CH₂-N⁺-(CH₂)₃-SO₃⁻
| |
CH₂   CH₃
|
CH₂=CH   [168]

CH₃
|
CH₂=CH-(CH₂)₈-C(=O)-N⟨ ⟩N⁺-(CH₂)₃-SO₃⁻   [164, 343]

ACRYLATES

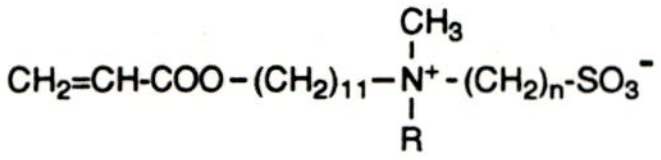

CH₃ and R on nitrogen:

$$CH_2=CH-COO-(CH_2)_{11}-\overset{CH_3}{\underset{R}{N^+}}-(CH_2)_n-SO_3^-$$

R:= -CH₃, C₄H₉   n:= 3,4

[227, 245, 331, 343, 406, 407]

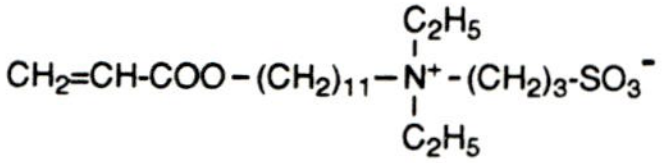

$$CH_2=CH-COO-(CH_2)_{11}-\overset{C_2H_5}{\underset{C_2H_5}{N^+}}-(CH_2)_3-SO_3^-$$

[331]

$$CH_3-(CH_2)_9-\overset{CH_3}{\underset{CH_2-CH_2-OOC-CH=CH_2}{N^+}}-(CH_2)_3-SO_3^-$$

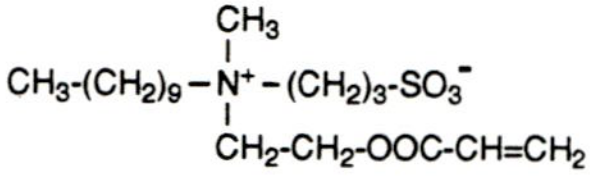

[227, 245, 331, 406]

METHACRYLATES

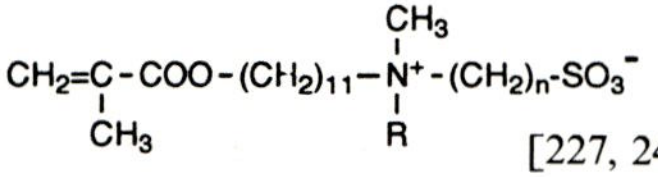

$$CH_2=\overset{CH_3}{\underset{\,}{C}}-COO-(CH_2)_{11}-\overset{CH_3}{\underset{R}{N^+}}-(CH_2)_n-SO_3^-$$

R:= -CH₃, C₄H₉   n:= 3,4

[227, 245, 267, 268, 299, 331, 343, 406, 407]

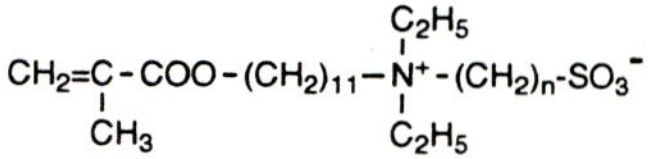

$$CH_2=\overset{CH_3}{\underset{\,}{C}}-COO-(CH_2)_{10}-OOC-\langle C_6H_4\rangle-N^+-(CH_2)_3-SO_3^-$$

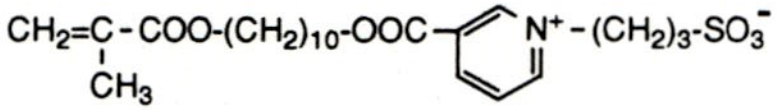

[227, 245, 267, 268, 343]

$$CH_2=\overset{CH_3}{\underset{\,}{C}}-COO-(CH_2)_{11}-\overset{C_2H_5}{\underset{C_2H_5}{N^+}}-(CH_2)_3-SO_3^-$$

[331]

$$CH_3-(CH_2)_9-\overset{CH_3}{\underset{CH_2-CH_2-OOC-\underset{CH_3}{C}=CH_2}{N^+}}-(CH_2)_3-SO_3^-$$

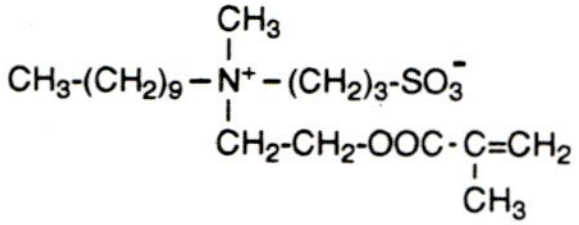

[167, 227, 232, 245, 331, 406, 407]

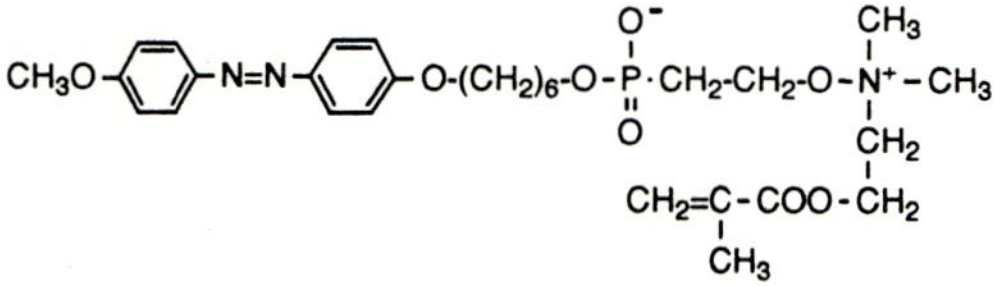

$$CH_3-(CH_2)_9-N-\overset{O}{C}-\langle C_6H_4\rangle-N^+-(CH_2)_3-SO_3^-$$
$$\underset{CH_3}{CH_2=C-COO-CH_2-CH_2}$$

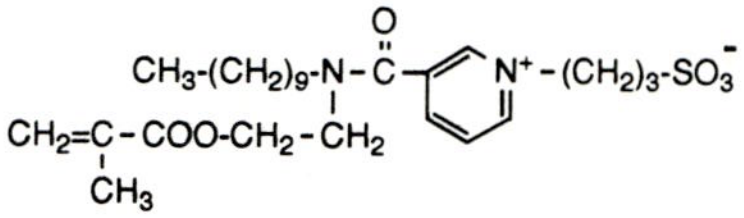

[227, 245, 248, 267]

$$CH_3O-\langle C_6H_4\rangle-N=N-\langle C_6H_4\rangle-O-(CH_2)_6-O-\overset{O^-}{\underset{O}{P}}=O-CH_2CH_2O-N^+-\underset{CH_3}{\overset{CH_3}{\underset{CH_2}{C}}}$$
$$\underset{CH_3}{CH_2=C-COO-CH_2}$$

[262]

Continued

ACRYLAMIDES AND METHACRYLAMIDES

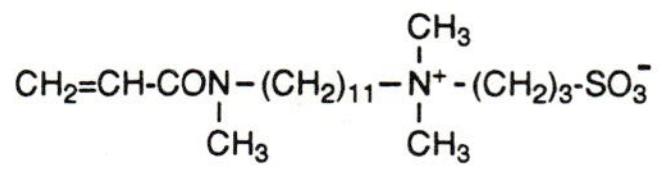

$$CH_2=CH-CON-(CH_2)_{11}-N^+-(CH_2)_3-SO_3^-$$
with CH_3 groups on the N of the amide and ammonium
[167, 331, 343]

$$CH_2=CH-CON-(CH_2)_{11}-N^+-(CH_2)_3-SO_3^-$$
with C_2H_5, C_2H_5 on ammonium and CH_3 on amide
[331]

$$CH_3(CH_2)_9-N^+-CH_2-CH_2-N^+-(CH_2)_3-SO_3^-$$
$C=O$, $CH_2=CH$; CH_3 groups
[167, 168, 227, 245, 343]

$$CH_3-(CH_2)_9-N^+-(CH_2)_3-SO_3^-$$
$CH_2-CH_2-NOC-CH=CH_2$; CH_3 groups
[167, 227, 245, 331]

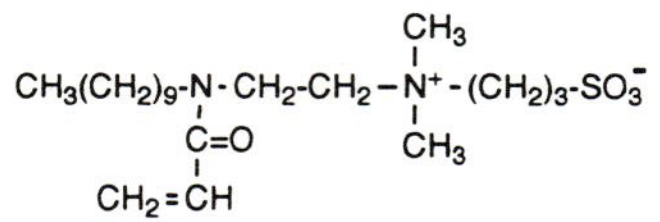

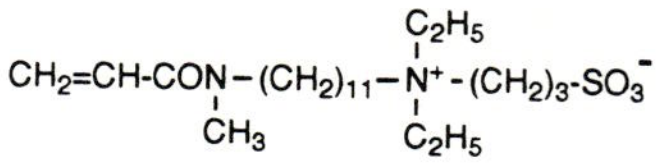

$$CH_2=C-CON-(CH_2)_{11}-N^+-(CH_2)_3-SO_3^-$$
CH_3, CH_3 groups
[331]

$$CH_2=CH-CONH-(CH_2)_{10}-NH-CO-\underset{}{\bigcirc}-N^+-(CH_2)_3-SO_3^-$$
[227, 245, 267, 268]

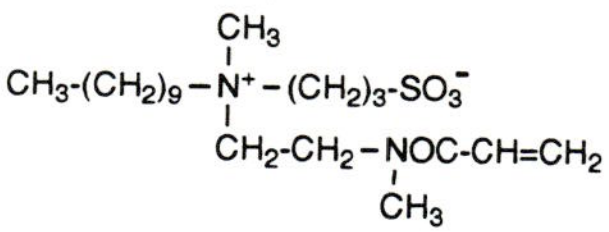

OTHERS

$$CH_3-(CH_2)_9-N^+-(CH_2)_3-SO_3^-$$
$CH_2-COO-CH=CH_2$; CH_3 group
[232]

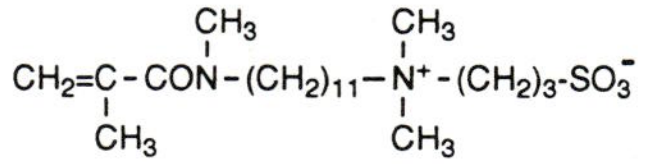

$$CH_3-(CH_2)_9-OOC-CH=CH-COO-CH_2-CH_2-N^+-(CH_2)_3-SO_3^-$$
CH_3, CH_3 groups
[168, 331]

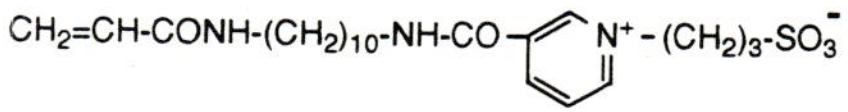

$$CH_3-(CH_2)_9-N^+-(CH_2)_3-SO_3^-$$
$CH_2-CH_2-O-CH=CH_2$; CH_3 group
[168, 331]

$$CH_3-(CH_2)_8-CH=CH-CH=CH-C-N-CH_2-CH_2-N^+-(CH_2)_3-SO_3^-$$
$O=$; CH_3, CH_3 groups
[168, 331]

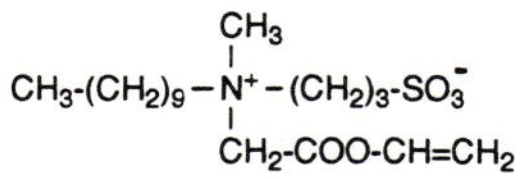

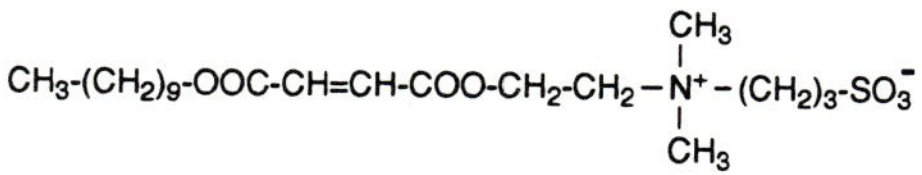

$$CH_3(CH_2)_9-N-CH_2-CH_2-N^+-(CH_2)_3-SO_3^-$$
$C=O$, $CH_2=CH-CH=CH$; CH_3 group
[168, 331]

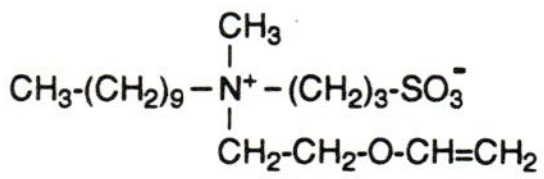

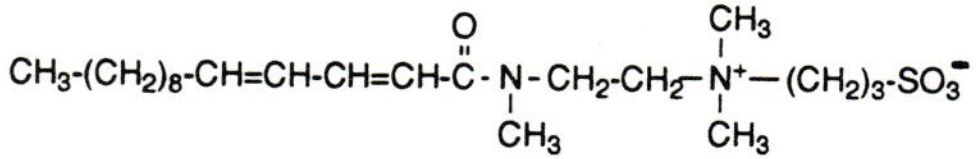

$$CH_3-(CH_2)_8-CH=CH-CH=CH-C-N^+-(CH_2)_3-SO_3^-$$
$O=$; piperidine ring with CH_3
[168, 331]

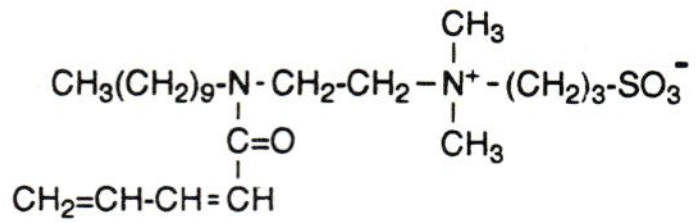

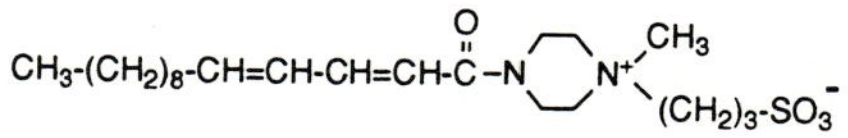

$$CH_3-(CH_2)_9-N-CO-CH=CH-\overset{}{\underset{}{C}}-N^+-(CH_2)_3-SO_3^-$$
[268]

$$CH_3-(CH_2)_9-OOC-\overset{C}{\underset{}{\diagdown}}-COO-CH_2-CH_2-N^+-(CH_2)_3-SO_3^-$$
CH_3, CH_3 groups
[168, 331]

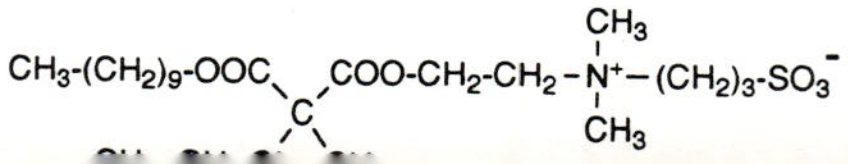

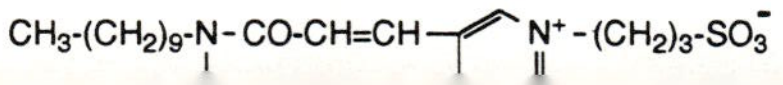

Table 5. Polymerizable surfactants with spacer groups

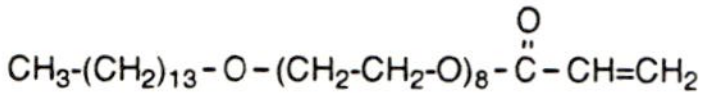

$CH_3-(CH_2)_{13}-O-(CH_2-CH_2-O)_{18}-\overset{O}{\overset{||}{C}}-C-CH=CH_2$ [231]

$CH_3-(CH_2)_n-O-(CH_2-CH_2-O)_x-\overset{O}{\overset{||}{C}}-C-CH=CH_2$ [251]

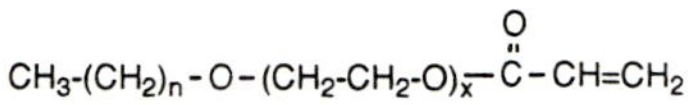

n:= 9,11   x:= 8
n:= 11-13;   x:=14,20,30
n:= 15-17   x:= 18
n:= 17   x:= 30,40

$CH_3-(CH_2)_n-O-(CH_2-CH_2-O)_x-\overset{O}{\overset{||}{C}}-\underset{CH_3}{C}=C=CH_2$ [102, 460]

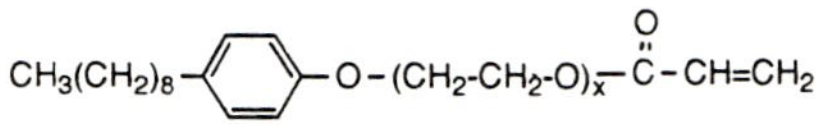

n:= 11   x:= 14,19
n:= 17   x:= 11, 23

$CH_3-(CH_2)_n-COO-(CH_2-CH_2-O)_x-\overset{O}{\overset{||}{C}}-\underset{CH_3}{C}=C=CH_2$ [223, 251]

n:= 9,11,13,15,17   x:= 8-9

$CH_3-(CH_2)_n-O-(CH_2-CH_2-O)_x-CH_2-\langle benzene\rangle-CH=CH_2$ [100, 102]

n:= 7   x:= 43
n:= 11   x:= 13
n:= 17   x:= 35

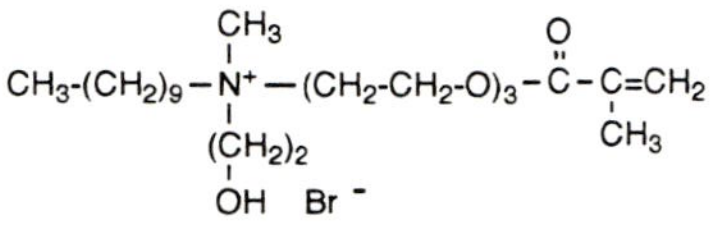

$CH_3(CH_2)_8-\langle benzene\rangle-O-(CH_2-CH_2-O)_x-\overset{O}{\overset{||}{C}}-C-CH=CH_2$ [157]

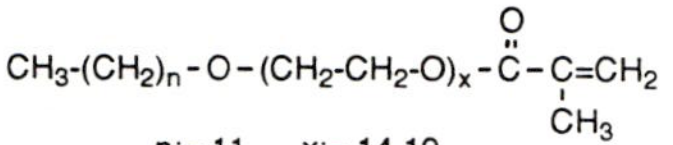

x:= 10,20,40

$CH_3-(CH_2)_9-N^+ \underset{\underset{OH \quad Br^-}{(-CH_2)_2}}{\overset{CH_3}{|}}-(CH_2-CH_2-O)_3-\overset{O}{\overset{||}{C}}-\underset{CH_3}{C}=CH_2$ [232]

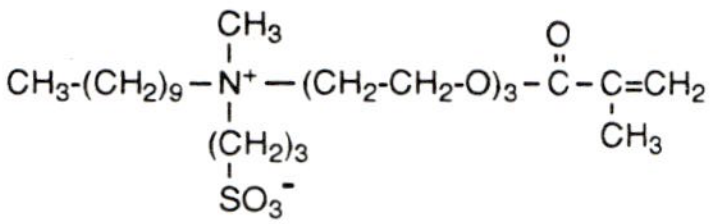

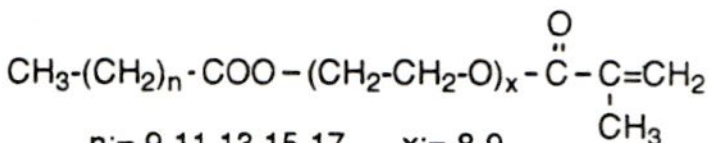

$CH_3-(CH_2)_9-N^+ \underset{\underset{SO_3^-}{(CH_2)_3}}{\overset{CH_3}{|}}-(CH_2-CH_2-O)_3-\overset{O}{\overset{||}{C}}-\underset{CH_3}{C}=CH_2$ [232]

$CH_3-(CH_2)_9-N^+ \underset{\underset{SO_3^-}{(CH_2)_3}}{\overset{CH_3}{|}}-CH_2-CHOH-CH_2-O-CH_2-CH=CH_2$ [164]

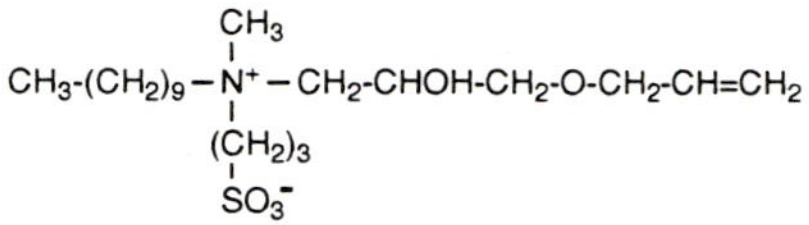

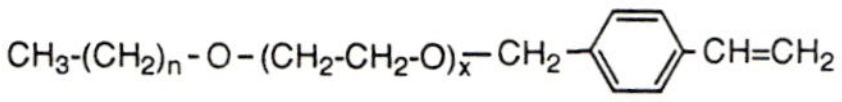

$CH_3(CH_2)_6-O-\langle benzene\rangle-\overset{O}{\underset{O}{\overset{||}{S}}}-N \underset{(CH_2)_3-N^+-(CH_2)_3-SO_3^-}{\overset{CH_3}{|} \quad \overset{CH_2-CH_2-COO(CH_2-CH_2-O)_4-C-CH=CH_2}{|}}$ [232]

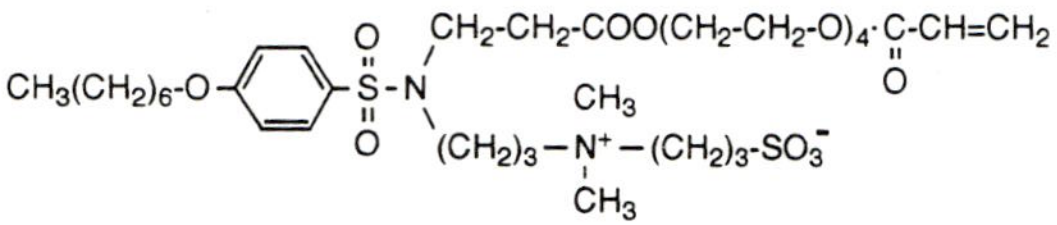

Within other applications, polysoaps have been studied as protective coatings [489, 490], photoswitchable systems [205], and proposed for chemical muscles [75].

8 Survey on Polymerizable Surfactants

As pointed out in Sect. 2.2, the polysoaps with the best defined molecular structure are produced by the polymerization of reactive surfactants. Because a known structure is a basic requirement, such reactive surfactants play a crucial role for model studies of polysoaps. The following tables 1–5 list a number of polymerizable surfactants to give an idea of their present variety. The tables are by no means comprehensive as, for example, patents are omitted. It should be realized as well that many potentially useful compound are reported in the literature (namely unsaturated long chain carboxylic acids) without considering their use as reactive surfactants so far.

9 Conclusions

Like all water-soluble polymers, polysoaps are intriguing substances from the scientific and the practical point of view. Having improved our knowledge of polysoaps considerably, the work of the past two decades has revealed some general relationships between the molecular architecture and key properties in aqueous solutions such as viscosity, surface activity and solubilization which cover much of the present data. These relationships provide useful guidelines for the application of polysoaps and for the synthesis of new types of polysoaps. With the uprise of water-based polymer technology, polysoaps and related systems may offer a great practical potential in the future. Nevertheless, a number of problems must still be resolved. As systematic structural investigations have just started, many novel aspects of polysoap systems are expected from future work.

References

1. Tanford C (1973) The hydrophobic effect. Wiley, New York
2. Evans DF (1988) Langmuir 4: 3
3. Blockzijl W, Engberts JBFN (1993) Angew Chem Int Eng Ed 32: 1545
4. Glass JE (ed) (1989) Polymers in aqueous media. Adv chemistry series 223, Am Chem Soc, Washington DC

5. El-Nokaly MA (ed) (1989) Polymer association structures, ACS Symposium Ser. 384, Am Chem Soc, Washington DC
6. Tomalia DA, Naylor AM, Goddard III WA (1990) Angew Chem Int Eng Ed 29: 138
7. Bader H, Dorn K, Hupfer H, Ringsdorf H (1985) Adv Polym Sci 64: 1
8. Ringsdorf H, Schlarb B, Venzmer J (1988) Angew Chem Int Eng Ed 27: 113
9. O'Brien DF, Liman U (1989) Angew Makromol Chem 166/167: 21
10. O'Brien DF, Ramaswami V (1989) in: Kroschwitz JI (ed) Encyclopedia of Polymer Science and Technology, Vol. 11, 2nd edn. Wiley-Interscience, New York, p 108
11. Paleos C (1990) J Macromol Sci Rev C30: 379
12. Tieke B (1985) Adv Polym Sci 71: 129
13. Embs F, Funhoff D, Laschewsky A, Licht U, Ohst H, Prass W, Ringsdorf H, Wegner G, Wehrmann R (1991) Adv Mater 3: 25
14. Arslanov VV (1992) Adv Coll Interface Sci 40: 307
15. Miyashita T (1993) Prog Polym Sci 18: 263
16. Shimomura M (1993) Prog Polym Sci 18: 295
17. Kawaguchi M (1993) Prog Polym Sci 18: 341
18. Lindmann B, Wennerström H (1980) Topics in Current Chem 87: 1
19. Hoffmann H (1984) Ber Bunsenges Phys Chem 88: 1078
20. Zana R (1986) J Chim Phys Chim Biol 83: 603
21. Falbe J (ed) (1987) Surfactants in Consumer Products. Springer, Berlin Heidelberg New York
22. Charvolin J, Sadoc JF (1988) J Phys Chem 92: 5787
23. Chevalier Y, Zemb T (1990) Rep Prog Phys 53: 279
24. Bekturov EA, Bakauova ZKh (1986) Synthetic Water-Soluble Polymers in Solution. Hütig & Wepf, Basel
25. Schmolka IR (1977) J Am Oil Chem Soc 59: 110
26. Hancock RI (1984) in: Tadros TF (ed) Surfactants. Academic Press, New York, p 287
27. Wanka G, Hoffmann H, Ulbricht W (1990) Colloid Polym Sci 268: 101
28. Brown W, Schillén K, Almgren M, Hvidt S, Bahadur P (1991) J Phys Chem 95: 1850
29. Deng Y, Ding J, Stubbersfield RB, Heatley F, Attwood D, Price C, Booth C (1992) Polymer 33: 1963
30. Hurter PN, Scheutjens JMHM, Hatton TA (1993) Macromolecules 26: 5592
31. Astafieva I, Zhang XF, Eisenberg A (1993) Macromolecules 26: 7339
32. Sawamoto M (1993) Trends in Polym Sci 1: 111
33. Kanaoka S, Omura T, Sawamoto M, Higashimura T (1992) Macromolecules 25: 6407
34. Rein D, Lamps JP, Rempp P, Lutz P, Papanagopoulos D, Tsitsilianis C (1993) Acta Polym 44: 225
35. Geeraert JM, Toncheva V, Goethals EJ (1993) Frontiers in Polymerization Symposium, Liège (Belgium) 6-8. Oct.
36. Seagusa T, Yamada A, Kobayashi S (1978) Polymer J 11: 53
37. Chujo Y, Shishino T, Tsukahara Y, Yamashita Y (1985) Polymer J 17: 133
38. Wesslén B, Wesslén KB (1992) J Polym Sci Polym Chem Ed A30: 355
39. Shoda SI, Masuda E, Furukawa M, Kobayashi S (1992) J Polym Sci Polym Chem Ed A30: 1489
40. Bayer U, Stadler R, Schmidt M (1993) Polym Prepr Am Chem Soc Div Polym Chem 34(2): 572
41. Kim YH, Webster OW (1990) J Am Chem Soc 112: 4592
42. Turro NJ, Barton JK, Tomalia DA (1991) Acc Chem Res 24: 332
43. Newkome GR, Moorefield CN, Baker GR, Saunders MJ, Grossmann SH (1991) Angew Chem 103: 1207 (Int Eng Ed 30: 1178)
44. Hawker CJ, Wooley KL, Frechet JMJ (1993) J Chem Soc Perkin Trans 1, 1287
45. Newkome GR, Moorefield CN (1994) Macromol Symp 77: 63
46. Strauss UP, Jackson EG (1951) J Polym Sci 5: 649
47. Jackson EG, Strauss UP (1951) J Polym Sci 5: 473
48. Layton LH, Jackson EG, Strauss UP (1952) J Polym Sci 9: 295
49. Strauss UP, Gershfeld NL (1954) J Phys Chem 58: 747
50. Medalia AI, Freedman HH, Sinha S (1959) J Polym Sci 40: 15
51. Sinha SK, Medalia AI (1957) J Am Chem Soc 79: 281
52. Woermann D, Wall FT (1960) J Phys Chem 64: 581
53. Ito K, Ono H, Yamashita Y (1964) J Coll Sci 19: 28
54. Dubin PL, Strauss UP (1975) in: Rembaum A, Sélégny E (eds) Polyelectrolytes and Their Applications. Reidel D Publishers, Dordrecht (NL), p 3

55. Strauss UP (1989) in: Glass JE (ed) Polymers in Aqueous Media. Adv Chemistry Series 223, Am Chem Soc, Washington DC, p 317
56. Kunitake T, Nakashima H, Takarabe K, Nagai M, Tsuge A, Yanagi H (1981) J Am Chem Soc 103: 5945
57. Soldi V, De Magalhaes Erismann N, Quina FH (1988) J Am Chem Soc 110: 5137
58. Nakaya T, Yasuzawa M, Imoto M (1989) Macromolecules 22: 3180
59. Kokkinia A, Paleos CM, Dais P (1989) Polym Prepr Am Chem Soc Div Polym Chem 30 (2): 448
60. Penzcek S, Pretula J (1993) Macromolecules 26: 2228
61. Larrabee jr CE, Sprague ED (1979) J Polym Sci Polym, Lett Ed 17: 749
62. Friberg SE, Thundatil R, Stoffer JO (1979) Science 205: 607
63. Thundatil R, Stoffer JO, Friberg SE (1980) J Polym Sci Polym Chem Ed A18: 2629
64. Sprague ED, Duecker DC, Larrabee jr CE (1981) J Am Chem Soc 103: 6797
65. Paleos CM, Stassinopoulou CI, Malliaris A (!983) J Phys Chem 87: 251
66. Arai K, Maseki Y, Ogiwara Y (1987) Makromol Chem Rapid Commun 8: 563
67. Arai K, Sugita J, Ogiwara Y (1987) Makromol Chem 188: 2511
68. Durairaj B, Blum FD (1990) Langmuir 5: 370
69. Gambogi RJ, Blum FD (1990) J Colloid Interface Sci 140: 525
70. Shibasaki Y, Fukuda K (1992) Colloids Surfaces 67: 195
71. Arai K, Okabe T (1992) Polymer J 24: 769
72. Boyer B, Lamaty G, Leydet A, Roque JP, Sama P (1992) New J Chem 16: 883
73. Arai K, Yamazaki K (1993) Polymer J 25: 1169
74. Denton JM, Duecker DC, Sprague ED (1993) J Phys Chem 97: 756
75. Strauss UP, Gershfeld NL, Crook EH (1956) J Phys Chem 60: 577
76. Hamid S, Sherrington DC (1987) Polymer 28: 332
77. a) Chang Y, McCormick CL (1993) Macromolecules 26: 6121
    b) McCormick CL, Chang Y (1994) Macromolecules 27: 2151
78. Anton P, Köberle P, Laschewsky A (1993) Makromol Chem 194: 1
79. Zuckerberg A, Diver A, Peeri Z, Gutnick DL, Rosenberg E (1979) Appl Env Microbiol 37: 414
80. Cooper DG, McDonald CR, Duff SJB, Kosaric N (1981) Appl Env Microbiol 40: 408 (1981)
81. Roe ET, Swern D (1953) J Am Chem Soc 75: 5479
82. Bistline jr RG, Stirton AJ, Weil JK, Port WS (1956) J Am Oil Chem Soc 33: 44
83. Freedman HH, Mason JP, Medalia AI (1958) J Org Chem 23: 76
84. Kammer U, Elias HG (1972) Kolloid Z u Z Polymere 250: 344
85. Salamone JC, Isreal SC, Snider B, Taylor P, Raia D (1972) Polym Prepr Am Chem Soc Div Polym Chem 13 (1): 271
86. Salamone JC, Israel SC, Taylor P, Snider B (1973) Polymer 14: 639
87. Finkelmann H, Rehage G (1984) Adv Polym Sci 60/61: 1
88. Hamid S, Sherrington DC (1986) J Chem Soc Chem Commun 936
89. Hamid S, Sherrington DC (1987) Polymer 28: 325
90. Paleos C, Malliaris A (1990) J Macromol Sci Rev C28: 403
91. Malyukova YeB, Naumova SV, Gritskova IA, Bondarev AN, Zubov VP (1991) Vysokomol Soyed A33: 1469 (English translation: Polym Sci USSR 33: 1361)
92. Yegorov VV, Zaitsev SYu, Zubov VP (1991) Vysokomol Soyed A33: 1587 (English translation: Polym Sci USSR 33: 1475)
93. Guyot A, Tauer K (1994) Adv Polym Sci 111: 43
94. Ringsdorf H, Thunig D (1977) Makromol Chem 178: 2205
95. Siol W (1979) Mizellbildende Monomere- Synthese und Untersuchung des Polymerisationsverhaltens. Thesis, Universität Mainz (Germany)
96. Batrakova EV, Orlov YuN, Egorov VV, Zubov VP, Kabanov VA (1985) Vysokomol Soyed B27: 87
97. Egorov V (1990) Makromol Chem Macromol Symp 31: 157
98. Yeoh KW, Chew CH, Gan LM, Koh LL, Teo HH (1989) J Macromol Sci Chem A26: 663
99. Yeoh KW, Chew CH, GAn LM, Koh LL (1990) J Macromol Sci Chem A27: 63
100. Ito K, Tanaka K, Tanaka H, Imai G, Kawaguchi S, Itsuno S (1991) Macromolecules 24: 2348
101. Chao D, Itsuno S, Ito K (1991) Polymer J 23: 1045
102. Ito K, Kobayashi H (1992) Polymer J 24: 199
103. Cochin D, Zana R, Candau F (1993) Polymer Int 30: 491
104. Cochin D, Zana R, Candau F (1993) Macromolecules 26: 5765
105. Dais P, Paleos CM, Nika G, Malliaris A (1993) Makromol Chem 194: 445
106. Paleos CM, Dais P, Malliaris A (1984) J Polym Sci Polym Chem Ed A22: 3383

107. Shapiro YuE, Orlov YuN, Egorov VV, Zubov VP (1984) Vestn Mosk Univ Ser 2: Khim 25: 595
108. Peiffer DG (1990) J Polym Sci Polym Chem Ed A28: 619
109. Arai K, Miyahara S (1990) Makromol Chem 191: 2647
110. Arai K, Miyahara S, Okabe T (1991) Makromol Chem 192: 2183
111. André B, Boyer B, Lamaty G, Leydet A, Roque JP (1991) Tetr Lett 32: 1881
112. Boyer B, Durand S, Lamaty G, Moussamou-Missima JM, Pavia AA, Pucci B, Roque JP, Rouvière J (1991) J Chem Soc Perkin Trans 2: 1311
113. Boyer B, Lamaty G, Moussamou-Missima JM, Pavia AA, Pucci B, Roque JP (1991) Tetr Lett 32: 1191
114. Nagai K, Satoh H, Kuramoto N (1992) Polymer 33: 5303
115. Tribet C, Gaboriaud R, Lelievre J (1992) Polymer Int 29: 79
116. Esumi K, Nakao T, Ito S (1993) J Coll Interface Sci 156: 256
117. Arai K, Yamazaki K (1993) Polymer J 25: 1169
118. Nagai K, Ohishi Y (1994) J Polym Sci Polym Chem Ed A32: 445
119. Sonessa AJ, Cullen W, Ander P (1980) Macromolecules 13: 195
120. Knapick EG, Hirsch JA, Ander P (1985) Macromolecules 18: 1015
121. Finkelmann H, Lühmann B, Rehage G (1982) Colloid Polym Sci 260: 56
122. Finkelmann H, Lühmann B, Rehage G (1984) Angew Makromol Chem 123/124: 217
123. Finkelmann H, Lühmann B, Rehage G (1985) Makromol Chem 186: 1059
124. Lühmann B, Finkelmann H (1987) Colloid Polym Sci 265: 506
125. Luca C, Avram E, Petrariu I (1988) J Macromol Sci Chem A25: 345
126. Finkelmann H, Jahns E (1989) in El-Nokaly MA (ed) Polymer Association Structures, ACS Symposium Ser. 384, Am Chem Soc, Washington DC, p 1
127. Paleos CM, Tsiourvas D, Anastassopoulu J, Theophanides T (1992) Polymer 33: 4047
128. Tsiourvas D, Paleos CM, Malliaris A (1993) J Polym Sci Polym Chem Ed A31: 387
129. a) Geckeler KE, Zhou R (1994) Macromol Rapid Commun 15: 55
     b) Palmer V, Zhou R, Geckeler KE (1994) Angew Makromol Chem 215: 175
130. Strauss UP, Assony SJ, Jackson EG, Layton LH (1952) J Polym Sci 9: 509
131. Jorgensen HE, Strauss UP (1961) J Phys Chem 65: 1873
132. Nakagawa T, Inoue H (1964) Kolloid Z u Z Polymere 195: 93
133. Inoue H (1964) Kolloid Z u Z Polymere 195: 102
134. Inoue H (1965) Kolloid Z u Z Polymere 196: 1
135. Klotz IM, Royer GP, Sloniewsky AR (1969) Biochem 8: 4752
136. Royer GP, Klotz IM (1969) J Am Chem Soc 91: 5885
137. Okube T, Ise N (1973) J Org Chem 38: 3120
138. Rodulfo T, Hamilton JA, Cordes EH (1974) J Org Chem 39: 2281
139. Kunitake T, Shinkai S, Hirotsu S (1975) J Polym Sci Polym Lett Ed C13: 377
140. Shinkai S, Kunitake T (1976) Biopolymers 15: 1129
141. Kunitake T, Shinkai S, Hirotsu S (1976) Biopolymers 15: 1143
142. Shinkai S, Ide T, Manabe O (1978) Bull Chem Soc Jpn 51: 3635
143. Tanchuk YuV, Pop GS (1978) Coll J USSR 40: 1209 (English translation 40: 1024)
144. Pop GS, Tanchuk YuV (1979) Ukr Khim Zh 45: 991
145. Sisido M, Akiyama K, Imanishi Y, Klotz IM (1984) Macromolecules 17: 198
146. Richlin J, Strauss UP (1986) Polymer Prepr Am Chem Soc Div Polym Chem 27(1): 425
147. Moriya M, Nishimura A, Hoaoda K, Takai M, Hidaka H (1986) J Am Oil Chem Soc 63: 263
148. Hidaka H, Onai T, Murata M, Fujiwara K, Takai M, Moriya M (1987) J Coll Interface Sci 119: 565
149. Wang KT, Iliopoulos I, Audebert R (1988) Polym Bull 20: 577
150. Tsiourvas D, Paleos CM, Dais P (1989) J Appl Polym Sci 38: 257
151. Iliopoulos I, Wang KT, Audebert R (1991) Langmuir 7: 617
152. Seo T, Take S, Miwa K, Hamada K, Iijima T (1991) Macromolecules 24: 4255
153. Kuo PL, Ni SC, Lai CC (1992) J Appl Polym Sci 45: 611
154. McCormick CL, Hoyle CE, Clark MD (1992) Polymer 33: 243
155. a) Rios HE, Rojas JS, Gamboa IC, Barraza RG (1993) J Coll Interface Sci 156: 388
     b) Rios HE, Aravena MH, Barraza RG (1994) J Colloid Interface Sci 165: 259
156. Salamone JC, Taylor P, Snider B, Israel SC (1975) J Polym Sci Polym Chem Ed A13: 161
157. Schulz DN, Kaladas JJ, Maurer JJ, Bock J, Pace SJ, Schulz WW (1987) Polymer 28: 2110
158. Tagagishi T, Nakata Y, Kuroki N (1974) J Polym Sci Polym Chem Ed A12: 807
159. a) Salamone JC, Thompson AM, Raheja MK, Su CH, Watterson AC (1988) Polym Prepr Am Chem Soc Div Polym Chem 29 (1): 281

   b) Salamone JC, Thompson AM,Su CH, Watterson AĆ (1989) Polym Prepr Am Chem Soc Div
       Polym Chem 30 (1): 333
160. Salamone JC, Thompson AM, Rice WC, Lai KT, Boden MW, Luo YM, Raheja MK, Su CH,
       Watterson AC (1989) in: Saegusa T, Higashimura T, Abe A (eds) Frontiers in Macromolecular
       Science, Blackwell Scientific, Oxford, p 227
161. Do CH, Butler GB (1989) Polym Prepr Am Chem Soc Div Polym Chem 30(2): 352
162. Peiffer DG (1990) Polymer 31: 2353
163. Yeoh KW, Chew CH, Gan LM, Koh LL, Ng SC (1990) J Macromol Sci Chem A27: 711
164. Anton P, Laschewsky A (1991) Makromol Chem Rapid Commun 12: 189
165. Peiffer DG (1991) Polymer 32: 134
166. Watterson AC, Hunter CR, Salamone JC (1992) Polym Prepr Am Chem Soc Div Polym Chem
       33(1): 1160
167. Köberle P, Laschewsky A, van den Boogaard D (1992) Polymer 33: 4029
168. Anton P, Laschewsky A (1993) Makromol Chem 194: 601
169. Pefferkorn E, Schmitt A, Varoqui R (1968) C R Acad Sc C267: 349
170. Varoqui R, Strauss UP (1968) J Phys Chem 72: 2507
171. Dubin PL, Strauss UP (1970) J Phys Chem 74: 2842
172. Schmitt A, Varoqui R, Skoulios A (1969) CR Acad Sci C268: 1469
173. Mathis A, Skoulious A, Varoqui R, Schmitt A (1974) Eur Polym J 10: 1011
174. Timofeevskii SL, Baikov VE, Panarin EF, Pautov VD (1994) Vysokomol Soyed A36: 15 (Eng.
       translation: Polym Sci USSR 36: 10
175. Strauss UP, Vesnaver G (1975) J Phys Chem 79: 1558 and 2426
176. Strauss UP, Schlesinger MS (1978) J Phys Chem 82: 571
177. Strauss UP, Schlesinger MS (1978) J Phys Chem 82: 1527
178. Mathis A, Schmitt A, Skoulios A, Varoqui R (1979) Eur Polym J 15: 255
179. Martin PJ, Morss LR, Strauss UP (1980) J Phys Chem 84: 577
180. Strauss UP, Barbieri BW (1982) Macromolecules 15: 1347
181. Barbieri BW, Strauss UP (1985) Macromolecules 18: 411
182. Varoqui R, Pefferkorn E (1985) in: Dubin P (ed) Microdomains in Polymer Solutions. Plenum
       Press, New York, p 225
183. Guillet JE, Rendall WA (1986) Macromolecules 19: 224
184. Hsu JL, Strauss UP (1987) J Phys Chem 91: 6238
185. Chu DY, Thomas JK (1987) Macromolecules 20: 213
186. Binana-Limbelé W, Zana R (1987) Macromolecules 20: 1331
187. McGlade MJ, Randall FJ, Tcheurekdjian N (1987) Macromolecules 20: 1782
188. Meier C (1989) Lyotrop flüssigkristalline Copolymere aus nicht-amphiphilen Monomeren.
       Thesis, Universität Freiburg (Germany)
189. McGlade MJ, Olufs JL (1988) Macromolecules 21: 2346
190. Shih LB, Mauer DH, Verbrugge CJ, Wu CF, Chang SL, Chen SH (1988) Polym Prepr Am
       Chem Soc Div Polym Chem 29(2): 451
191. Shih LB, Shen EY, Chen SH (1988) Macromolecules 21: 1387
192. Shih LB, Mauer DH, Verbrugge CJ, Wu CF, Chang SL, Chen SH (1988) Macromolecules
       21: 3235
193. Morishima Y, Kobayashi T, Nozakura S (1989) Polymer J 21: 267
194. Weinert RJ, Piirma I (1989) J Appl Polym Sci 38: 1295
195. Hamad E, Qutubuddin S (1990) Macromolecules 23: 4185
196. Binana-Limbelé W, Zana R (1990) Macromolecules 23: 2731
197. Chu DY, Thomas JK (1991) Macromolecules 24: 2212
198. Miyashita T, Arito Y, Matsuda M (1991) Macromolecules 24: 872
199. Yang YJ, Engberts JFBN (1991) Rec Trav Chim Pays-Bas 110: 384
200. Yang YJ, Engberts JFBN (1991) J Org Chem 56: 4300
201. Ito K, Hashimura K, Itsuno S, Yamada E (1991) Macromolecules 24: 3977
202. Morishima Y, Tominaga Y, Kamachi M, Okada T, Hirata Y, Mataga N (1991) J Phys Chem
       95: 6027
203. Morishima Y, Tsuji M, Kamachi M, Hatada K (1992) Macromolecules 25: 4406
204. Seki M, Morishima Y, Kamachi M (1992) Macromolecules 25: 6540
205. Koňák C, Kopečková P, Kopeček J (1992) Macromolecules 25: 5451
206. Benrraou M, Zana R, Varoqui R, Pefferkorn E (1992) J Phys Chem 96: 1468
207. Hamad E, Qutubuddin S (1992) J Chem Phys 96: 6222
208. Maltesh C, Xu Q, Somasundaran P, Benton WJ, Nguyen H (1992) Langmuir 8: 1511

209. Ueda T, Oshida H, Kurita K, Ishihara K, Nakabayashi N (1992) Polymer J 24: 1259
210. Sakurai I, Kawamura Y, Suetsugu T, Nakaya T (1992) Macromolecules 25: 7256
211. Nowakowska M, Foyle VP, Guillet JE (1993) J Am Chem Soc 115: 5975
212. Varadaraj R, Bock J, Brons N, Pace S (1993) J Phys Chem 97: 12991
213. Kuo PL, Hung MN (1993) J Appl Polym Sci 48: 1571
214. Betton F, Theretz A, Elaissari A, Pichot C (1993) Colloids Surfaces B1: 97
215. Zdanowicz VS, Strauss UP (1993) Macromolecules 26: 4770
216. Watterson AC, Haralabakopoulos A, Salamone JC (1993) Polym Prepr Am Chem Soc Div Polym Chem 34(1): 610
217. Morishima Y (1994) Trends in Polym Sci 2: 31
218. Xie X, Hogen-Esch TE (1994) Polym Prepr Am Chem Soc Polym Chem Div 35(1) 498
219. Nagai K, Fujii I, Kuramoto N (1992) Polymer 33: 3060
220. Hill A, Candau F, Selb J (1993) Macromolecules 26: 4521
221. Xin Y, Hu Y, Zeldin M, Fife WK (1993) Macromolecules 26: 4670
222. Yassini M, Hogen-Esch TE (1994) Polym Prepr Am Chem Soc Polym Chem Div 35(1) 478
223. Hamid S, Sherrington DC (1984) Brit Polym J 16: 39
224. Paleos CM, Malliaris A, Dais P (1987) Polym Prepr Am Chem Soc Polym Chem Div 28(2): 434
225. Paleos CM, Margomenou-Leonidopoulou G, Malliaris A (1988) Mol Cryst Liq Cryst 161: 385
226. Michas J, Paleos CM, Dais P (1989) Liq Cryst 5: 1737
227. Laschewsky A, Zerbe I (1991) Polymer 32: 2070
228. Ishihara K, Tsuji T, Sakai Y, Nakabayashi N (1993) J Polym Sci Polym Chem Ed A32: 859
229. Wagner D (1985) Über den Einfluß der Polymerfixierung von Tensiden auf das Aggregationsverhalten in Wäßriger Lösung. Thesis, TU Clausthal (Germany)
230. Charreyre MT, Boullanger P, Pichot C, Delair T, Mandrand B, Llauro MF (1993) Makromol Chem 194: 117
231. Jahns E, Finkelmann H (1987) Colloid Polym Sci 265: 304
232. Laschewsky A (1991) Colloid Polym Sci 269: 785
233. Elbert R, Laschewsky A, Ringsdorf H (1985) J Am Chem Soc 107: 4134
234. Laschewsky A, Ringsdorf H, Schmidt G, Schneider J (1987) J Am Chem Soc 109: 788
235. Ringsdorf H, Schmidt G, Schneider J (1987) Thin Solid Films 152: 207
236. Percec V, Pugh C (1989) in McArdle CB (ed) Side Chain Liquid Crystals. Blackie & Son Ltd, Glasgow, p 30
237. Zentel R (1989) in: Allen G, Bevington JC (eds) Comprehensive Polymer Science Vol 5. Pergamon Press, Oxford (UK), p 723
238. Kobayashi K, Sumitomo H (1980) Macromolecules 13: 234
239. Abid SK, Hamid SM, Sherrington DC (1987) J Coll Interface Sci 120: 245
240. Gallot B, Marchin B (1989) Liq Crystals 5: 1729
241. Barakat Y, Gendy TS, Mohammed AI, Youssef AFM (1989) Brit Polym J 21: 383
242. Barakat Y, Gendy TS, Basily IK, Mohammed AI (1989) Brit Polym J 21: 451
243. Barakat Y, Basily IK, Mohammed AI, Youssef AFM (1989) Brit Polym J 21: 459
244. Gallot B (1991) Mol Cryst Liq Cryst 203: 137
245. Laschewsky A, Zerbe I (1991) Polymer 32: 2081
246. Frankel DA, O'Brien DF (1991) J Am Chem Soc 113: 7436
247. Fuhrhop JH, Blumtritt R, Lehmann C, Luger P (1991) J Am Chem Soc 113: 7436
248. Ahlheim M, Hallensleben ML (1992) Makromol Chem 193: 779
249. Abid SK, Sherrington DC (1993) Eur Polym J 29: 153
250. Frankel DA, O'Brien DF (1994) Macromol Symp 77: 141
251. Ferguson P, Sherrington DC, Gough A (1993) Polymer 34: 3281
252. Shimizu T, Kwak JC, Minakata A (1985) J Polym Sci Polym Phys Ed B23: 1139
253. Kopolow S, Hogen-Esch TE, Smid J (1973) Macromolecules 3: 133
254. Roland B, Smid J (1983) J Am Chem Soc 105: 5269
255. Roland B, Kimura K, Smid J (1984) J Coll Interface Sci 97: 392
256. Menger FM, Tsuno T (1990) J Am Chem Soc 112: 6723
257. Kaneka M, Yamada A (1984) Adv Polym Sci 55: 2
258. Morishima Y (1992) Adv Polym Sci 104: 51
259. Morishima Y, Tominaga Y, Nomura S, Kamachi M, Okada T (1992) J Phys Chem 96: 1990
260. Morishima Y, Tsuji M, Seki M, Kamachi M (1993) Macromolecules 26: 3299
261. Husson F, Herz J, Luzzati V, Wyart J (1961) CR Acad Sci C252: 3290
262. Sugiyama K, Shiraishi K (1989) Makromol Chem 190: 2381
263. Matsuo T, Sakamoto T, Takuma K, Sakura K, Ohsako T (1981) J Phys Chem 85: 1277

82 A. Laschewsky

264. Tundo P, Kippenberger DJ, Politi MJ, Klahn P, Fendler JH (1982) J Am Chem Soc 104: 5352
265. Nakahira T, Grätzel M (1985) Makromol Chem Rapid Comm 6: 341
266. Okahata Y, Ariga K, Seki T (1986) J Chem Soc Chem Commun 73
267. Anton P, Köberle P, Laschewsky A (1992) Prog Colloid Polym Sci 89: 56
268. Anton P, Heinze J, Laschewsky A (1993) Langmuir 9: 77
269. Anton P (1993) Redoxaktive Polyseifen: Wege zu schaltbaren Polymeren. Thesis, Universität Mainz (Germany)
270. Collard DM, Fox MA (1991) J Am Chem Soc 113: 9414
271. Coche-Guerente L, Deronzier A, Galland B, Labbe P, Moutet JC, Reverdy G (1991) J Chem Soc Chem Commun 386
272. Chen SA, Hua MY (1993) Macromolecules 26: 7108
273. Shi S, Wudl F (1990) Macromolecules 23: 2119
274. Rau IU, Rehahn M (1994) Acta Polymer 45: 3
275. Salamone JC, Israel SC, Taylor P, Snider B (1974) J Polym Sci Symp 45: 65
276. Ghesquiere D, Chachaty C, Ban B, Loucheux C (1976) Makromol Chem 177: 1601
277. Ikeda T, Tazuke S (1983) Makromol Chem Rapid Comm 4: 459
278. Ikeda T, Tazuke S, Suzuki Y (1984) Makromol Chem 185: 869
279. Nagai K, Ohishi Y, Inaba H, Kudo S (1985) J Polym Sci Polym Chem Ed A23: 1221
280. Nagai K, Ohishi Y (1987) J Polym Sci Polym Ed A25: 1
281. Schambil F, Schwuger MJ (1987) in: Falbe J (ed) Surfactants in Consumer Products, Springer, Berlin Heidelberg New York, p 133
282. Wang KT, Iliopoulos I, Audebert R (1989) Polym Prepr Am Chem Soc Div Polym Chem 30 (2): 377
283. Yeoh KW, Chew CH, Gan LM, Koh LL (1989) Polym Bull 22: 123
284. Gan LM, Yeoh KW, Chew CH, Koh LL, Tan TL (1991) J Appl Polym Sci 42: 225
285. Magny B, Iliopoulos I, Audebert R (1991) Polymer Comm 32: 456
286. Yegorov VV, Ksenofontova OB (1991) Vysokomol Soyed A33: 1780 (english translation: Polym Sci USSR 33: 1665)
287. Seo T, Take S, Akimoto T, Hamada K, Iijima T (1991) Macromolecules 24: 4801
288. Cochin D, Candau F, Zana R, Talmon Y (1992) Macromolecules 25: 4220
289. Yasuda Y, Rindo K, Tsushima R, Aoki S (1993) Makromol Chem 194: 485
290. Yasuda Y, Rindo K, Tsushima R, Aoki S (1993) Makromol Chem 194: 1893
291. Yasuda Y, Rindo K, Aoki S (1993) Polym J 25: 1203
292. Cochin D, Candau F, Zana R (1993) Macromolecules 26: 5755
293. Nakaya T, Nishio K, Memita M, Imoto M (1993) Makromol Chem Rapid Comm 14: 77
294. Yassini M, Hogen-Esch TE (1992) Polym Prepr Am Chem Soc Polym Chem Div 33(1): 933
295. Seery TAP, Yassini M, Hogen-Esch TE, Amis EJ (1992) Macromolecules 25: 4784
296. Zhang YX, Da AH, Butler GB, Hogen-Esch TE (1992) J Polym Sci Polym Chem Ed A30: 1383
297. Hwang FS, Hogen-Esch TE (1993) Macromolecules 26: 3156
298. Ringsdorf H, Simon J, Winnik FM (1992) Macromolecules 25: 5353
299. Köberle P, Laschewsky A, Lomax TD (1991) Makromol Chem Rapid Commun 12: 427
300. Finkelmann H, Schafheutle M (1986) Colloid Polym Sci. (1986) 264: 786
301. Klein J, Kunz M, Kowalczyk J (1990) Makromol Chem 191: 517
302. Fuhrhop JH, Spiroski D, Schnieder P (1991) React Polym 15: 215
303. Morishima Y, Itoh Y, Nozakura S (1981) Makromol Chem 182: 3135
304. Katchalsky A, Miller I (1951) J Phys Chem 55: 1182
305. Ohno N, Nitta K, Makino S, Sugai S (1973) J Polym Sci Polym Phys Ed B11: 413
306. Okuda T, Ohno N, Nitta K, Sugai S (1973) J Polym Sci Polym Phys Ed B11: 749
307. Vallin D, Huguet J, Vert M (1980) Polymer J 12: 113
308. Couvercelle JP, Huguet J, Vert M (1993) Macromolecules 26: 5015
309. De Winter W, Alaerts L (1981) in: Abstracts of the 27th IUPAC Symposium on Macromolecules Vol 1, Strasbourg, p 316
310. Plaisance M, Ter-Minassian-Saraga L (1976) J Coll Interface Sci 56: 33
311. Plaisance M, Ter-Minassian-Saraga L (1977) J Coll Interface Sci 59: 113
312. Bansleben DA, Vogl O (1985) J Polym Sci Polym Chem Ed A23: 703
313. Tyminski PN, Ponticello IS, O'Brien DF (1987) J Am Chem Soc 109: 6541
314. Tauer K, Goebel KH, Kosmella S, Stähler K, Neelsen J (1990) Makromol Chem Macromol Symp 31: 107
315. Cornforth JW, D'Arcy Hart P, Rees RJW, Stock JA (1951) Nature 168: 150
316. Anton P, Laschewsky A, Eur Polym J, in press

317. Kunitake T, Shinkai S, Hirotsu S (1977) J Org Chem 42: 306
318. Laughlin RG (1978) in: Brown HG (ed) Advances in Liquid Crystals Vol 3. Academic Press, New York, p 41 and ibid p 99
319. Chung DC, Kostelnik RJ, Elias HG (1977) Makromol Chem 178: 691
320. Wessling RA, Pickelman DM (1981) J Disp Techn 2: 281
321. Elias HG, Chung DC, Donkai N, Hellmann GP (1987) Makromol Chem 188: 537
322. Nagai K, Elias HG (1987) Makromol Chem 188: 1095
323. Berg JM, Claesson PM (1994) J Coll Interface Sci 163: 289
324. Cederstav AK, Novak BM (1992) Polym Prepr Am Chem Soc Div Polym Chem 33(1): 1090
325. Seery TAP, Wallow TI, Novak BM (1993) Polym Prepr Am Chem Soc Div Polym Chem 34(2): 727
326. Chaturvedi V, Tanaka S, Keriyama K (1993) Macromolecules 26: 2607
327. Gieselman MB, Reynolds JR (1993) Macromolecules 26: 5633
328. Watterson AC, Twu YK, Salamone JC (1993) Polym Prepr Am Chem Soc Div Polym Chem 34(2): 402
329. Cleij (1989) Enantioselective Ester Cleavage by Histidine Containing Oligopeptides at Micellar Interfaces. An Enzyme Model. Thesis, Rijksuniversiteit Utrecht (NL)
330. Mao G, Tsao Y, Tirrell M, Davis T, Hessel V, Ringsdorf H (1993) Langmuir 9: 3461
331. a) Köberle P, Laschewsky A (1994) Macromolecules 27: 2165
    b) Köberle P, Laschewsky A Macromol Symp (in press)
332. Okubo T, Ise N (1977) Adv Polym Sci 25: 135
333. Ford WT, Badley RD, Chandran RS, Babu SH, Hassanein M, Srinivasan S, Turk H, Yu H, Zhu W (1992) in: Daniels ES, Sudol ED, El-Aasser MS (eds) Polymer Latexes, ACS Symposium Ser 492. Am Chem Soc, Washington DC, p 422
334. Batz HG (1977) Adv Polym Sci 23: 25
335. Ferruti P, Danusso F, Franchi G, Polentarutti N, Garattini S (1973) J Med Chem 16: 496
336. Pratten MK, Lloyd JB, Hörpel G, Ringsdorf H (1985) Makromol Chem 186: 725
337. Charreyre MT, Boullanger P, Declair T, Mandrand B, Pichot C (1993) Colloid Polym Sci 271: 668
338. Tanchuk YuV, Kotenko SI, Ladtschenko SYu, Yablonko BM (1988) UKr Chim Zh 54: 990
339. Fuoss RM, Strauss UP (1948) J Polym Sci 3: 246
340. Skolnick J, Fixman M (1977) Macromolecules 10: 944
341. Odijk T (1977) J Polym Sci Polym Phys Ed B15: 477
342. Mandel M (1988) in: Kroschwitz JI (ed) Encyclopedia of Polymer Science and Technology 2nd ed Vol 11. Wiley-Interscience, New York, p 739
343. Anton P, Laschewsky A (in press) Colloid Polym Sci
344. Förster S, Schmidt M, Antonietti M (1990) Polymer 31: 781
345. Schmidt M (1991) Macromolecules 24: 5361
346. Förster S, Schmidt M, Antonietti M (1992) J Phys Chem 96: 4008
347. Salamone JC, Israel SC, Taylor P, Snider B (1973) Polym Prepr Am Chem Soc Div Polym Chem 14(2): 778
348. Lunkenheimer K, Miller R (1987) J Coll Interface Sci 120: 176
349. Nitsch W, Kremnitz W, Schweyer G (1987) Ber Bunsenges Phys Chem 91: 218
350. Okubo T (1988) J Coll Interface Sci 125: 386
351. Kamide K, Saito M, Akedo T (1992) Polym Int 27: 35
352. Glatzhofer DT, Cho G, Lai CL, O'Rear EA, Fung BM (1993) Langmuir 9: 2949
353. Kuo PL, Hung MN, Lin YH (1993) J Appl Polym Sci 47: 1295
354. Gu T, Sjöblom J (1992) Colloids Surfaces 64: 39
355. Morishima Y, Nozakura S (1986) J Polym Sci Polym Symp 74: 1
356. Salamone JC, Rice WC (1988) in: Kroschwitz JI (ed) Encyclopeida of Polymer Science and Technology, Vol 11, 2nd edn. Wiley-Interscience, New York, p 514
357. Laschewsky A, unpublished
358. McBain MEL, Hutchinson E (1955) in: Solubilization and Related Phenomena. Academic Press, New York
359. Elworthy PH, A.T. Florence AT, MacFarlane CB (1968) in: Solubilization by Surface-Active Agents. Chapman and Hall, London
360. Hoffmann H (1994) Adv Mater 6: 116
361. Boyer B, Lamaty G, Leydet A, Roque JP, Sama P (1992) New J Chem 16: 887
362. Grieser F, Drummond CJ (1988) J Phys Chem 92: 5580
363. Dong DC, Winnik MA (1984) Can J Chem 62: 2560

364. Mukerjee P, Ko JS (1992) J Phys Chem 96: 6090
365. Winnik FM (1993) Chem Rev 93: 587
366. Ephardt H, Fromherz P (1993) J Phys Chem 97: 4540
367. Gao Z, Kwak JCT, Labonté R, Marangoni DG, Wasylishen RE (1990) Colloids Surfaces 45: 269
368. Hoshino T, Imamura Y (1990) Bull Chem Soc Jpn (1990) 63: 502
369. Fujiwara H, Kanzoki K, Kimura A, Tanka K, Da YZ (1992) J Chem Soc Chem Commun 736
370. Zana R, Lang J (1990) Colloids Surfaces 48: 153
371. Kalyanasundaram K, Thomas JK (1977) J Am Chem Soc 99: 2039
372. Street jr KW, Acree jr WE (1986) Analyst 111: 1197
373. Gao Z, Wasylishen RE, Kwak JCT (1989) Macromolecules 22: 2544
374. Kiefer M (1990) Synthese und Untersuchungen zum Hämolyse und Solubilisierungsverhalten nichtionogener amphiphiler Seitenkettenpolymere. Thesis, Universität Freiburg (Germany)
375. Cochin D (1991) Polymérisation de tensioactifs sous forme micellaire. Thesis, Université Louis Pasteur, Strasbourg (France)
376. Yang YJ, Engberts JBFN (1992) Eur Polym J 28: 881
377. Esumi K, Watanabe N, Meguro K (1989) Langmuir 5: 1420
378. Esumi K, Watanabe N, Meguro K (1991) Langmuir 7: 1775
379. Owenson B, Pratt LR (1984) J Phys Chem 88: 2905
380. Cabane B, Duplessix R, Zemb T (1985) J Physique 46: 2161
381. Gruen DWR (1985) Progr Colloid Polym Sci 70: 6
382. Chevalier Y, Cachaty C (1985) J Phys Chem 89: 875
383. Jansson M, Li P, Stilbs P (1987) J Phys chem 91: 5279
384. Jansson M, Li P, Stilbs P (1991) J Coll Interface Sci 142: 593
385. Bacaloglu R, Blasko A, Bunton CA, Cerichelli G, Shirazi A (1991) Langmuir 7: 1107
386. Ebelhäuser R, Spieß HW (1987) Makromol Chem 188: 2935
387. Fahmy T, Wesser J, Spieß HW (1989) Angew Makromol Chem 166/167: 39
388. Disch S, Schnepp W, Schmidt C (1992) Makromolekulares Kolloquium Freiburg (Germany)
389. Schnepp W (1994) Deuterium-NMR-Untersuchungen an deuterierten lyotropen Flüssigkristallen. Thesis, Universität Freiburg (Germany)
390. Aniansson EAG, Wall SN, Almgren M, Hoffmann H, Kielmann I, Ulbricht W, Zana R, Lang J, Tondre C (1976) J Phys Chem 80: 905
391. Magny B, Iliopoulos I, Audebert R (1991) Polym Prepr Am Chem Soc Div Polym Chem 32(1): 577
392. Richtering W, Löffler R, Burchard W (1992) Macromolecules 25: 3642
393. Elias HG (1973) J Macromol Sci Chem A7: 601
394. Laschewsky A, Schmidt M, unpublished
395. Bachmann L, Dasch W, Kutter P (1981) Ber Bunsenges Phys Chem 85: 883
396. Bellare JR, Kaneko T, Evans DF (1988) Langmuir 4: 1066
397. Vinson PK, Bellare JR, Davis HT, Miller WG, Scriven LE (1991) J Coll Interface Sci 142: 74
398. Herz J, Husson F, Luzzati V (1961) C R Acad Sc C252: 3462
399. Herz J, Reiss-Husson F, Rempp P, Luzzati V (1964) J Polym Sci C4: 1275
400. Gallot B, Douy A (1987) Mol Cryst Liq Cryst 153: 367
401. Gallot B, Douy A (1989) Markromol Chem Macromol Symp 24: 321
402. Gallot B, Marchin B (1989) Liq Crystals 5: 1719
403. Löffler R, Finkelmann H (1990) Makromol Chem Rapid Commun 11: 321
404. Naitoh K, Ishii Y, Tsujii K (1991) J Phys Chem 95: 7915
405. Friberg S, Yu B, Ahmed AU, Campbell GA (1993) Colloids Surfaces 69: 239
406. Köberle P, Laschewsky A, Tsukruk V (1992) Makromol Chem193: 1815
407. Tsukruk V, Mischenko N, Köberle P, Laschewsky A (1992) Makromol Chem 193: 1829
408. Hessel V, Ringsdorf H, Festtag R, Wendorff JH (1993) Makromol Chem Rapid Commun 14: 707
409. Geetha B, Mandel AB, Ramasami T (1993) Macromolecules 26: 4083
410. Fuoss RM, Edelson D (1951) J Am Chem Soc 73: 269
411. Menger FM, Wrenn S (1974) J Phys Chem 78: 1387
412. Fuhrhop JH, Fritsch D (1986) Acc Chem Res 19: 130
413. Menger FM, Littau CA (1991) J Am Chem Soc 113: 1451
414. Rosen MJ (1993) Chemtech 30
415. Menger FM, Littau CA (1993) J Am Chem Soc 115: 10083
416. Robinson DI, Sherrington DC, Suckling CJ (1984) Tetr Lett 40: 785

417. Hamid SM, Sherrington DC, Suckling DJ (1986) Colloid Polym. Sci 264: 883
418. Murakami Y, Nakano A, Akiyoshi K, Fukuya K (1981) J Chem Soc Perkin Trans 1: 2800
419. Menger FM, Takeshita M, Chow JF (1981) J Am Chem Soc 103: 5938
420. Menger FM, Angel DeGreiff AJ, Jaeger DA (1984) J Chem Soc Chem Commun 545
421. Menger FM (1986) Top Curr Chem 136: 2
422. Keller-Griffith R, Ringsdorf H, Vierengel A (1986) Colloid Polym Sci 264: 924
423. Murakami Y, Nakano A, Miyata R, Matsuda Y (1979) J Chem Soc Perkin Trans 1: 1669
424. Hessel V, Ringsdorf H, Laversanne R, Nallet F (1993) Rec Trav Chim Pays-Bas 112: 339
425. Hessel V a) (1990) Multipolare Polymerisierbare Amphiphile: Photoreaktionen zur Stabilisier-
     ung von Zylindermizellen. Diplomarbeit, Universität Mainz (Germany)
     b) (1993) Multipolare Amphiphile. Thesis, Universität Mainz (Germany)
426. Cipiciani A, Fracassini MC, Germani R, Savelli G, Bunton CA (1987) J Chem Soc Perkin
     Trans 2: 547
427. Masuyama A, Yokota M, Zhu YP, Kida T, Nakatsuji Y (1994) J Chem Soc Perkin Trans 2: 547
428. Zana R, Yiv S, Kale KM (1980) J Coll Interface Sci 77: 456
429. McKenzie DC, Bunton CA, Nicoli DF, Savelli G (1987) J Phys Chem 91: 5709
430. Ikeda K, Yasuda M, Ishikawa M, Esumi K, Meguro K, Binana-Limbele W, Zana R (1989)
     Colloid Polym Sci 267: 825
431. Ikeda K, Ishikawa M, Yasuda M, Esumi K, Meguro K, Binana-Limbele W, Zana R (1989) Bull
     Chem Soc Jpn 62: 1032
432. Yasuda M, Ikeda K, Esumi K, Meguro K (1989) Bull Chem Soc Jpn 62: 3648
433. a) Zhu YP, Masuyama A, Okahara M (1990) J Am Oil Chem Soc 67: 459
     b) Zhu YP, Masuyama A, Okahara M (1990) J Am Oil Chem Soc 68: 268
434. Zhu YP, Masuyama A, Kobata Y, Nakatsuji Y, Okahara M, Rosen MJ (1993) J Coll Interface
     Sci 158: 40
435. Alami E, Beinert G, Marie P, Zana R (1993) Langmuir 9: 1465
436. Forrest BJ, Hecker De Carvalho L, Reeves LW (1981) J Am Chem Soc 103: 245
437. Wong RC, Ikeda K, Meguro K, Söderman O, Olsson U, Lindman B (1989) J Phys Chem 93:
     4861
438. Ikeda K, Khan A, Meguro K, Lindmann B (1989) J Coll Interface Sci 133: 192
439. Ishikawa M, Matsumura KI, Esumi K, Meguro K, Binana-Limbele W, Zana R (1992) J Coll
     Interface Sci 151: 70
440. Meguro K, Ikeda K, Otsugi A, Taya M, Yasuda M, Esumi K (1987) J Coll Interface Sci 118:
     372
441. Deinega YuF, Ul'berg ZR, Marotchko LG, Rudi VP, Denisenko VP (1974) Kolloidn Zh 36:
     649
442. Deviniski F, Masarova L, Lacko I (1985) J Coll Interface Sci 105: 235
443. Devinski F, Lacko I, Bittererova F, Tomeckova L (1986) J Coll Interface Sci 114: 314
444. Rozycka-Roszak B, Witek S, Przestalski S (1989) J Coll Interface Sci 131: 181
445. Zana R, Benrraou M, Rueff R (1991) Langmuir 7: 1072
446. Parreira HC, Lukenbach ER, Lindemann MKO (1979) J Am Oil Chem Soc 56: 1015
447. Ul'berg ZR, Podol'skaya VI (1978) Kolloidn Zh 40: 292
448. Devinski F, Lacko I, Imam T (1991) J Coll Interface Sci 143: 336
449. Zhou GB, Smid J (1993) Langmuir 9: 2907
450. Zhou GB, Smid J (1993) Polymer 34: 5128
451. Lühmann B (1985) Über den Einfluß einer Polymerfixierung auf das lyotrope Mesophasenver-
     halten nichtionischer Amphiphile. Thesis, TU Clausthal 1985
452. Kuo PL, Chen CJ (1993) J Polym Sci Polym Chem Ed A31: 99 (1993)
453. Hyde AJ, Robb DJM (1963) J Phys Chem 67: 2089
454. Greene BW, Saunders FL (1970) J Coll Interface Sci 33: 393
455. De Winter W, Mariën A (1984) Makromol Chem Rapid Commun 5: 593
456. Gan LM, Chew CH (1984) J Disp Sci Tech 5: 179
457. Chew CH, Gan LM (1985) J Polym Sci Polym Chem Ed A23: 2225
458. Chen S, Chang HS (1985) J Polym Sci Polym Chem Ed A23: 2615
459. Malyukova EB, Nesmelova SV, Gritskova IA, Pravednikov AN, Egorov VV, Zubov VP,
     Kabanov VA (1985) Dokl Akad Nauk SSSR 284: 1420
460. Ito K, Yokoyama S, Arakawa F (1986) Polym Bull 16: 345
461. Tsaur SL, Fitch RM (1987) J Coll Interface Sci 115: 450
462. Tsaur SL, Fitch RM (1987) J Coll Interface Sci 115: 463
463. Hassanein M, Ford WT (1988) Macromolecules 21: 525

464. Choubal M, Ford WT (1989) J Polym Sci Polym Chem Ed A27: 1873
465. Yamaguchi K, Watanabe S, Nakahama S (1989) Makromol Chem 190: 1195
466. Guillaume JL, Pichot C, Guillot J (1990) J Polym Sci Polym Chem Ed A28: 137
467. Raj WRP, Sasthav M, Cheung HM (1992) Langmuir 8: 1931
468. Watanabe S, Ozaki H, Mitsuhashi K, Nakahama S, Yamaguchi K (1989) Makromol Chem 193: 2781
469. Babu SH, Ford WT (1992) J Polym Sci Polym Chem Ed A30: 1917
470. Urquiola MB, Dimonie VL, Sudol ED, El-Aasser MS (1992) J Polym Sci Polym Chem Ed A30: 2619
471. Urquiola MB, Dimonie VL, Sudol ED, El-Aasser MS (1992) J Polym Sci Polym Chem Ed A30: 2631
472. Urquiola MB, Sudol ED, Dimonie VL, El-Aasser MS (1993) J Polym Sci Polym Chem Ed A31: 1403
473. Uyama H, Honda Y, Kobayashi S (1993) J Polym Sci Polym Chem Ed A31: 123
474. Kobayashi S, Uyama H, Lee SW, Matsumoto Y (1993) J Polym Sci Polym Chem Ed A31: 3133
475. Arai K (1993) Makromol Chem 194: 1975
476. Nagai K, Satoh H, Kuramoto N (1993) Polymer 34: 4969
477. Raj WRP, Sasthav M, Cheung HM (1993) J Appl Polym Sci 49: 1453
478. Aoyagi T, Terashima O, Nagase Y, Matsui K (1991) Polymer 32: 2106
479. Holt P, Thadani C (1973) Makromol Chem 169: 55
480. Aoyagi T, Terashima O, Suzuki N, Matsui K, Nagase Y (1990) J Controlled Release 13: 63
481. Pucci B, Polidori A, Rakotomanomana N, Chorro M, Pavia AA (1993) Tetr Lett 34: 4185
482. Shinkai S (1982) Prog Polym Sci 8: 1
483. Bunton CA, Nome F, Quina FH, Romsted LS (1991) Acc Chem Res 24: 357
484. Shimidzu T (1977) Adv Polym Sci 23: 55
485. Engberts JFBN (1992) Pure Appl Chem 64: 1653
486. Flach HN, Grassert I, Oehme G (1994) Macromol Chem Phys 195: 3289
487. Suh J, Scarpa IS, Klotz IM (1976) J Am Chem Soc 98: 7060
488. Boutevin B, Robin JJ, Boyer B, Roque JP, Senhaji O (1994) Macromol Chem Phys 195: 129
489. Torstensson M, Hult A (1992) Polym Bull 29: 549
490. Schmidt DL, Coburn CE, Dekoven BM, Potter GE, Meyers GF, Fischer DA (1994) Nature 368: 39
491. a) Senhaji O (1993) Synthèse de tensioactifs méthacryliques et étude physico-chimique de leurs télomères et polymères. Thesis, Université de Montpellier (France)
     b) Boutevin B, Robin JJ, Roque JP, Senhaji O (1994) Macromol Chem Phys 195: 2177
492. Yasuda Y, Rindo K, Aoki S (1992) Makromol Chem 193: 2875
493. Gan LM, Chew CH, Yeoh KW, Koh LL (1990) J Coll Interface Sci 137: 597
494. Imae T, Chew CH, Gan LM (1991) Colloids Surfaces 61: 75
495. Greene BW, Sheetz DP, Filer TD (1970) J Coll Interface Sci 32: 96
496. Greene BW, Sheetz DP (1970) J Coll Interface Sci 32: 96
497. Fang TR, Zhu XB (1991) Polym Bull 25: 467
498. Rehse H, Ritter H (1988) Makromol Chem 189: 529
499. Keller-Griffith R (1986) Strukturvariation von Amphiphilen: Multipolare und polymerisierbare Tenside. Thesis, Universität Mainz (Germany)
500. León A, Abuin E, Lissi E, Gargallo A, Radic D (1987) J Coll Interface Sci 115: 529
501. León A, Gargallo A, Radic D, Horta A (1991) Polymer 32: 761
502. Von Berlepsch H, Strey R (1993) Ber Bunsenges Phys Chem 97: 1403
503. Fife WK, Ranganathan P, Zeldin M (1989) Polym Prep Am Chem Soc Polym Chem Div 30 (2): 123
504. Fife WK, Ranganathan P, Zeldin M (1990) J Org Chem 55: 5610
505. Paleos CM, Voliotis S, Margomenou-Leonidopoulou G, Dais P (1980) J Polym Sci Polym Chem Ed A18: 3463
505. Yegorov VV, Batrakova YeV, Zubov VP (1990) Vysokomol Soyed A32: 927 (english translation: Polym Sci USSR 32: 861)
506. Wibbing M (1986) Monomere und polymere multipolare Amphiphile - Synthese und Phasenverhalten in wäßrigen Lösungen. Diplomarbeit, Universität Mainz (Germany)
507. Richtering W, Löffler R, Burchard W (1992) J Phys Chem 96: 3883

Editor: Prof. Ringsdorf
Received: August 1994

Aromatic and Heterocyclic Amines in Polymer Stabilization

Jan Pospíšil
Institute of Macromolecular Chemistry, Academy of Sciences of the Czech Republic, Prague, Czech Republic

Amines constitute a family of stabilizers protecting plastics and coatings against free radical oxidative degradation and diene based elastomers against both ozonation and oxidation. The respective antioxidant, antifatigue, photo-antioxidant and antiozonant protection of polymers is dependent on the structure of the particular amine. Mechanisms of specific contributions of aromatic and heterocyclic amines have been interpreted in detail. The data are based on up-to-date knowledge of the inherent chemical efficiency of individual aminic types and the physical factors governing the final effect. Application of polymer stabilizers is mandatory from the technical point of view. The public concern dealing with the potential impact on the environment is increasing. Therefore, the final part of the paper is devoted to environmental aspects encountering amine stabilizers.

List of Symbols and Abbreviations   89

Introduction   91

1 Degradation of Polymers by Environmental Deteriogens   91

2 Structure of Aminic Stabilizers   94

3 Antioxidant, Anti-Flex-Crack and Photo-Antioxidant Activities
   of Amines 100
   3.1 Aromatic Amines 100
        3.1.1 Antioxidant Activity 100
              3.1.1.1 Aminyl Radicals 101
              3.1.1.2 Nitroxides 105
              3.1.1.3 Quinone Imines and Nitrones 110
        3.1.2 Anti-Flex-Crack Activity 116
   3.2 Heterocyclic Amines 117
        3.2.1 Derivatives of Indole and Carbazole 117
        3.2.2 Derivatives of Dihydroquinoline and Dihydroacridine . . . 118
        3.2.3 Phenothiazines 122
        3.2.4 Anilino Derivatives of Heterocyclic Amines 123
   3.3 Hindered Amines 124
        3.3.1 Activity Mechanisms of Hindered Amine Stabilizers 126
              3.3.1.1 Free-Radical Species Generated from Hindered
                      Amines 127

3.3.1.2 Free-Radical Scavenging by Hindered Amines . . .   130
      3.3.1.2.1 Scavenging of Alkylradicals
            by Nitroxides   130
      3.3.1.2.2 Scavenging of O-Centred Radicals
            by Hindered Amines   132
      3.3.1.2.3 Regeneration of Nitroxides from
            O-Alkylhydroxylamines   133
3.3.1.3 Deactivation of Organic Peroxides   136
3.3.1.4 Hydrogen Abstraction Activity of Nitroxides   138
3.3.1.5 Activity of Hydroxylamines   139
3.3.1.6 Interference of Hindered Amines with Singlet
      Oxygen, Excited Chromophores and Charge-
      Transfer-Complexes   139
3.3.1.7 Complexation of Metal Ions   141
3.3.1.8 Efficiency of Mixtures of HAS Transformation
      Products   142
3.3.1.9 Depletion of Hindered Amines   142
3.3.2 Hindered Amines in Contact with Acid Species   146

4   Antiradiant Activity of Amines   149

5   Antiozonant Activity of Amines   151

6   Blends of Amines with Other Polymer Stabilizers   156

7   Bifunctional Stabilizers Containing Amine Moiety   163

8   Physical Factors Affecting Stabilization of Polymers   165
  8.1   Physically Persistent Amines   167
  8.2   Properties of Macromolecular Amines   171

**9   Ecological Problems Encountering Application
   of Aminic Stabilizers**   174

10   Conclusions   177

11   Appendix   178

12   References   183

List of Symbols and Abbreviations

(Abbreviations of polymers according to [1, 2])

ABS	Acrylonitrile-butadiene-styrene polymer
AES	Acrylonitrile-EPDM-styrene polymer
AF	Antifatigue (anti-flex-crack) agent
AO	Antioxidant
AOZ	Antiozonant
AR	Antiradiant
BR	Butadiene rubber
BQ	Benzoquinone
BQDI	Benzoquinone diimine
BQMI	Benzoquinone monoimine
CB AO	Chain-breaking antioxidant
CR	Polychloroprene
CTC	Charge-transfer-complex
DABCO	1,4-Diazabicyclo[2.2.2]octane
DHI	1,2-Dihydro-3H-indole
DHQ	1,2-Dihydro-2,2,4-trimethylquinoline
DPA	Diphenylamine
EPDM	Ethylene-propylene-diene terpolymer
EPR	Elastomeric poly(ethylene-co-propylene)
EPR	Electron paramagnetic resonance spectroscopy
FR	Fire retardant
HA	Heat Aging
HAS	Hindered amine stabilizer
HD AO	Hydroperoxide decomposing AO
HD PE	High-density PE
HMW	High-molecular weight
HS	Heat stabilizer
IIR	Butylrubber, poly(isoprene-co-isobutylene)
IP	Induction period
IR	Synthetic isoprene rubber
LS	Light stabilizer
LD PE	Low-density PE
L-LDPE	Linear LD PE
LMW	Low-molecular weight
LTHA	Long-term heat aging
LTHS	Long-term heat stabilizer
MD	Metal deactivator
NBR	Nitrile rubber, poly(butadiene-co-acrylonitrile)
NA	Naphthylamine
NR	Natural rubber

PC	Polycarbonate
PD	1,4-Phenylenediamine
PE	Polyethylene
PNA	Phenylnaphthylamine
PO	Polyolefin
POP	Poly(oxyphenylene)
PP	Polypropylene
PS	Polystyrene
PT	Phenothiazine
PUR	Polyurethane
PVC	Poly(vinyl chloride)
QI	Quinone imine
QM	Quinone-methide
SAN	Poly(styrene-*co*-acrylonitrile)
SBR	Styrene butadiene rubber
TEMPO	2,2,6,6-Tetramethylpiperidine-1-oxyl
TMIO	1,1,3,3-Tetramethylisoindoline-2-oxyl
TLC	Thin layer chromatography
UV	Ultraviolet
UVA	Ultraviolet light absorber
X-LPE	Crosslinkable PE

Introduction

Regarding the enhanced use of plastics, rubbers and coatings in environments having increased chemical and physical aggressivity, the research and development are aimed at innovative technologies for processing and end-use stabilizations. Amine stabilizers have a broad application spectrum due to their structural diversity. It is understandable that they continuously attract scientific attention. The up-to-date knowledge of the chemical phenomena controlling activity mechanisms of amines ("inherent chemical efficiency") [3, 4], of transformations under conditions of polymer degradation [5, 6] and of physical factors governing the efficiency [7, 8] constitute a base for understanding the role of amines in polymer stabilization. The mechanistic knowledge has been profitably exploited to meet increasing demands of polymer end-use customers.

Principal structures of aminic stabilizers, their involvement in individual degradation processes of polymers, behaviour in mixtures with other polymer additives and an outline of environmental impacts due to the amines are included. The most relevant literature sources published by the first quarter of 1994 are reported. Some earlier data has to be mentioned as a reminder of the original ideas and to improve the interpretation of results. Where relevant, recent comprehensive reviews are cited. Principal types of commercial stabilizers are included in Appendix.

1 Degradation of Polymers by Environmental Deteriogens

Regular components of the troposphere, oxygen and actinic solar radiation (UV-A radiation 320–400 nm and a part of the UV-B radiation 290–320 nm having wavelengths higher than 300 nm) are, together with heat and mechanical stress, the principal deteriogens attacking polymers. Some defect structures and free radicals arising during fabrication and processing of polymers increase together with metallic impurities and residues of polymerization catalysts increase the degradation potency from the very beginning of the end-use of polymers. Tropospheric pollutants (nitrogen oxides, acid rain, ozone) enhance the adverse effect of the environment.

Nitrogen dioxide is one of the most dangerous atmospheric photo-pollutants. Its background is 0.02 to 0.03 p.p.b. (parts per billion). NO_2 acts as a free radical initiator accelerating the rate of hydrocarbon oxidation [9]. The attack by NO_2 results in gas fading of light polymers and shading of coloured polymers. Moreover, NO_2 depletes aminic chain-breaking antioxidants (CB AO). Oxidation of NO_2 and SO_2 emissions in the atmosphere results in acid depositions returning to the earth as acid rain, fog or dry particles. These acid pollutants adversely affect activity of amine stabilizers.

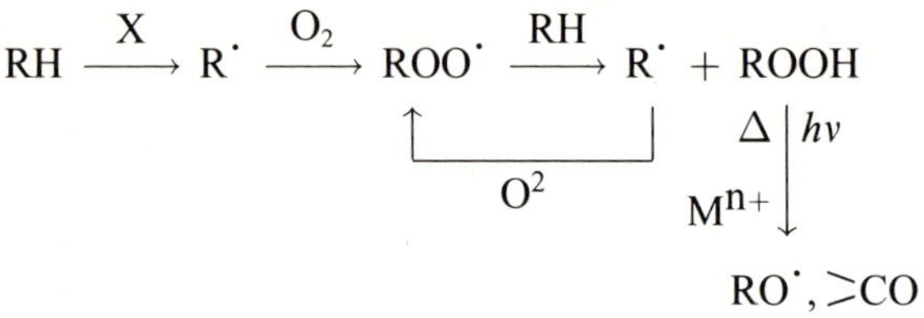

$$RH \xrightarrow{X} R^{\cdot} \xrightarrow{O_2} ROO^{\cdot} \xrightarrow{RH} R^{\cdot} + ROOH$$

with O_2 feedback and $\Delta \mid h\nu$, M^{n+} pathway to $RO^{\cdot}, {>}CO$

Scheme 1

Most oxidative mechanisms characteristic of the degradation of polymers during processing, storage and long-term end-use involve free radicals induced thermally, catalytically, mechanochemically or by radiation. Macroalkyls $R^{\cdot}$ are formed in hydrocarbon polymers RH by breaking C–H and C–C bonds during processing, mechanodegradation and by high energy radiation [10, 11]. In the presence of oxygen, $R^{\cdot}$ are quickly oxidized into alkylperoxyls $ROO^{\cdot}$ and the chain oxidation governs polymer degradation. Mechanistic pathways of the oxidative process have been described in detail recently [12, 13]. Free radicals $R^{\cdot}$, $RO^{\cdot}$ and $ROO^{\cdot}$, alkylhydroperoxides ROOH and carbonyl groups formed by disproportionation or β-scission of $RO^{\cdot}$ are species arising in the *hydroperoxide pathway*, the principal process accounting for autoxidation and photo-oxidation of polymers (Scheme 1).

ROOHs are the most dangerous chromophores in polymer stabilization and are homolyzed thermally by UV radiation and ions of transition metals.

Hydroperoxides **1** and cyclic peroxides **2** are formed in diene based rubbers by oxidation with singlet oxygen $^1O_2$ [14].

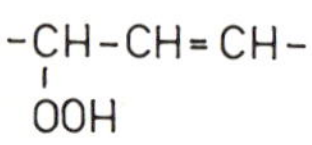

$$-\underset{\underset{OOH}{|}}{CH}-CH=CH- \qquad\qquad -\underset{\underset{OOH}{|}}{CH}-\overset{\overset{CH-CH}{\diagup}}{CH}\underset{O\,-\,O}{\diagdown}\overset{\diagup}{CH}=\underset{\underset{OOH}{|}}{CH}-$$

$$\qquad\qquad 1 \qquad\qquad\qquad\qquad\qquad\qquad 2$$

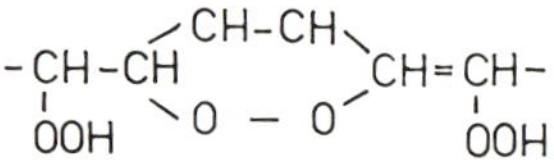

The *peracid pathway* (Scheme 2) is of minor importance for the oxidative transformation of polymers [15]. It must, however, be considered in the integral mechanism due to high reactivity of involved species with aminic stabilizers. Aldehydes formed as intermediates in the autoxidation are potential chain carriers and a source of acyl and acylperoxyradicals and peroxyacids [16].

Acyl radicals may also arise in Norrish I photolysis of carbonyl groups containing polymers [14].

Thermo- and mechano-initiated oxidation of vulcanized rubbers during cyclic dynamic flexure accounts for mechanical defects on the surface and in the bulk of

$$-CH_2CHO \xrightarrow{ROO^{\cdot}} -CH_2C^{\cdot}(O) \xrightarrow{O_2} -CH_2C(O)OO^{\cdot} \xrightarrow{RH} -CH_2C(O)OOH$$

Scheme 2

rubbers. The process has been recognized as flex-cracking or fatigue of rubbers. It has been accompanied by C-centred radicals formed by splitting of the rubber chains and by S-centred radicals arising in the restructuralization of the poly-sulfidic network [5]. The aging of diene polymers is controlled by free-radical oxidation, either photothermal in the presence of light or thermal in the dark.

Ozonation contributes to polymer degradation by ionic and free-radical mechanisms. Formation of ozone due to anthropogenic activities and photochemical processes in the troposphere closely correlates with formation of NO_2 [17]. Tropospheric ozone is a dangerous photo-oxidant in the urban atmosphere. Besides the harmful effect on human health and plants, it actively attacks unsaturated rubbers [3]. Ozone enhances external cracking during fatigue and accelerates the crack growth in rubber vulcanizates subjected to constant or repeated loading. Ozone is responsible together with oxygen for fleck-cracking phenomena [10, 19]. The macroscopically observable effect has also been called *ozone cracking*. Its physical mechanism is rather well described.

The ozone concentration in the troposphere during the daytime is typically about 1 pphm (parts per hundred million parts of air by volume) [20]. Values up to 100 pphm were measured in some photochemical smog areas. The molecular mechanism of the ozone aging of diene based elastomers was studied in detail and is well understood [19, 21]. Products or intermediates different from those arising in autoxidation or photo-oxidation of polymers were identified: ozonides (**3**), zwitterions (**4**), diperoxides (**5**), polyperoxides (**6**), polymeric ozonides (**7**) and terminal aldehydes (**8**). Reactivity of aminic antiozonants (AOZ) with these species accounts for the protection of rubbers against atmospheric O_3. AOZ must also possess antioxidant properties, because the free radical processes are concerted with ozonation due to the permanent presence of oxygen.

$$-\overset{\displaystyle O - O}{\underset{\displaystyle O}{\underset{|}{CH}}}\overset{}{\underset{|}{CH}}- \qquad -\overset{+}{C}H-O\overset{-}{O} \qquad -\overset{\displaystyle O-O}{\underset{\displaystyle O-O}{CH}}\overset{}{CH}-$$

$$\text{3} \qquad\qquad\qquad \text{4} \qquad\qquad\qquad \text{5}$$

$$\{CH(R)OO\}_n \qquad \{CH(R)OCH(R^1)OO\} \qquad -CHO$$

$$\text{6} \qquad\qquad\qquad \text{7} \qquad\qquad\qquad \text{8}$$

The effect of ozone on saturated polymers, polyolefins (PO) and polystyrene (PS) in particular, is underrated. Ozone is involved in the initiation phase of their oxidation. Its attack results in formation of $R^{\cdot}$, $RO^{\cdot}$, $ROO^{\cdot}$, ROOH and $>CO$ species in PO, and free radicals and polymeric peroxides in PS. Moreover, ozone decomposes hydroperoxides into oxygen-centred radicals $RO^{\cdot}$, $ROO^{\cdot}$ and $HO^{\cdot}$.

The polymer lifetime is also influenced indirectly by stratospheric ozone and its photochemistry. Terrestrial fluxes of the solar UV radiation are changing with the stratospheric ozone concentration. Stratospheric ozone depletion accounting for global ozone losses of $2.7 \pm 1.4\%$ per decade may have increased to about

4% per decade outside of the tropics [22]. This is manifested by a danger of increasing the amount of the UV-B radiation. At the current depletion rates, this could happen in 30 to 50 years. Recently [23], an increase in UV-B radiation of 1.9% per year has been measured during the summer time in the Toronto area since 1989 and directly correlated with the stratospheric ozone loss. Higher rates of UV degradation of plastics in near equatorial regions were attributed to higher UV radiation doses and explained as a consequence of the thinner stratospheric ozone layer above this region. This changing situation is reflected in requirements for more efficient photostabilizers. This may be reached by some bifunctional aminic photoantioxidants having UV absorbing and radical scavenging capacity.

The practical use of various goods fabricated from polymers is very diverse as regards the severity of environmental attack. New functional moieties are formed in polymers during aging. Some of them are true sensitizers or initiators of further degradation, having irreversibly physical consequences, including discolouring, loss of gloss, chalking, surface cracking, crazing, changes of solubility, mechanical and electrical properties. Most technical polymers cannot therefore be used without stabilization. Amines having a proper chemical structure are suitable for protecting polymers against all tentatively described oxidative processes and react with the intermediates and products formed.

2 Structures of Aminic Stabilizers

Two groups of amines rank among polymer stabilizers. The first one consists of aromatic and nonhindered heterocyclic amines, the classical antidegradants applied mainly in rubbers and to a lower extent in PO as antioxidants (AO), heat stabilizers (HS), antifatigue (anti-flex-crack) agents (AF) and antiozonants (AOZ) [3, 21, 24, 25]. The secondary amino group in α-position to the aromatic nucleus and/or and endocyclic amino group represent the key functionality assuring the inherent chemical efficiency. Few changes have been made in the development of structures of commercial amines during the last decade.

The second group consists of hindered heterocyclic amines (hindered amine stabilizers, HAS) introduced on the market in the 1970s rank among the currently most extensively studied additives. HAS have been used almost exclusively in PO and coatings as light stabilizers (LS, photo-antioxidants) or HS. There is a continuous trend of improving some ancillary properties of HAS by modification of their molecular architecture [26]. The mechanistic knowledge has been fully exploited in commercial development optimizing the inherent chemical efficiency, resistance against harmful influences of impurities in polymers, physical persistence and compatibility with the host polymer.

Aromatic amines have structures of substituted diphenylamines (**9**, DPA), N-phenyl-2-naphthylamines (**10**, PNA) or 1-naphthyl isomers, and N, N'-disubstituted 1,4-phenylenediamines (**11, 12**, PD), the almost universal and largest scale rubber antidegradant. Terpolymer **13** is an example of a polymeric amine stabilizer.

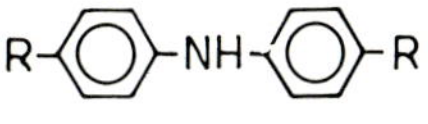

$R-\langle\bigcirc\rangle-NH-\langle\bigcirc\rangle-R$

9

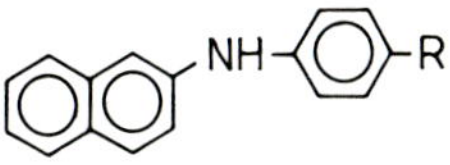

$\langle\bigcirc\bigcirc\rangle^{NH-\langle\bigcirc\rangle-R}$

10

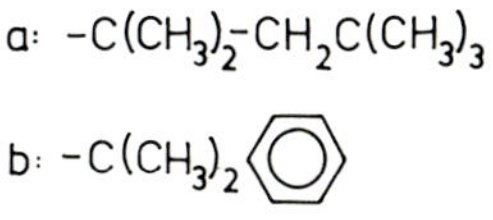

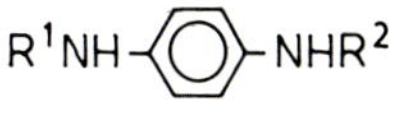

a: $-C(CH_3)_2-CH_2C(CH_3)_3$

b: $-C(CH_3)_2\langle\bigcirc\rangle$

$R^1NH-\langle\bigcirc\rangle-NHR^2$

11

$A\searrow_N\swarrow A$ over $N\bigcirc N$ over A

12

a: $R^1 = R^2 = -CH(CH_3)C_6H_{13}$
b: $R^1 = -CH(CH_3)_2$, $R^2 = -C_6H_5$
c: $R^1 = -CH(CH_3)CH_2CH(CH_3)_2$
   $R^2 = -C_6H_5$
d: $R^1 = R^2 = $ aryl

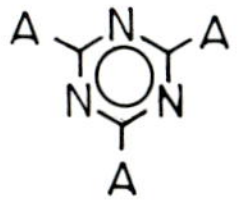

a: $A = -NH-\langle\bigcirc\rangle-NHR$

b: $A = -N=\langle\bigcirc\rangle=NR$

$R = -CH(CH_3)(CH_2)_3CH(CH_3)_2$

$\left[CHCH_2\atop CN\right]_m\left[CH_2CH=CHCH_2\right]_n\left[C\atop{C=O\atop NH\langle\bigcirc\rangle NH\langle\bigcirc\rangle}\right]\!{}^{CH_3}\!-CH_2\right]_o$

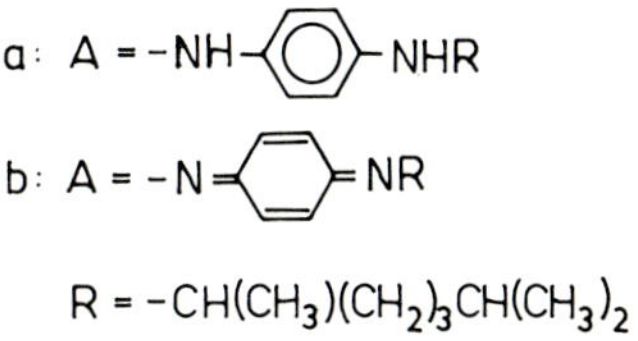

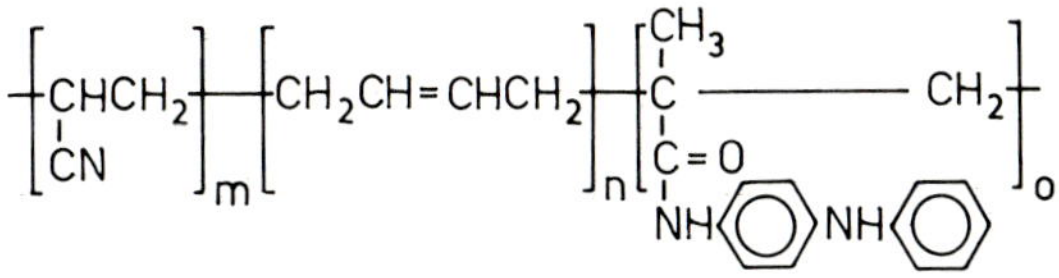

13

$N-N-CH_2N\phi_2$ over $S{\overset{\displaystyle\diagup}{\diagdown}}S{\diagup}S$ over $CH_2N\phi_2$

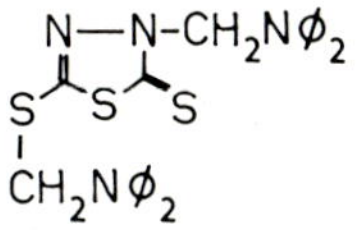

14

DPA are excellent AO and HS in various rubbers and plastics, lubricating greases and oils [27]. They have some antifatigue effect and provide no antiozonant protection [1]. Their photostabilization activity is rather low. DPA may be used in radiation resistant polymer compositions for cable coatings in nuclear power plants [28]. Various 4,4′-disubstituted DPA contain a *tert*-alkyl (e.g. 1,1,3,3-tetramethylbutyl **9a**, 1,1-dimethylethyl or their combinations), other alkyls or aralkylgroups (e.g. α,α-dimethylbenzyl, **9b**) [1, 3, 24]. A tertiary amine 3-[(diphenylamino)methyl]-5-[(diphenylaminomethyl)thio]-1,3,4-thiadiazolidine-2 (**14**) was proposed as AO for rubbers [29].

Both phenyl-1- and 2-naphthylamines are highly effective discolouring AO imparting a good fatigue resistance to natural rubber (NR), but a low resistance to styrene-butadiene rubber (SBR) vulcanizates. PNA have no antiozone protection

capacity [1]. They lost their importance for rubbers because of toxicological considerations. PNA are, however, of continuous interest in stabilization of synthetic
ester lubricating oils and greases. Persistent **10** have bulky substituents on the
phenyl moiety, like 4,4'-methylene-bis[*N*-(4-*tert*-octylphenyl-1-naphthylamine] or
substituted naphthalenediamines [30].

N,N'-Disubstituted PD **11** are discolouring stabilizers protecting vulcanized
NR, SBR, butyl rubber (IIR), nitrile rubber (NBR) or ethylene-propylene-diene
terpolymer (EPDM) against oxidative degradation, dynamic fatigue, heat and
ozone degradation [1, 21, 31]. PD function as AO/HS in plastics, greases or
lubricating oils and Diesel fuels and were proposed as stabilizers for very
advanced applications in electrophotographic photoreceptors or materials for
high-speed recording [28, 32, 33]. The properties are influenced by the nature
of *N,N'*-substituents [1, 3, 21, 24, 31]. Derivatives having at least one aromatic
substituent (phenyl, naphthyl) possess antifatigue and some antiozonant activities. *N,N'*-Diaryl-PD (**11d**) are therefore excellent AO/AF. The presence of
N-*sec*-alkyl group slightly reduces the antifatigue effect but enhances the
antiozonant effect. Bis-*sec*-alkyl-PD (e.g. **11a**) are strong AOZ. However, the un-
symmetrically substituted *N*-*sec*-alkyl-*N'*-aryl-PD **11b, c** provide the best
balanced antioxidant/antifatigue/antiozonant protection under dynamic stress and
aggressive environment. Great effort was spent on the improvement of industrial syntheses and optimal exploitation of substituent effects in PD [34, 35].
Various heterocyclic amines, like aminopyrroles, isoindolidones, pyridazines
or carbazoles having endocyclic secondary amino groups or nitrogen heterocycles with exocyclic amino substituents protect organic substrates against
oxidative attacks. Only some heterocycles attracted commercial and scientific interest: 2,2-disubstituted-1,2-dihydro-3-oxo(or phenylimino)-3*H*-indoles (**15**, DHI),
6-substituted 1,2-dihydro-2,2,4-trimethylquinolines (**16a, b**, DHQ), various other
6-substituted DHQ, like 6-hydroxy-, 6-anilino-or 6,6-alkylidenebis-DHQ (derivatives having substituents in position 8 are ineffective stabilizers), 1,2,3,4-
tetrahydro-2,2,4-trimethylquinolines (**17**), 2,2,4-trimethyldecahydroquinoline (**18**),
3,3-dialkyldecahydrochinoxaline-2-one (**19**), substituted phenothiazines (**20**),
oligomeric 1,2-dihydro-2,2,4-trimethylquinoline (**21**, the second largest scale
rubber antidegradant), condensation products of DPA with acetone (**22**) or
acetone and formaldehyde (**23**), containing bound-in 9,10-dihydroacridine and/or
carbazole moieties [36–39].

15   16   17

a: X = O   a: R = H

b: X = Nϕ b: R = OC$_2$H$_5$

All heterocyclic amines **15** to **23** are more or less discolouring stabilizers. DHI (**15**) is HS for PO. Monomeric DHQ **16** are very good AO in vulcanized rubbers and protect against heat aging (HA) and fatigue. They also stabilize oils and fats against oxidation. Oligomeric DHQ **21** are AO and HS with little antifatigue and no antiozonant activity. Oligomers **22** and **23** are excellent AO for rubber vulcanizates [1].

Partially hindered heterocyclic amines fused with alicyclic groups as in **18** and **19** impart an excellent light stability effect to PO. The effect is weaker than that of HAS [39]. This may be ascribed to the presence of the α-CH bond in **18** and **19**.

PT **20** contains a sulfur atom in the heterocyclic system, contributing by its hydroperoxide decomposing (HD) potency to the integral antioxidant effect. A similar effect may be expected in **14**, in derivatives DHQ containing dithiolothione or thiadiazolidinethione moieties [29].

The lack of hydrogens on both α-carbons to the nitrogen atom in the heterocycle is the main structural feature of amines ranking in the HAS category. This structural feature is reflected in the impossibility of forming nitrone or quinone imine (QI) structures during sacrificial transformations of HAS without a deeper destruction of the heterocycle. HAS generally have structures of secondary ($>$NH) and tertiary amines ($>$NR, R is mostly methyl; other alkyls or aralkyls are used in development HAS). The more recent HAS belong among hindered hydroxylamines ($>$NOH), O-alkylhydroxylamines ($>$NOR) and acylamines ($>$NCOCH$_3$).

Derivatives of oxazolidine, imidazolidine, 1,4-dihydropyridine, piperidine, piperazine or hexahydropyrimidine are examples of tested structures. Only two of these structures have been commercialized [13, 40–46]. Most HAS are derivatives of 2, 2, 6, 6-tetramethylpiperidine (for example, structures **24**, **25**, **28**–**39**). 3,3,5,5-Tetramethylpiperazin-2-ones (e.g. **27**) represent another excellent commercialized structure. HAS derivative of 1,4-dihydro-2, 6-dimethylpyridine (**26**, R = 1,2,2,6,6-pentamethylpiperidyl) is of interest for chlorinated polymers.

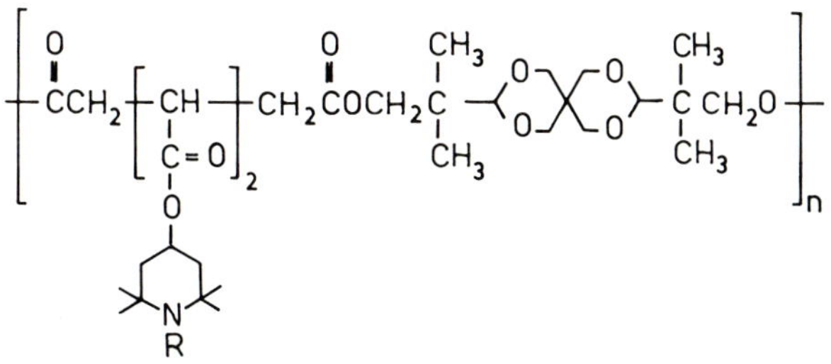

33

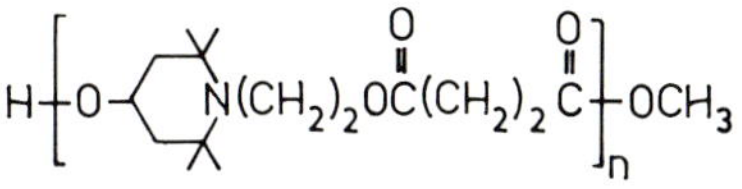

34

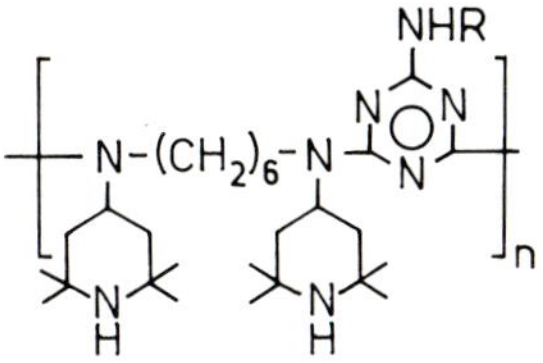

35

a : R = -C(CH$_3$)$_2$CH$_2$C(CH$_3$)$_3$

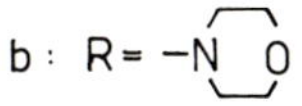

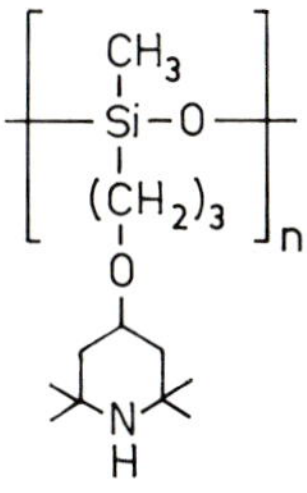

b : R = -N O

36

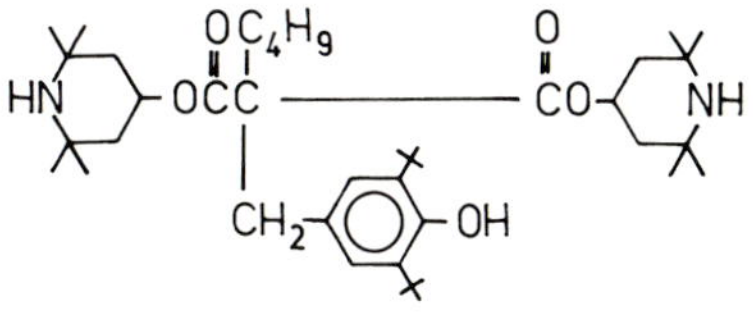

37

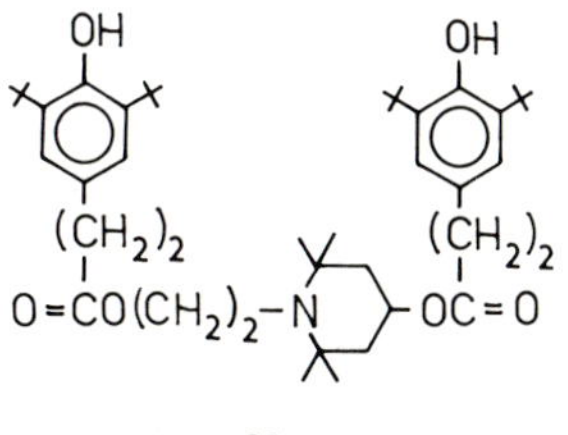

38

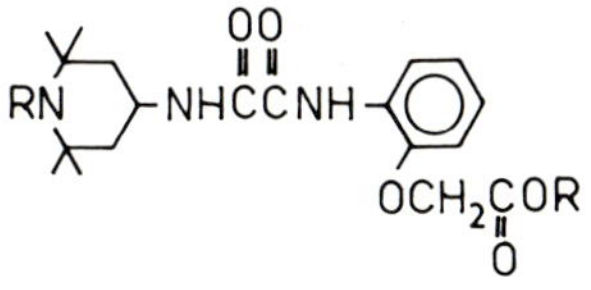

39

Great attention has been paid to high-molecular weight (HMW) HAS like **31**, **32** and oligomeric HAS, e.g. **33–36**. The molecular structure of some HAS is rather complicated, as in bifunctional stabilizers **37–39**.

3 Antioxidant, Anti-Flex-Crack and Photo-Antioxidant Activities of Amines

Interference with free-radical and peroxidic species resulting in polymer autoxidation, fatigue and photo-oxidation belong to the principal processes of polymer stabilization. AO, AF and radical scavenging/hydroperoxide decomposing LS (photo-antioxidants) fulfil the requirements by scavenging of R˙ immediately after their formation, by scavenging of RO˙, ROO˙ and RC(O)OO˙ radicals and preventing homolysis of ROOH [25]. Secondary aromatic and heterocyclic amines belong to the category of CB AO. The simultaneous antifatigue activity of some of them (PNA, PD, DHQ) arises exclusively from products of their transformation [3–5,15]. HAS possess photo-antioxidant activity reflected in their light and heat stabilizing properties. Mechanistic features of the most important structural types of amines are outlined in the following sub-chapters.

3.1 Aromatic Amines

Aromatic amines are listed among rubber processing chemicals, rank among the oldest polymer antidegradants and hold the leading position in production of all additives. They are used mostly in the stabilization of rubber vulcanizates and protect them against crazing, surface cracking, loss of mechanical properties and drastic shortening of their useful lifetime. Discolouration of polymers and staining are inherent disadvantages arising from quinone iminoide transformation products. Smaller amounts of aromatic amines are used in combination with phenolic AO in stabilization of plastics in applications where discolouration is not a problem [47]. Aromatic amines are superior AO to phenols in PO and carbon black filled rubbers. Some derivatives of DPA or PD retard photoprocesses in PO. Derivatives of PD are considered as weak metal deactivators (MD) in catalyzed oxidations [48]. The inherent chemical efficiency of amines and the strength of the N–H bond are affected by their structure and polar effects of substituents [3, 4]. Primary aromatic amines may contaminate commercial PD. They do not substantially reduce their antioxidant activity. However, the cure process is unfavourably influenced and some condensation reactions resulting in more complicated QI are triggered.

3.1.1 Antioxidant Activity

Hydrogen transfer from the secondary amine $>$NH to the ROO˙ radical resulting in formation of an aminyl radical $>$N˙ (Eq. 1) is the key and rate controlling step of the CB AO activity of aromatic amines [3, 4, 49]. Kinetic deuterium isotope effect supports the process according to Eq. (1):

$$>\text{NH} + \text{ROO}^{\cdot} \longrightarrow >\text{N}^{\cdot} + \text{ROOH} \tag{1}$$

The rate constants of the reaction at Eq. (1) are in the range 10^5–10^7 dm^3 mol^{-1} s^{-1} [50]. Electron donating substituents attached to the aromatic ring enhance the reactivity with ROO$^\bullet$ [51]. Dissociation energies D_{N-H} are in the range 340–380 kJ mol^{-1}. Amines having higher strength of the $>$N–H bond have only low reactivity in Eq. (1). Those having D_{N-H} lower than 340 kJ mol^{-1} may be easily depleted by direct oxidation with molecular oxygen:

$$>NH + O_2 \longrightarrow >N^\bullet + HOO^\bullet . \qquad (2)$$

Rate constants in Eq. (2) are $\sim 10^{-5}$ dm^3 mol^{-1} s^{-1} for monoamines **9** and **10** and $\sim 10^{-2}$ dm^3 mol^{-1} s^{-1} for PD **11** [50]. At a proper substitution, the amount of amines oxidized according to Eq. (2) is negligible in comparison with the process in Eq. (1). A considerable charge separation was observed in the transition state of Eq. (1) [51]. Charge-transfer-complexes (CTC) have been envisaged for aromatic monoamines (**40**) and diamines (**41**) as the primary step preceding formation of the aminyl:

$$\left[ROO^- \; \bigcirc\!\!-N^{+\bullet}H\!-\!\!\bigcirc \right] \qquad \left[ROO^- \; R^1NH\!-\!\!\bigcirc\!\!-N^{+\bullet}HR^1 \right]$$

$$\textbf{40} \qquad\qquad\qquad\qquad \textbf{41}$$

The antioxidant activity of aromatic amines gets stronger with increasing conjugation of the system in the series DHQ$<$DPA$<$PNA$<$PD.

3.1.1.1 Aminyl Radicals

Formation of the aminyl $>$N$^\bullet$ according to Eq. (1) is the first step of the service or sacrificial transformation of amines in the antioxidant mechanism [3, 4, 6, 21, 49]. Aminyls also result in Eq. (2) by oxidation with $^1O_2$, HO$^\bullet$ radicals, hydroperoxides, ions of transition metals or NO$_2$ and by excited chromophores (Eq. 3). A part of these reactivities reduces the efficiency of the main pathway (Eq. 1) and has been included among activity depleting processes.

$$>NH \xrightarrow[\text{ROOH, M}^{n+}\text{, NO}_2]{\text{HO}^\bullet\text{, ROO}^\bullet\text{, O}_2\text{,}\,^1O_2} >N^\bullet \longrightarrow \text{Products} \qquad (3)$$

Aromatic amines and derived nitroxides are efficient chemical and physical quenchers of $^1O_2$ [52]. The reactivity with ROOH (Eq. 4) is in general lower than that of aliphatic amines:

$$>NH + ROOH \longrightarrow >N^\bullet + RO^\bullet + H_2O \qquad (4)$$

Formation of hydroxylamine $>$NOH was also considered after reaction of DPA or PD with ROOH via a complex [$>$N$^{\bullet+}$H, RO$^\bullet$, HO$^\bullet$].

"Products" arising in Eq. (3) are derived from the respective aminyl, independently of the way of its generation. Some products possess free-radical scavenging activity [4–6].

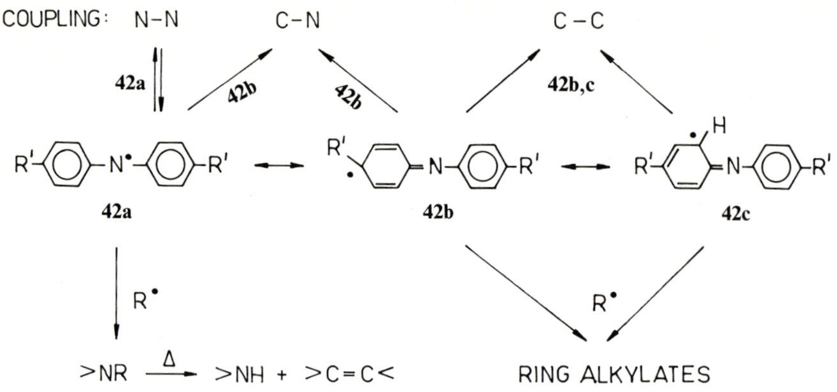

COUPLING: N–N   C–N   C–C

Scheme 3

Aminyls were evidenced by ESR spectroscopy or pulse radiolysis. They are short-living radicals having properties of weak oxidation agents. The reactivity of aminyls decreases with increasing conjugation of the parent amine system, i.e. from DPA to PNA and PD series. Reactive aminyls derived from DPA behave as N-centred (**42a**) or C-centred free radicals (**42b, c**) and undergo bimolecular N–N, C–N and C–C couplings [3] accounting for dimeric and trimeric transformation products (Scheme 3, $R' =$ H, *tert*-alkyl or aralkyl). The existence of mesomeric forms **42a–42c** augments the variability of the coupling products [3, 4]. Rate constants for coupling reactions (without any specification of products formed) have been reported to be in the order 10^6–10^9 dm^3 mol^{-1} s^{-1} [50]. The coupling processes are diffusion controlled, limited in viscous media and may take place in the polymer matrix only at sites with an increased aggregation of amines.

The N-centred radical **42a** participates in a reversible N–N coupling/dissociation process [3]. Tetraphenylhydrazine **43** was formed in the yield of 40% from **42a** generated from DPA with α, α-diphenyl-β-picrylhydrazyl. Various dimeric C–N and C–C coupling products, e.g. **44**, **45** were isolated [3–5, 53].

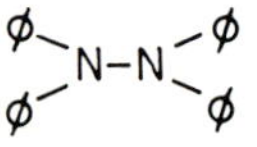

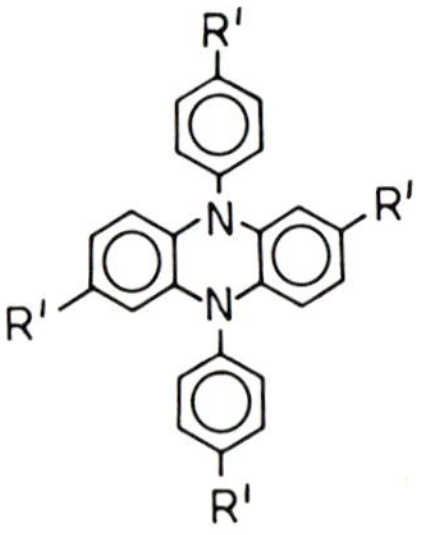

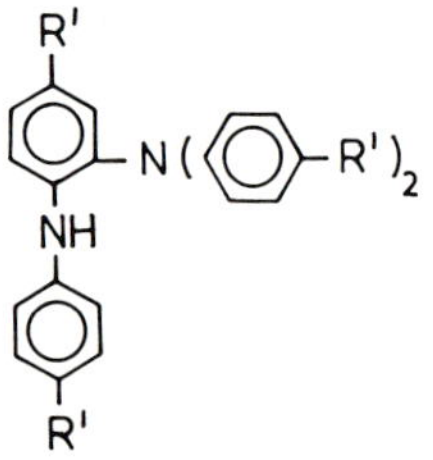

The ratio of the N–N, N–C$_{para}$ and N–C$_{ortho}$ couplings was estimated to be 1:0.15:0.03 [54]. Oligomeric condensates having six or seven DPA units were formed among the products.

It was suggested that part of the aminyls of the DPA series participates in an amine regeneration via reactivity with ROO·:

$$>N· + ROO· \longrightarrow >NH + O_2 + Olefin \tag{5}$$

Aminyls formed in DPA doped octadecene or generated in model experiments by thermolysis of 1,4-diphenyl-1,4-bis(2-naphthyl)-2-tetrazene facilitated deciphering the mechanism and products of transformation of PNA in rubber vulcanizates and synthetic oils [55]. N–N, C–C (e.g. **46**) or C–N coupling products (e.g. **47**) and more complicated trimers and tetramers are formed from 1- or 2-PNA [3, 5, 53, 56]. 7-Phenyl-dibenzo[**c**, **g**]carbazole (**48**) was formed as a trace product.

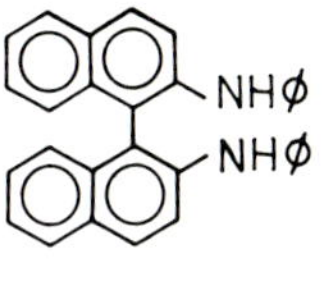

46 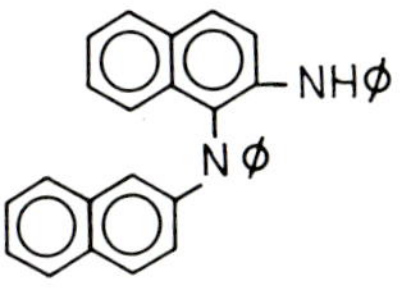   **47** 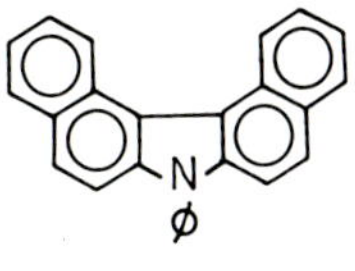   **48**

Mechanistic studies reveal transformations of N, N'-disubstituted PD **11** in autoxidized and photo-oxidized polymers and model hydrocarbons [57, 58]. Wurster's ion radicals **49**, associated with ROO· in CTC **41** are formed in the first step after electron transfer from **11** [3, 24]. **49** is formed regardless of the oxidizing agent and is stabilized via electron delocalization in mesomeric forms [59]:

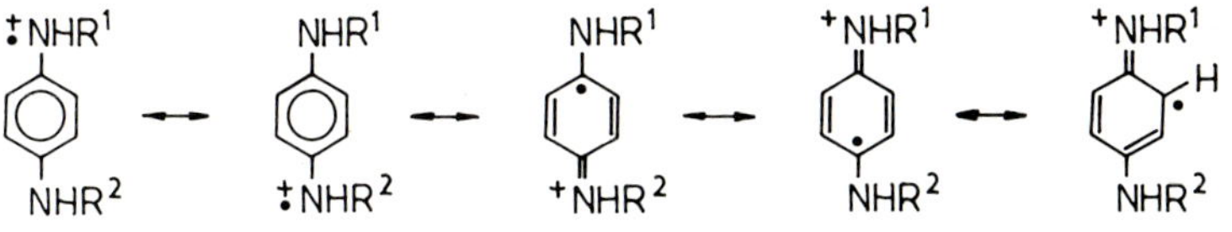

49

Processes similar to those exemplified for DPA in Scheme 3 are envisaged for mesomeric aminyls **50a, b, c** of the PD series. Coupling products **51** or **52** are examples.

Disproportionation of **50a** results in regeneration of **11** and formation of benzoquinonediimine (BQDI) **53** (Scheme 4). Coupling products of **50** having preserved the secondary amino-group as in **46**, **47**, **51** or **52** are strong CB AO [53, 56] and function according to Eq. (1). The respective new aminyls account for the formation of oligomeric coupling products and augment the product complexity. Because of the high reactivity, aromatic aminyls were suspected of abstracting hydrogen from C–H bonds in hydrocarbon polymers (Eq. 6). Model experiments in ethylbenzene reveal rate constants in the range 10 to 10^{-1} dm^3 mol^{-1} s^{-1} [50].

$$>N· + RH \longrightarrow >NH + R· \tag{6}$$

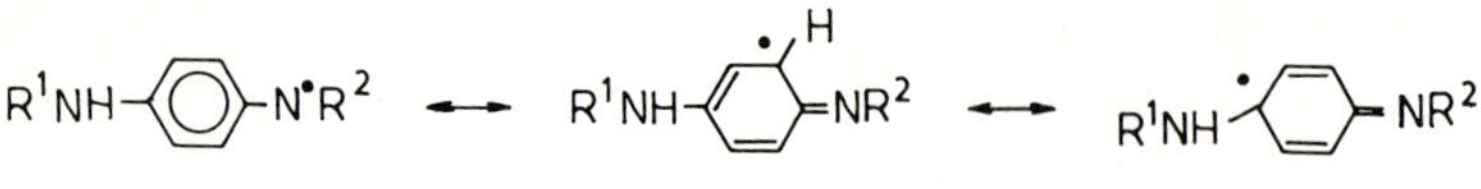

50a   **50b**   **50c**

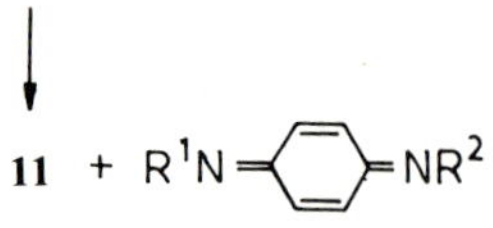

11 + R^{1}N=⟨⟩=NR^2

53

Scheme 4

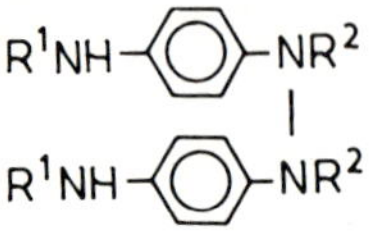

51 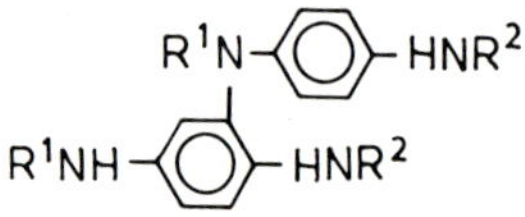   **52**

Fortunately, aminyls derived from efficient antidegradants of the DPA and PNA series do not react in polymeric systems according to Eq. (6).

Reactivity of aminyls with ROOH, according to Eq. (7), was considered to diminish the antioxidant effect of amines. CTC **40** and **41** were evidenced as intermediates in Eq. (7), and the full reversibility of the process represented by Eq. (1) was thus confirmed.

$$>N^• + ROOH \longrightarrow >NH + ROO^• \tag{7}$$

The rate constants reported in the literature for Eq. (7) are in a very broad range 10^2–10^5 dm^3 mol^{-1} s^{-1} [50]. Other unspecified reactions were included during kinetic measurements because the amines used in kinetic experiments were not the optimal models of efficient AO. The practical validity of the data is therefore vague. The reaction at Eq. (7) may be omitted in the PD series and is negligible for polymers doped with DPA and PNA.

Scavenging of R$^•$ and ROO$^•$ by aromatic aminyls is an important contribution to the stabilization mechanism. Scavenging of R$^•$ by N-centred radicals **42a** is involved in the regeneration of the $>$NH function. An olefin is released after thermolysis of the N-alkylate $>$NR. C-Centred radicals **42b, c** yield alkylates irreversibly. The process was proved using low-molecular weight (LMW) R$^•$. Formation of polymer-bound species after scavenging of macroalkyls may be extrapolated from model experiments. Trapping of R$^•$ was also confirmed

54 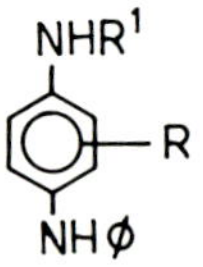   **55**

for aminyls of the PNA and PD series. Thermolabile *N*-alkylate **54** and stable *C*-alkylate **55** are examples of products formed from **11b, c** [4].

3.1.1.2 Nitroxides

Reactivity of aromatic aminyls of the DPA (**42**) and PNA series in their N-centred forms with ROO˙ results in a nitroxide **56** or its mesomeric nitrone form [60] (Scheme 5). The kinetics of the formation of the nitroxide in DPA doped hydrocarbons was monitored by ESR spectroscopy [61].

Tertiary ROO˙ are more efficient than the primary and secondary ones in the $>$NO˙ formation [51, 60]. In the latter two cases, the caged $>$NO˙ and RO˙ radicals (**61**) may disproportionate into a hydroxylamine and a carbonyl compound as a competitive process to $>$NO˙ formation (Scheme 6). This is not possible with tertiary ROO˙.

The rate constant for the reaction of the aminyl with ROO˙ was reported 6×10^8 dm^3 mol^{-1} s^{-1} [50]. Although the rate is very high in the DPA series, it is lower than the rate of initiation of the chain oxidation. It is so because ROO˙ attacks not only the N-centred radical **42a**, but also the rings of the C-centred radicals **42b, c** [60]. The $>$NO˙ formation has been diminished by the presence of organic acids. The nitroxide arises in the yield about 80% also in ozonation of DPA or its 4,4'-dialkylderivatives [62].

Various concentration levels of nitroxides are observed in the oxidized amine doped substrate. The higher the reactivity of the formed $>$NO˙, the lower is its detectable concentration. The reactivity of aminyls increases in the series DPA < PNA < PD.

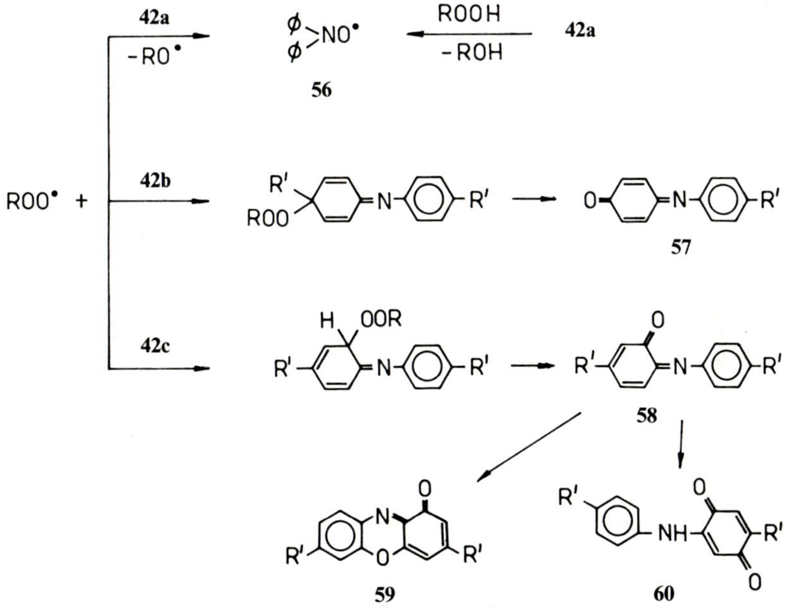

Scheme 5

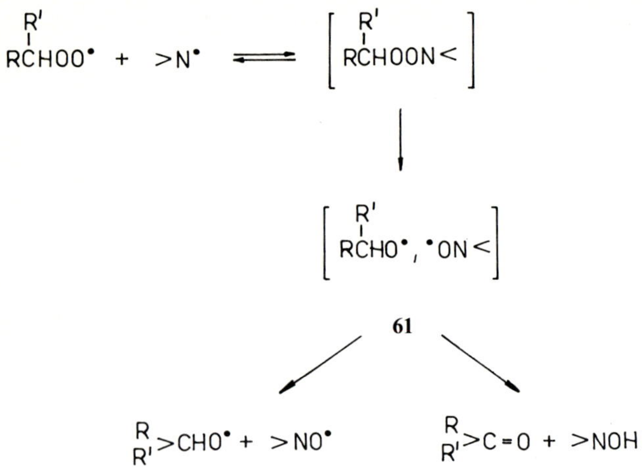

$$\underset{\substack{R' \\ |}}{RCHOO^\bullet} + >N^\bullet \;\rightleftharpoons\; \left[\underset{\substack{R' \\ |}}{RCHOON<} \right]$$

$$\left[\underset{\substack{R' \\ |}}{RCHO^\bullet}, \; {}^\bullet ON< \right]$$

61

$$\underset{R'}{\overset{R}{>}}CHO^\bullet + >NO^\bullet \qquad\qquad \underset{R'}{\overset{R}{>}}C=O + >NOH$$

Scheme 6

Thermodynamic and kinetic data indicate strong solvation of the nitroxides by a polar environment [63]. This may influence redistribution of $>NO^\bullet$ in the oxidized polymer matrix. Some mechanistic conclusions obtained in the DPA series were extrapolated to other aromatic amines.

Antioxidant properties of aromatic nitroxides have been reported in polymers and model systems [3, 24]. The antioxidant activity is lower than that of the parent amines or derived hydroxylamines.

Diarylnitroxides are strong abstractors of hydrogen atoms (Eq. 8). The respective hydroxylamine is formed from $>NO^\bullet$. Nitroxides oxidise phenols via the respective phenoxyls. For example, 2,6-di-*tert*-butylphenol was converted in 61% yield into 2,6-di-*tert*-butyl-1,4-benzoquinone [64].

$$>NO^\bullet + XH \longrightarrow >NOH + X^\bullet \tag{8}$$

Diarylnitroxide **56** reacts in mesomeric forms **56a** to **56d** [65].

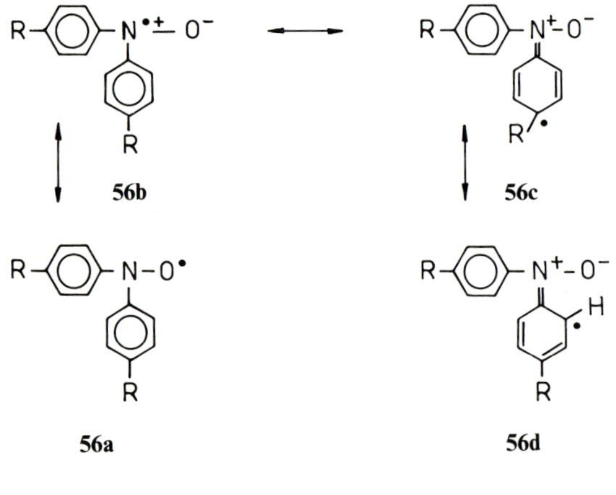

56b

56c

56a

56d

This enables trapping both alkyl and ROO˙ radicals (Scheme 7). The reaction with ROO˙ is much slower:

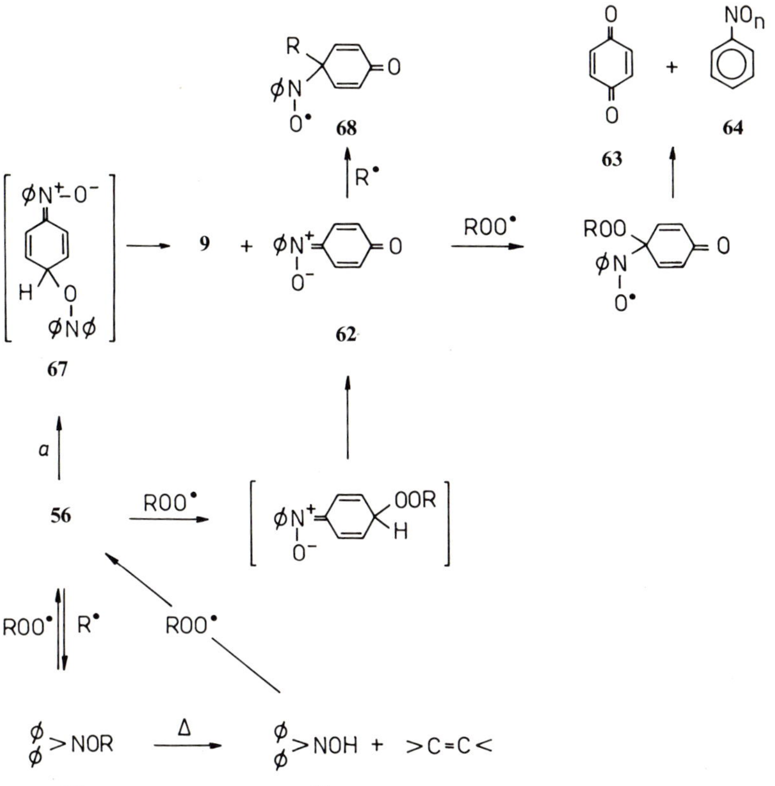

Scheme 7

The nitroxide **56** traps in its mesomeric form **56c** radicals ROO˙. Quinone imine N-oxide **62**, is formed via alkylperoxycyclohexadieneimine intermediate. **62** is destroyed by the further attack of ROO˙. Benzoquinone (**63**) and nitroso- (**64**, $n=1$) and nitrobenzene (**64**, $n=2$) are formed in the ultimate phase of the lifetime of **62** [65]. An alternative pathway for oxidation of **56** with ROO˙ in cumene autoxidation suggests formation of an olefin and hydroxylamine **66** as CB antioxidant species [66]:

$$56 + ROO˙ \longrightarrow {>}C = C{<} + \textbf{66} \tag{9}$$

Scavenging of R˙ by **56** results in O-alkyldiphenylhydroxylamine ${>}NOR$ (**65**) [66]. ${>}NO˙$ is consumed in an inert atmosphere by the rate equal to that of R˙ formation. This reactivity contributes to the antifatigue effect of aromatic amines only in the oxygen-deficient environment. After oxygen was admitted into the system, $>$NO$^•$ reappears, presumably by the following reaction [66]:

$$65 + ROO^• \longrightarrow ROOH + 56 + {>}C = C{<} \tag{10}$$

Reactivity with ROO$^•$ was also used for explanation of the nitroxide regeneration from **65** and, consequently, of the high stoichiometric factor of DPA in autoxidation of hydrocarbons [66]. Thermolysis of $>$NOR **65** at temperatures exceeding 100 °C was proposed as another process regenerating the antioxidant activity [65]. Hydroxylamines **66**, strong CB AO, are formed. $>$NOH **66** are reoxidized to **56** either by scavenging ROO$^•$ (Scheme 7) or by oxidation with ROOH via CTC [$>$N$^{•+}$OH, HO$^-$, RO$^•$].

Diphenylhydroxylamine **66** is not an ideal AO. It disproportionates slowly into **56** and the parent amine and is oxidized easily to **56** by air oxygen [67]. Unlike **66**, good experience was obtained with dibenzylhydroxylamine, the relevant bis(4-alkoxyderivatives), amide **69** or its ester analogue [68, 69].

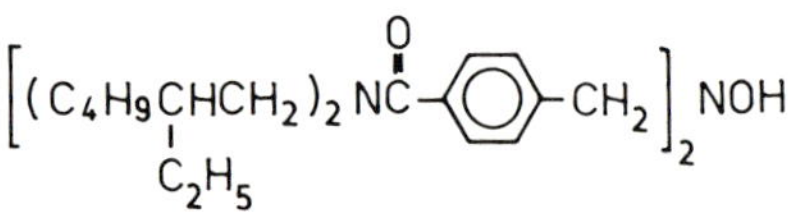

$$\left[(C_4H_9CHCH_2)_2 \overset{\overset{O}{\|}}{N}C{-}\langle\bigcirc\rangle{-}CH_2 \right]_2 NOH$$
$$C_2H_5$$

69

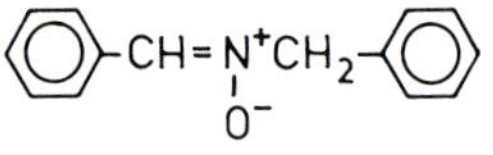

$$\langle\bigcirc\rangle{-}CH{=}\overset{+}{N}CH_2{-}\langle\bigcirc\rangle$$
$$O^-$$

70

$$\langle\bigcirc\rangle{-}\underset{R}{CH}{-}\underset{O^•}{NCH_2}{-}\langle\bigcirc\rangle$$

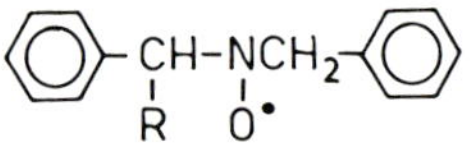

71

These compounds stabilize PO against thermal and light induced degradation and form efficient combinations with phenolic AO. According to an early observation [70], nitrone **70** may be formed from dibenzylhydroxylamine and serves as an alkyl radical scavenger generating a "secondary" nitroxide **71**.

Regeneration of the radical scavenging activity via formation of the nitroxide **56** from $>$NOR **65** or $>$NOH **66** (Scheme 7) has only minor importance for aromatic amines. A great proportion of the present nitroxides is lost in side reactions involving their mesomeric forms, resulting in **63** and **64**. Moreover, trapping of R$^•$ in air atmosphere is a very vague process. Unfortunately, the participation of the couple $>$NO$/>$NOR in a cyclic trapping of R$^•$ and ROO$^•$ (Scheme 7) has been repeatedly reported in the literature even for systems and conditions where the formation of $>$NOR is practically impossible.

It is a pity that the probability of the transient formation of **65** is so low. Synthetically prepared stabilizers of this class, like O-methyldibenzylhydroxylamine [71] have an appreciable effect in PO protection.

Pathway a in Scheme 7 represents self-reactions involved in disproportionation of **56**, resulting in the parent amine **9** and N-phenyl-1,4-benzoquinonemonoimine-N-oxide (**62**) [61]. The disproportionation proceeds via the coupling product **67**.

The principal attack of ROO˙ on N-phenyl-2-naphthylnitroxide is on the naphthyl ring of the PNA system [60]. QI **72** and **73** and naphthoquinones **74** and **75** are formed [3, 5, 53, 56]. The disproportionation of phenylnaphthylnitroxide into **10** and 2-phenylamino 1,2-naphthoquinone-N-oxide **76** proceeds similarly as for **56** in Scheme 7. Analogous transformations were observed with the isomeric N-phenyl-1-naphthylnitroxide [53, 56].

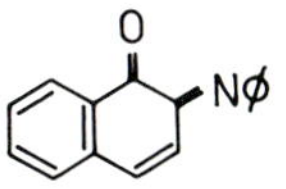

72 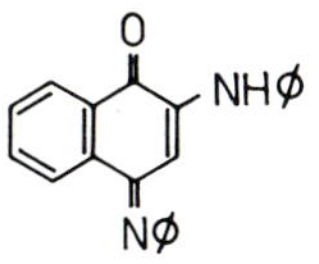　　　　　　　**73** 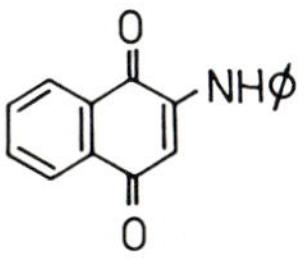　　　　　　　**74**

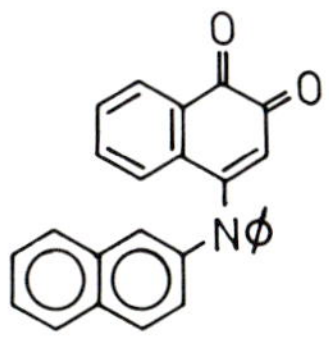

75 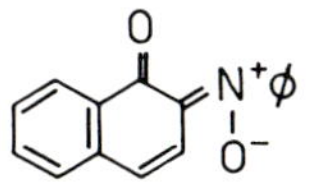　　　　　　　**76**

Mechanistic data obtained with the diarylnitroxide **56** have also mostly been extrapolated for nitroxides derived from heterocyclic amines DHQ, DHI, PT and HAS. The relative participation of the relevant nitroxides in the three principal pathways (Scheme 7), i.e. scavenging of R˙ and ROO˙ and disproportionation, has been determined by the chemical structure of the particular nitroxide and its stability reflected in the tendency to react in mesomeric forms. The adjacent aromatic moiety enhances participation of the mesomeric structures. The stability of the nitroxides diminishes in the series from HAS to DHQ or DHI and DPA and is the lowest in the PD series [4]. As a consequence, the nitroxides derived from PD exist only in the form of bisnitrones (**81** in Scheme 9). The ROO˙ scavenging ability is involved only in the aromatic (DPA, PNA) and heterocyclic (DHQ, DHI) amines and cannot be included in the reactivity of HAS derived nitroxides. On the contrary, the most efficient R˙ scavenging is in the HAS series.

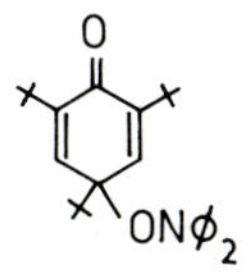

77

Other radical reactions that R˙ and ROO˙ scavenging encounter diarylnitroxides **56**. Mechanistic studies revealed formation of **77** with 2,4,6-tri-*tert*-butylphenoxyl [72]. Recombination with thiyl- and sulfinyl radicals arising during rubber vulcanization, fatigue and restructuring of rubber vulcanizates or oxidative splitting of sulfidic crosslinks in vulcanizates results in $>$NOSR or $>$NOS(O)R, hydrolyzing into amine **9** and sulfenic (**78**, n $=2$) or sulfonic (**78**, n $=3$) acids [4]. The process may be considered as regeneration of the aminic function:

$$\begin{array}{c} RS^{\cdot} \\ 56 \; \longrightarrow \; \begin{array}{|c} \hline \end{array} \end{array}$$

$$56 \; \longrightarrow \; \left[\begin{array}{l} RS^{\cdot} \longrightarrow \; >NOSR \\ \\ RS^{\cdot}(O) \longrightarrow \; >NOS(O)R \end{array} \right] \xrightarrow{H_2O} \; \mathbf{9} + RS(O)_n H \qquad (11)$$

$$\underset{\mathbf{78}}{}$$

Nitroxide **56** and analogous nitroxides generated from 4,4′-disubstituted DPA, PNA, DHQ or DHI should be categorized as the "primary" ones. ESR signals of $>$NO$^{\cdot}$ observed in flex-cracked polymers may also arise from "secondary" nitroxides formed after R$^{\cdot}$ trapping by nitrones [4, 73] (see Sect. 3.1.1.3, Scheme 11). These nitroxides are most probably those detected as the transformation species of PD.

3.1.1.3 Quinone Imines and Nitrones

Reaction of mesomeric C-radicals **42b** and **42c** with ROO$^{\cdot}$ is a competitive process to the nitroxide formation and accounts for coloured benzoquinone monoimines (BQMI), e.g. 1-phenyl-1,4-benzoquinone-4-imine (**57**, R′ $=$ H) or 1,2-benzoquinone imine **58**, (R′ $=$ *tert*-octyl or an aralkyl) [5]. QI **58** transforms further into 3,7-di-*tert*-octyl-1*H*-phenoxazone-1 (**59**) or 2-(4′-*tert*-octylphenylamino)-5-*tert*-octylcyclohexa-2,5-diene-1,4-dione (**60**, R′ $=$*tert*-octyl). The ratio of the $>$NO$^{\cdot}$ and BQMI formed according to Scheme 5 is influenced by the structure of DPA [60]. Bulky substituents in positions 4,4′ hinder BQMI formation. BQMI are also generated from DPA by oxidation with $^1O_2$ in a methylene blue sensitized reaction [74] or by oxidation of 4-aminophenols. The reaction proceeds for example with the bifunctional AO 2-(3,5-di-*tert*-butyl-4-hydroxyanilino)-4,6-bis(octylthio)-1,3,5-triazine [5]. The respective BQMI is formed transiently, and 2,6-di-*tert*-butylbenzoquinone is released after its hydrolysis. QI **72** or **73** and naphthoquinones **74** and **75** are formed in the PNA series [3, 56].

BQDI **53** is the principal product of both sacrificial and depleting transformations of PD in model hydrocarbons, PO and rubbers and is formed via oxidation with ROO$^{\cdot}$, oxygen, ozone, ozonides, organic peroxides or ions of transition metals. BQDI have therefore been found in PD doped polymers degraded under various conditions [3–5, 58]. They are formed in high preparative yield by oxidation with ferricyanide and also result in interactions of PD with some rubber chemicals, e.g. with benzothiazolyl-2-sulfenemorpholide.

QI constitute inherent impurities in commercial PD. For example, 1-phenylamino-4-*sec*-alkylamino-3,6-bis(4-phenylaminophenylimino)-1,4-cyclohexadiene

(**79**, R^1 = isopropyl or 1,3-dimethylbutyl) is present in amounts up to 3–5% in commercial **11b, c**. **79** does not diminish, however, their antidegradant activity [75].

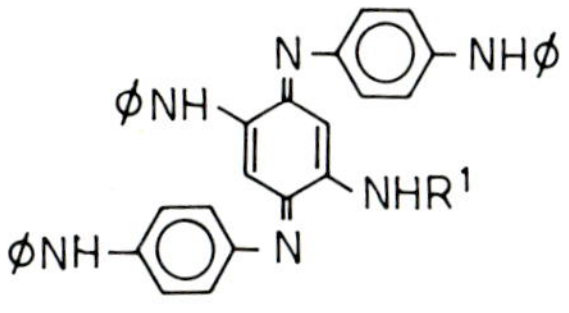

79

Regardless of which oxidation agent is used, Wurster's cation-radicals **49** are expected as intermediates. BQDI **53** is formed via disproportionation of aminyls **50a** (Scheme 4) or their oxidation with ROO˙ (Scheme 8) [3–5, 76]. The respective theoretically possible bisnitroxide **80** is unstable and transforms in situ into bisnitrone **81** (azomethine N-oxide, N,N'-disubstituted 1,4-quinonediimine-N,N'-dioxide). Consequently, no ESR signals of $>$NO˙ were detected in oxidized PD [60].

The chemistry or photochemistry of BQDI **53** [1,4-benzoquinone-bis(alkyl or aryl)imine] or BQMI **57** [1,4-benzoquinone-4-(alkyl or aryl)imine] is characteristic of the cross-conjugated systems having exocyclic $>$C=N– bonds [77, 78]. Both QI exist in the form of syn and anti geometric isomers. Model experiments revealed that QI retard thermooxidation of hydrocarbons and polymers and photo-oxidation of hydrocarbons [3, 4, 58, 75]. They are weaker AO than the parent PD and their effect is more pronounced in unsaturated substrates. For example, the Bandrowski's base **79** has a good antioxidant activity in

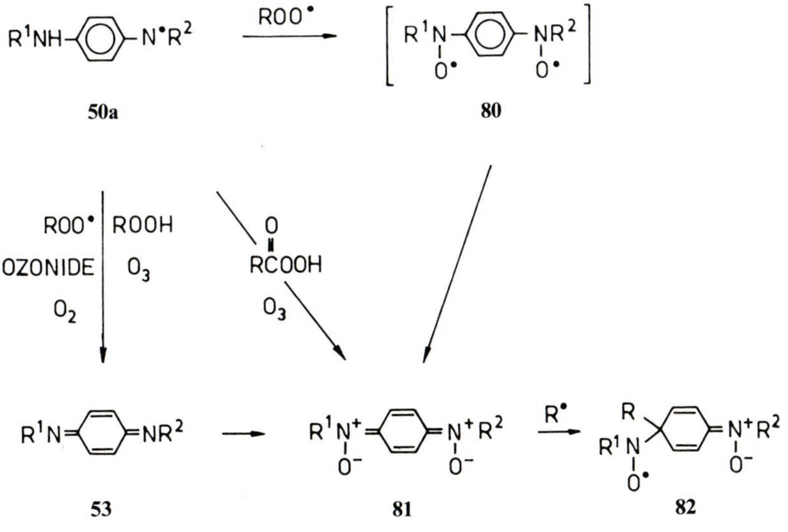

Scheme 8

2,6,10,15,19,23-hexamethyltetracosa-2,6,10,14,18,22-hexaene (squalene) and NR vulcanizate[4, 75].

Both **53** and **81** outlined in Scheme 8 contribute to the stabilizing mechanism by trapping R$^{\cdot}$ [4, 5]. The bisnitrone **81** is formed in high yields by oxidation of PD with organic peracids [79]. This transformation may proceed in photo-oxidized PO doped with PD. **81** was also found among products of ozonation of **11a** [19, 80]. **81** decomposes by irradiation at 300–450 nm into **62**, azobenzene, nitrosobenzene **64** (n = 1) and benzoquinone **63**.

BQDI **53** may induce undesirable reduction of the processing safety and efficiency of crosslinking in rubber vulcanization [3].

The character of the *N,N'*-substituents affects the reactivity of **53** [4, 58]. In the presence of a weak organic acid or silica gel, a simple hydrolysis takes place if one of the N-substituents is a secondary alkyl. For example, **53**, $R^1 = sec$-alkyl, R^2 = phenyl are deaminated into BQMI **57** (Eq. 12). Benzoquinone **63** may arize in the ultimate phase. Reduced forms of both BQDI **53** and BQMI **57**, i.e. PD **11** and 4-hydroxy-DPA **82**, are present in the respective reaction mixture together with traces of other products. The redox generation of amines from QI contributes to the observed antidegradant effect of BQDI [4].

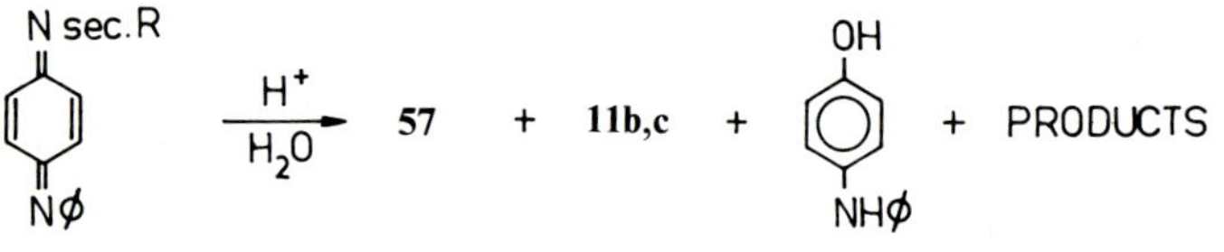

$$\text{53} \quad \xrightarrow[\text{H}_2\text{O}]{\text{H}^+} \quad \textbf{57} \quad + \quad \textbf{11b,c} \quad + \quad \text{83} \quad + \quad \text{PRODUCTS} \qquad (12)$$

N,N'-Diphenyl-1,4-BQDI **53**, R^1, R^2 = phenyl, hydrolyses very slowly. A mixture of products is formed via substitution, disproportionation, and cyclization. Indoamines and indoanilines, Bandrowski's bases like azophenin **84**, its hydrolyzed analogue **85**, polynuclear heterocycles 2,10-dihydro-10-phenyl-3-(phenylamino)-2-(phenylimino)phenazine (**86**) or 1,8-diphenylfluorindine **87** were isolated [4, 58].

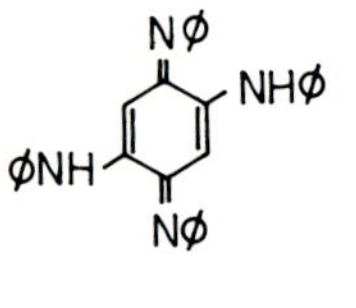

84

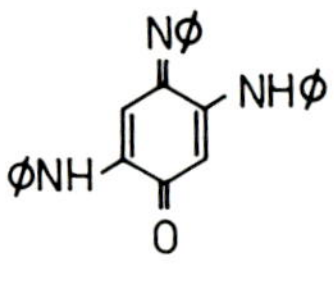

85

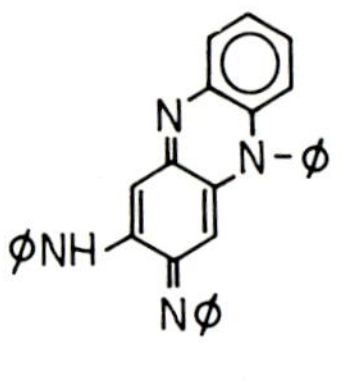

86

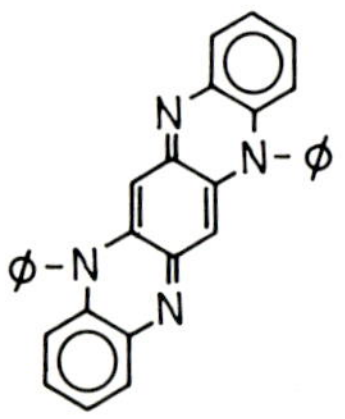

87

Deeply coloured polynuclear amino-imines having structural moieties close to that of iminoblue, quinoneimine red, nigraniline, pernigraniline and polyanilines are also formed by oligomerization [81, 82]. QI have a strong absorption in the visible part of the light between 440 and 480 nm [77] and are responsible for discolouration of the host polymer at concentrations as low as 15 p.p.m. Most QI also have staining properties. Although coloured products arise in the service transformation of amines, their formation is not directly correlated to the integral efficiency of amines. Discolouration due to QI formation is restricted to some extent in DPA **9a, b** having bulky substituents in positions 4,4′- or in PD **11b, c** bearing at least one N-sec-alkyl [4].

Chemical reactivity of QI is reflected in the complexity of chemical processes taking part in aging of unsaturated vulcanized rubbers [3, 5, 24, 58]. QI are oxidants and are transformed in high yields into the respective reduced formes, e.g. **11** or **83**. From the point of view of the polymer stabilization, the process should be understood as regeneration of the AO/AF species. This was observed in the reaction with 2-mercaptobenzimidazole and other thiols during the vulcanization process [4, 77]. The majority of ArSH was bound to the arylene ring of the reduction product, e.g. in **88**.

88   **89**

90

Another example of the reductive regeneration of PD was observed in acid catalyzed reaction of N-isopropyl-N′-phenyl-1,4-BQDI with 2,6-dialkylphenols. They were oxidized into the respective 3,5,3′,5′-tetraalkyl-4,4′-biphenyldiols, 3,5,3′,5′-tetraalkyl-4,4′-diphenoquinones and 2,6-dialkylbenzoquinones, i.e. species scavenging either ROO˙ or R˙ [3, 4]. BQDI was reduced into **11b** and partially bound in the form of 2,6-dialkyl-4-(4-phenylaminophenylimino)-2,5-cyclohexadiene-1-one **89** and a cyclic product **90** (R = isopropyl).

Formation of **89** has been favoured on the account of **90** with the increasing steric hindrance of the phenolic moiety. **89** was the main product with 2,6-di-tert-butylphenol. It effectively protects squalene against autoxidation [3, 4]. The cyclic condensate **90** thermolyses and releases PD. The reactivity of BQDI with 2, 6-dialkylphenols slows down the consumption of PD, but cannot prevent it.

QI are photochemically reduced into **11** or **83** in the presence of suitable donors of hydrogen [83].

Reactivity of QI with compounded NR at vulcanization temperatures results in two concerted processes [4, 24]. More than 30% of the QI was reduced into CB AO **11** or **83**, another part was bound into the polymeric substrate. The mechanism was proved using 2-methyl-3-pentene as a model hydrocarbon. The bound species have structures **54**, **91** or **92** (R = alkyl or a polymeric residue). Reactivity of QI with R˙ seems to be responsible for the observed result.

NHφ

OR

91

NHφ

—R

OH

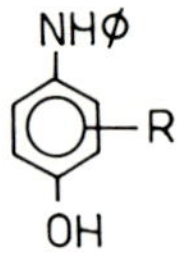

92

Scavenging of R˙ may explain retardation of autoxidation of polymers by QI, as a supporting antioxidant mechanism [75]. To explain the reactivity with alkyls, reactions of 1-cyano-1-methylethyl with PD **11b, d**, the derived BQDI **53** (R^1 = isopropyl, R^2 = phenyl and R^1, R^2 = phenyl) and BQMI **57** (R' = H) were studied in the presence of catalytic amounts of organic acids [4]. Four sites are available for a nucleophilic attack on the N,N'-disubstituted BQDI: positions 2 and 3 on the central ring and both nitrogens. The process is accompanied by PD regeneration.

An appreciable reaction rate was not observed with the BQDI used alone. The reaction was successful when a mixture of BQDI with PD was used, ac-

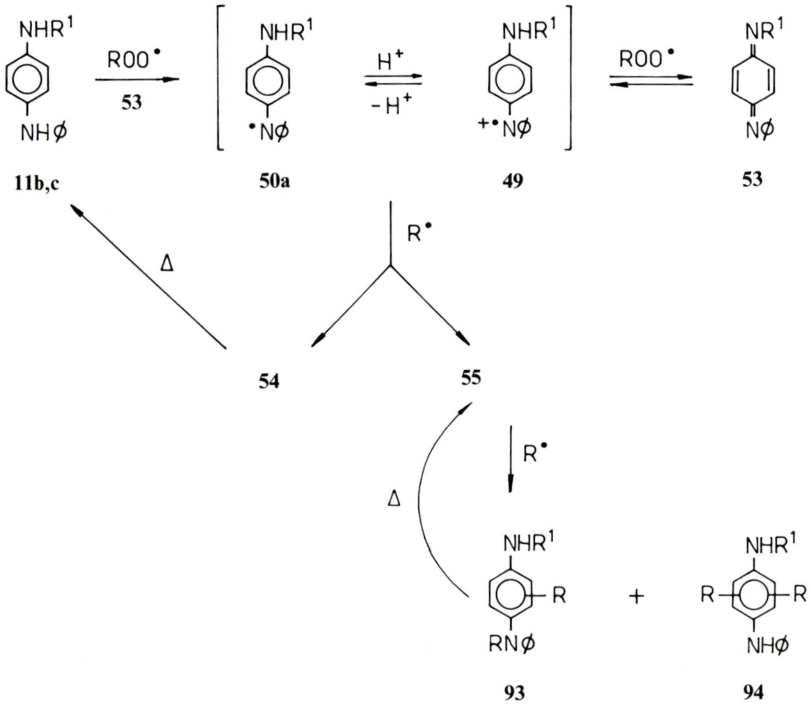

Scheme 9

counting for activation of the reactivity with R˙ by in situ formation of either an aminyl **50a** or cation-radical **49** [4] (Scheme 9). The mechanism encountered both hydrolysable and hydrolysis resistant BQDI.

N-Alkylates **54** or **93** and ring-alkylate **55** and dialkylate **94** were isolated. Polymer-incorporated species arise with macroalkyls. N-Alkylated compounds like **54** or **93** are thermolabile [4]. The alkyl group is split-off over 120 °C or in the presence of an acid catalyst in the form of an olefin. The respective parent **11** and ring-substituted **55** are regenerated.

A reaction pathway similar to that in Scheme 9 may be drawn for the BQMI **57**, R' = H. O-Alkylate **91**, alkylated aminophenol **92** and an analogous dialkylate were isolated [4]. BQMI **57** reacts with R˙ directly. The presence of a redox couple with 4-hydroxy-DPA **83** is not necessary. When using the couple, the reactivity increases and is higher than that of the sole BQMI. A reactivity series of various quinones with R˙ may be drawn [4] (Scheme 10).

The discussed reactivities may be extrapolated for QI derived from other aromatic and heterocyclic amines.

The scavenging of R˙ by $>$NO˙ or QI results in a partial regeneration of the sec-amine moiety and accounts for the integral high antioxidant effect [4, 58].

Nitrones formed from aromatic amines, e.g. **62**, **81** or hydroxylamines, e.g. **70**, and from partially hindered heterocyclic amines **16–19** react in canonical forms and are efficient spin traps [73, 84].

Scavenging of alkyls accounts for nitroxides (e.g. **68**, **71** or **82** [3, 4, 73]. They should be considered as "secondary" $>$NO˙ and are also observable in reactions of amines of the PD series not providing "primary" $>$NO˙ **80** due to their instability. The reaction of nitrones with O-centred radicals (RO˙, ROO˙) results in α-alkoxy- (or α-alkylperoxy) nitroxides (the reaction with ROO˙ is envisaged in Scheme 11). Primary amines convert nitrone into a derivative of hydroxylamine (Scheme 11). This reaction may be expected with impurities contaminating commercial PD.

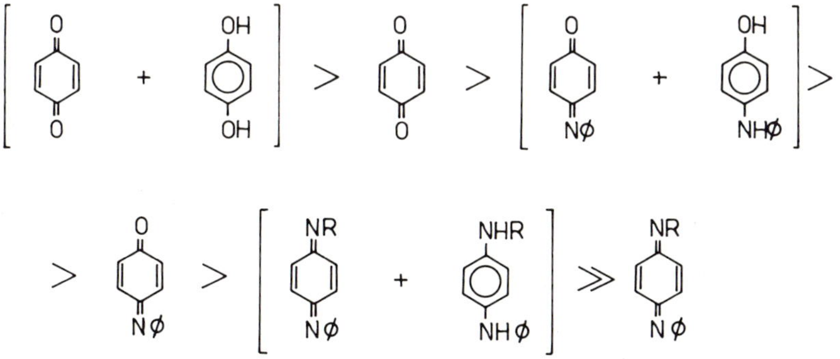

Scheme 10

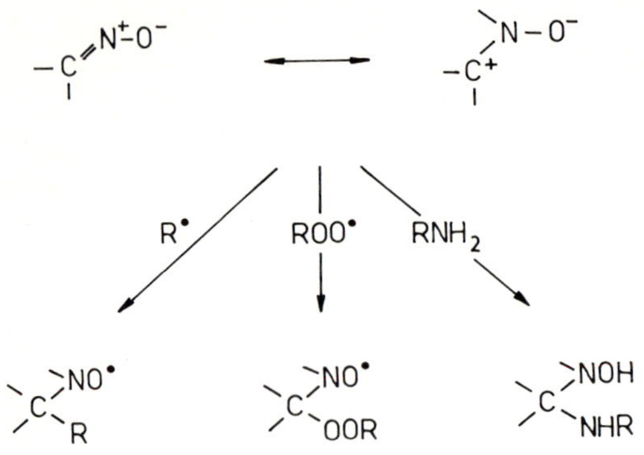

$$-C \overset{N^{\pm}O^-}{\underset{|}{\diagup}} \quad\longleftrightarrow\quad \overset{}{\underset{-C^+}{\diagup}} N-O^-$$

R· | ROO· RNH₂

$$\overset{}{\underset{/}{C}} \overset{NO·}{\underset{R}{\diagdown}} \qquad \overset{}{\underset{/}{C}} \overset{NO·}{\underset{OOR}{\diagdown}} \qquad \overset{}{\underset{/}{C}} \overset{NOH}{\underset{NHR}{\diagdown}}$$

Scheme 11

3.1.2 Anti-Flex-Crack Activity

Scavenging of alkylradicals immediately after their formation and before self-transformations or trapping by oxygen is a step contributing to the antifatigue activity [6]. Although derivatives of DPA, PNA, PD and DHQ have been categorized as AF agents, they are in their parent structures actually fully inert towards the nucleophilic attack by R·. Only some "products" formed according to (Eq. 3) in both the sacrificial and depleting transformations of amines are able to trap R·. The antioxidative pathway (Eq. 1) is the most favourable for the generation of "products" from the point of view of the integral stabilizing potency. This envisages a close relationship between antioxidant and antifatigue activities, accounting for alternative scavenging of ROO· and R·. Model experiments performed with combinations of a ROO· scavenger (DPA, PNA) and a R· scavenger (substituted benzoquinone, α,α-diphenyl-β-picrylhydrazyl, nitroxide of a hindered piperidine) [85] confirm the idea. The maximum effect was observed in O_2 deficient atmosphere and with a molar ratio ROO· to R· scavengers 3:1. No synergism was observed in a blend of PD **11** with a R· scavenger. Aminyls, nitroxides, nitrones, quinones and quinone imines [3–6, 49, 86, 87] are the true scavengers of R· in amine doped substrates. These combinations of ROO·/R· scavengers are formed in the polymer matrix in situ.

The analysis of the anti-flex-crack activity of various aminic AO revealed that the moiety ΦNHΦ-subst. present in the DPA, PNA and N,N'-diaryl stabilizers is the principal structural precursor of an efficient AF [6, 31]. The mechanism of R· trapping by the derived aminyls is shown in Schemes 3, 7–9 and 11.

The high efficiency of aromatic amines in rubbers, and R·/ROO· scavenging capacities due to some amine transformation products, encouraged an extended use of amines in stabilization of PO [47, 88] and, moreover, in stabilization of vinylic monomers [89]. For example, combinations of PD with the respective BQDI, i.e. the ROO·/R· scavenging system, are effective stabilizers against premature polymerization [90]. A similar activity is performed by **11** in vinylic monomers in the presence of traces of oxygen or **20** with benzoquinone as an oxidant [6, 91].

The R˙ trapping also represents alternatives to the formation of so called "unextractable" nitrogen observed in aged vulcanizates, and formed as a result of reactivity with vulcanizing ingredients, aldehydic end groups or zwitterions (for mechanistic details see Sect. 5).

3.2 Heterocyclic Amines

Antidegradants having a nonhindered or partially hindered endocyclic amino group and hydrogen atom on α-carbon to the amino group, like five-membered **15** or six-membered heterocycles **16–19** and the relevant oligomers **21–23** are included. All of them are CB AO with some anti-flex-crack potency. Contribution to the photo-protecting effects increases with the steric hindrance of the amino group.

3.2.1 Derivatives of Indole and Carbazole

Five-membered heterocyclic amines have been attracting attention for a long time. Their antioxidant effect was confirmed in model experiments. Derivatives of indole and carbazole are HS for PVC and copolymers of vinyl chloride [92]. Some indoles are medical or biological AO protecting against chemical carcinogenesis or chemically induced hepatoxicity [93]. 6-Hydroxycarbazole suppresses oxidation of methyl linoleate, its sacrificial transformation yields a QI. 2,2-Disubstituted 1,2-dihydro-3-oxo(or phenylimino)-3H-indoles (**15a**, **b**, R = Me, Et) have a potential importance as HS for PO [36].

A nitroxide is formed from **15** via oxidation with ROO˙. Conjugation with the adjacent aromatic ring enhances its reactivity in the mesomeric C-radical form. Stability of the indolidone nitroxide is somewhat lower than that of nitroxides derived from HAS. HCl and strong organic acids catalyze its disproportionation into the parent amine and quinoneimine-N-oxide [94]. Coupling with alkyl or 4-hydroxybenzyl radicals results in O-alkyl-(or 4-hydroxybenzyl)hydroxylamines **95** [95]. Electrophilic radicals like RO˙, ROO˙, RC(O)OO˙, $>$N˙ or phenoxyls recombine with positions 5 and 7 of the mesomeric nitrone form and a mixture of isomeric substituted nitroxides arises. For example, *tert*-butoxy or *tert*-butylperoxy radicals yield a mixture of 5-*tert*-butoxy (**96**), 7-*tert*-butoxy and 5,7-di-*tert*-butoxy derivatives. Isomeric 5- and 7-acetyl and 5,7-diacetyl derivatives were prepared with lead tetraacetate. Reaction with diphenylaminyl generated in situ provided a 5-diphenylamino derivative **97** of the respective hydroxylamine [96]. Oxidation of the indolidone nitroxide with *tert*-butylperoxyl gives two isomeric quinoneimine-N-oxides **98** and **99**. **96** and **97** are similarly oxidized into **98** (R^1 = tBuO) and **98** (R^1 = diphenylamino), respectively.

95   96   97

98   99

DHI **15** have been proposed as HS for PO. Therefore information dealing with the thermal stability of the parent amine and of the derived nitroxide and *O*-alkylhydroxylamine at processing temperatures is of interest. The oxoderivative **15a** is more stable at 200 °C than the phenyliminoderivative **15b** [36]. The nitroxide derived from **15a** (R = Me or Et) does not decompose significantly at 200 °C. The nitroxide derived from **15b** yielded at the same temperature low amounts (less than 2%) of indolenine-*N*-oxide (**100**) and indolenine (**101**). **95** (R' = Me) was transformed at 200 °C into the parent amine **15a** [36].

100   101

3.2.2 Derivatives of Dihydroquinoline and Dihydroacridine

Low molecular weight and oligomeric DHQ **16** and **21**, respectively, have a broad application spectrum. They have been used for a long time as HS and AO with some anti-flex-crack effect in the rubber industry [1] and as HS in PO [97]. Their contribution to polymer photostability is rather poor. Some antiozonant efficiency has been admitted for 6-ethoxy-DHQ (**16b**) [1]. The substance **16b** holds a specific position as a harmless AO for other sensitive organic materials including pharmaceuticals, lipids, fish meat and oils [98] and has been used in controlling the scalding of pears during storage caused by oxidation of α-farnesene in skin tissue of pears [99]. Most mechanistic studies were performed with **16b** and the results may be extrapolated for other derivatives of DHQ and the oligomer **21**.

Abstraction of the hydrogen atom from the $>$NH group according to Eq. (1) and formation of aminyl **102** is the primary step of the antioxidant action of **16b**. Aminyl **102** was generated for model studies by photolysis of **16b** [100]. Aminyl reacts in mesomeric forms **102a–102c**.

102a   102b   102c

High density of the unpaired electron in position 8 was determined by ESR spectroscopy [98]. Self-termination of **102** involves N–C coupling yielding a dimer **103** [101]. The rate constant of the decay of **102** via dimerization was $5 \pm 3 \times 10^6 \, \mathrm{dm^3 \, mol^{-1} \, s^{-1}}$. The steric hindrance of position 8 by methyl group reduces the rate constant to $4 \pm 2 \times 10^2 \, \mathrm{dm^3 \, mol^{-1} \, s^{-1}}$, blocks the 1,8-dimerization and, consequently, reduces the antioxidant effect of the 6-ethoxy-8-methyl-DHQ [98]. Theoretically possible N–N and C–C coupling products of **102** were not observed [101]. N–C-Dimerization was also observed with 6-hydroxy-DHQ [102].

103   104   105

106   107

The dimer **103** was also formed by air oxidation of **16b** [101].

Using monoelectron oxidizing agents simulating reactions with ROO˙, it was found that the coupling of **102** is accompanied by oxidation to QI **104–106**. **105** is the principal product, formed after elimination of the ethoxy group from the aminyl **102** [101]. **105** was identified as the only product formed from 6-hydroxy-DHQ at 60 °C in ethylbenzene or at 150 °C in *n*-decane [103]. At higher concentrations of the oxidation agent, the amount of **106** increases due to increased formation of the dimer **103** and its oxidation [101]. All arising QI discolour the polymer matrix doped with **16**. 6-Ethoxy-2,4-dimethylquinoline **107** results as a minority product after demethylation of **16b** [101]. Formation of organic peracids has been confirmed in photo-oxidation of PO [104]. *m*-Chloroperbenzoic acid, used as a model peracid, transformed the parent **16b** into the respective nitroxide **108** [101]. The nitroxide **108** was also formed by oxidation of **102** with oxygen and was detected in autoxidized fish oil doped with **16b** [105] and in squalene oxidized with ozonized or NO_2 containing air. The same agent oxidized the dimer **103** and QI **105** into 2,2,4-trimethyl-6-quinolone-*N*-oxide **109** [98].

| 108a | 108b | 109 |

| 110 | 111 | 112 |

Nitroxide **108a** reacts in its mesomeric form **108b** and is destroyed in a weak acid environment into a mixture of the parent **16b** and nitrone **109** [101]. Using a low concentration of ROO˙, **16b** was transformed into the dimer **103** as the main product. QI **104** and **105** were present only in small amounts [106]. The amount of QI increased with the concentration of ROO˙ present in the system. Compounds **103**–**105** were also detected in autoxidized squalene doped with **16b**. This confirms the transformation pathway of **16b** based on transitional formation of aminyls **102**. It should be noted that QI **104** and **105** were also identified as metabolites of **16b** in rats [101]. This indicates a paralelism of processes involving **16b** in vivo and in vitro.

Transformation products of **16b** contribute to some extent to the antioxidant effect. The dimer **103** is a weak AO in squalene [106] and in fish oil [98]. The nitroxide **108** stabilizes fish oil, if present at a relatively high concentration (0.1%) [105]. Its effect is not significant at lower concentrations. QI **105** was a weak retarder in oxidized fish oil [98] and decane [103]. It may be anticipated that other QI arising from **16** contribute similarly to the stabilization effect. A transient formation of the respective hydroxylamines is theoretically possible after hydrogen abstraction from a donor by **108** according to Eq. (8) or after thermolysis of the respective *O*-alkylhydroxylamine (an analogy to the process shown in Scheme 7). Their contribution to the stabilization effect is certainly not important, as revealed by a model study with ⊃NOH derived from **16b**. It is very unstable and disproportionates to the parent **16b** and nitroxide **108** [5, 101].

Both **102** and **108** were confirmed as (macro)alkyl radical scavengers in an oxygen deficient environment. **102** traps 1-cyano-1-methylethyl used in excess [107]. The respective thermolabile *N*-alkylate **110** (R = EtO, 20% yield) and a stable 8-alkylate **111** (R = EtO, 50% yield) are formed. 6-Ethoxy-8-alkoxy-2,2,4-trimethyl-1,2-dihydroquinoline **112** (R = EtO, R^1 = 1-cyano-1-methylethyl) is formed as the main product from **108**, probably by isomerization of the O-substituted hydroxylamine generated in the first stage from **108**.

The chemistry of the service transformation of **16b** during antioxidant and anti-flex-crack activity in elastomers and model results obtained with oils and liquid hydrocarbons may be exploited for explanation of the activity of **16a**. The derived aminyl **102** (R = H) is unstable [108] and terminates by coupling at 20 °C with the rate $4.4 \times 10^7 \, dm^3 \, mol^{-1} \, s^{-1}$. The respective nitroxide **108** (R = H) arises after ROO˙ scavenging by **16a** or **102** (R = H). Reactions analogous to that involving the respective radicals derived from **16b** may be expected. Thin layer chromatographic analysis (TLC) reveals formation of deeply coloured QI **104**–**106** (R = H) after oxidation of **16a** with silver oxide. Oxidative scission of bonds connecting the individual construction units in the oligomer **21** results in formation of QI analogous to those arising from **16a**.

1,2,3,4-Tetrahydro-2,2,4-trimethylquinoline (**17**) and its derivatives, e.g. the respective methylenebis (tetrahydrotrimethylquinoline) or N-substituted derivatives [109] are categorized as partially hindered heterocyclic amines having properties of HS and LS for rubbers, lubricants or hydraulic fluids. Because of the chemical similarity with **16**, an analogy to the activity mechanism may be drawn. The respective aminyl was formed by photolysis of **17** [110] and a nitroxide was generated by oxidation with hydrogen peroxide.

A fully hydrogenated derivative of DHQ, decahydroquinoline (**18**) and its derivatives may be considered as a link between partially hindered heterocyclic amines and HAS. This imparts a significantly enhanced light stabilizing effect of **18** in comparison with DHQ. Compound **18** reacts as cis and trans geometric isomers [38]. The light stabilizing activity of the bicyclic amine **18** appears to depend on the degree of the hindrance of the nitrogen atom and the way in which the two rings are fused together. This conclusion is based on the fact that the trans isomer is a better LS in PP than the cis isomer, as a consequence of the higher stability of the nitroxide derived from the former isomer.

Nitroxides derived from amines having one hydrogen atom on α-carbon atom to nitrogen oxidize with ROO˙ to nitrones [111] and/or disproportionate easily into the respective hydroxylamine and nitrone [112]. The cis and trans nitroxides derived from **18** differ in their disproportionation rates. The disproportionation proceeds via an intermediately formed dimer:

$$(13)$$

Decahydroquinoxaline **19** (R = benzyl) is another excellent LS and AO for PP
[39]. The photostabilizing effect is lower than that of HAS [113], most probably
due to the presence of the α-CH bond to the amino group enabling formation
of mesomeric structures of the derived aminyl. The respective nitroxide may
disproportionate similarly as shown with **18** in Eq. (13). A stable nitroxide **113**
was prepared from **19** and perbenzoic acid [39]. It is not clear if **113** may also
form with ROO·.

113

9,10-Dihydroacridines (acridanes) are attracting attention as AO for PO [114].
Oligomeric amine/aldehyde or amine/ketone condensation products having structures **22** or **23** are AO for NR, isoprene rubber (IR), butadiene rubber (BR), SBR,
NBR or EPDM [1]. Mechanisms analogous to that shown for other non-hindered
nitrogen heterocycles may be envisaged. Oxidation of the derived aminyls and
nitroxides accounts for discolouring QI.

3.2.3 Phenothiazines

Phenothiazine (PT), its derivatives (**20**, R = *tert*-alkyl, aralkyl) [115], angular PT,
like benzo[*a*]phenothiazine **114** or its analogues [116] are inhibitors of polymerization of vinylic monomers, strong AO in petroleum lubricants, greases, functional fluids, fuels for internal combustion engines, sensitive organic materials
like vitamin A, rubbers and ethylene homo- and copolymers, PS or PUR [1, 53,
116, 117]. The antioxidant activity of PT in PE was appreciably enhanced by
combination with thiosynergists. Copolymers of polymerizable derivatives of PT
have been proposed for stabilization of plastic films and polymers used for lenses
and optical devices or hydrogels [118].

The broad application spectrum of PT has enhanced interest in elucidation
of the chemistry of PT during stabilization of polymers and sensitive organic
substrates. It is obvious that the antioxidant efficiency has been explained in terms
of ROO· trapping according to Eq. (1). Inhibition of the vinylic polymerization,
where trapping of alkyls is envisaged, is understood as a process taking place with
R·-reactive transformation products of PT (the respective cation-radical, aminyls,
nitroxides or QI) [117]. Aminyl and nitroxide radicals are formed from PT during
photochemically induced oxidation with *tert*-butylperoxyls [119].

A detailed mechanistic study performed in synthetic ester oils with 3,7-di-*tert*-
octyl-PT (**20**, R = *tert*-octyl) revealed [53] the primary formation of the >N· and
its involvement in C–N coupling accounting for dimeric and trimeric products.

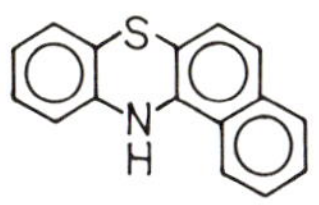

114

115

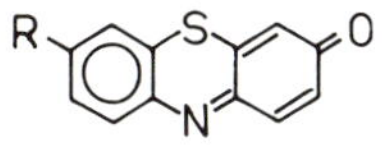

116

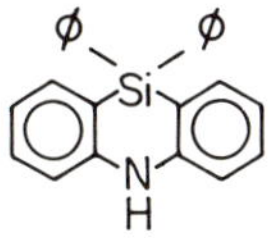

117

Compound **115** (R = *tert*-octyl) is one example. Because of the presence of the exocyclic aminogroup in **115** and analogous trimers, a residual antioxidant activity is expected. Oxidation of the aminyl results in QI derived from both the parent PT and the coupling products [5, 53, 56], e.g. **116**. Coloured oxidation products having structures of endocyclic QI were also formed from other derivatives of PT [116]. The respective sulfoxides were formed in trace amounts among degradation products of **20** (R = *tert*-octyl). They result from reactivity of **20** with ROOH.

Diphenylphenazasiline **117**, a silicone analogue of PT, is a strong and thermoresistant AO, applicable at temperatures over 200 °C [120].

3.2.4 Anilino Derivatives of Heterocyclic Amines

Exocyclic anilino groups bound to a nitrogen containing heterocycle should be considered as a chain breaking moiety reacting analogically to substituted DPA according to Eq. (1). It is therefore not surprising that electron-donor substituents in para-position of the anilino moiety enhance the antioxidant efficiency in PP as demonstrated with 3-anilino-1,5-diphenylpyrazoles **118** or 1-aryl-2,2-diacylhydrazines **119** [121].

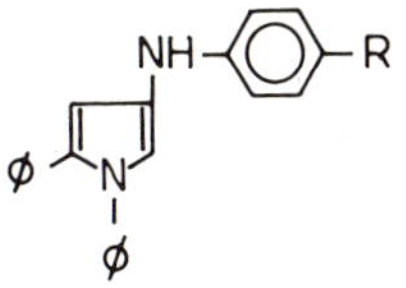

118

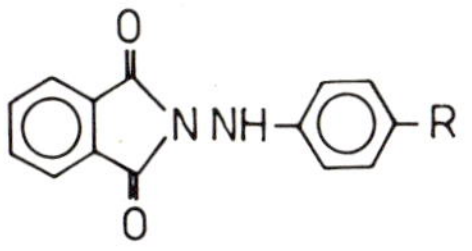

119

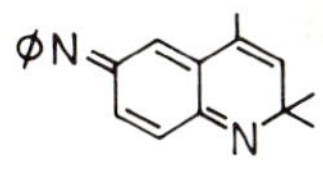

120

The aminyl radical generated in **118** after hydrogen abstraction either dimerizes via C–N coupling or oxidizes to the respective nitroxide or nitrone. Scavengers of R˙ are thus generated.

6-Anilino-2,2,3-trimethyl-1,2-dihydroquinoline (16, R = -NHC$_6$H$_5$) is an interesting stabilizer [37]. Its sacrificial transformation yields QI 120. An exocyclic aminyl is envisaged as the primary transformation product. QI 120 hydrolyzes slowly in a weak acid environment into 105.

3.3 Hindered Amines

Hindered amine stabilizers (HAS) rank among the most recent development in stabilizers for plastics and coatings. They protect polymers by a combined mechanism involving free radical scavenging, hydroperoxide deactivation and formation of CTC, as the principal activities. HAS have been recognized to have a superior performance with most polymers and outperform the other classes of LS [13, 26, 122]. The stabilizing effect was confirmed in many model studies [123–126]. The efficiency is not dependent on the thickness of the polymer. Therefore HAS are suited for stabilization of thick as well as of ultrathin materials (monofilaments, foils, tapes). Protection against photo-oxidation is the main application field [13]. Therefore, HAS have been generally called Hindered Amine Light Stabilizers (HALS) and categorized as photo-antioxidants. This property was extrapolated from the activity in photo-oxidized polymers [13, 15, 125, 127, 128] and in PO autoxidized at moderate temperatures [26, 129]. The photo-antioxidant activity assumes interference with oxidative processes induced by light or energetic ionization radiation and continuing in the dark [130]. The photo-antioxidant effect of HAS has certainly been connected with the efficiency of the derived nitroxides. Nitroxides generated from binuclear HAS like 28 were reported to be more active than those derived from mononuclear HAS [108]. The distance between $\geq$NH (or $\geq$NO˙) moieties influences the effect as well: HAS, having a shorter alkylidene bridge connecting two piperidine nuclei, is of higher activity. This favours binuclear HAS analogous to 28 but having only two methylene groups in the bridge.

The over-accelerated testing conditions in rating stabilizer properties may account for an incorrect interpretation of their activity [129, 131]. This problem also encountered the rating of HAS as HS in oven tests. New findings obtained in temperature ranges below 130 °C, i.e. closer to the realistic long-term heat aging (LTHA) conditions of polymers, revealed an outstanding performance of commercial HMW and oligomeric HAS as long-term HS (LTHS) [26, 122, 129, 132–136]. The term Hindered Amine Thermal Stabilizer (HATS) was introduced for this major recent development in HAS application [129]. This effect is appreciated in stabilization of PP fibres, tapes and thick sections, HDPE plaques and thick sections, LDPE and L-LDPE films. Excellent contribution to the long-term heat stability was observed only in solid PO and is characteristically exclusive for HMW HAS like 32 and oligomers like 34 and 35 [129, 133, 137]. The stability effect is dependent on the concentration of HAS. Of both the oligomers, compound 35 imparted a better protection. The conventional thermostabilizers (hindered phenols and hydrolysis resistant phosphites) should generally be present in formulations with HAS [122, 129, 133]. Oven tests used for rating the heat stability of PE and PP doped with HAS should be performed at 100 and 120 °C respectively. Tests of the LTHS in PO indicate a gradual loss of the efficiency of **32, 34** or **35a, b** with increasing temperature from 110 to 150 °C. This was confirmed when HAS was used either as a single stabilizer or in combination with phenol and phosphite. Some differences exist between PP and PE with respect to the relative response to HAS and phenolic AO [134]. At 140 °C, the hindered phenol Irganox 1076 and **34** show a comparable effect in PP. In LDPE, HAS was found to be better HS at 90 and 100 °C than the phenol. However, it was concluded from comprehensive experiments that application of oligomeric HAS as LTHS is not the solution to every LTHA problem. Over 120 °C, phenols and thiosynergists are more efficient LTHS than HAS. Only below 120 °C can HAS offer both superior LTHS and LS to PO.

All HAS commercially available so far ($>$NH, $>$NR, $>$NOR) fail as melt (processing) stabilizers when used as a sole additive [129, 133, 138]. Some melt stabilizing efficiency was reported for HAS derived nitroxide ($>$NO$^{\cdot}$) [139]. It is not surprising that HAS analogous hydroxylamine ($>$NOH) acts as a melt stabilizer ($>$NOH is a classical CB AO).

The lack of the melt processing activity of HAS means that they do not contribute to the conventional melt processing formulations consisting of hindered phenols and aromatic phosphites. A properly selected combination of processing stabilizers and HAS has, however, a favourable effect on the LTHS and light stability of the polymer during the longterm application [129, 135].

Stabilization of PP, HDPE, LDPE and LDPE/L-LDPE blends is the main application field of HAS [12]. The polymer matrix influences, to some extent, the efficiency. For example, the light stabilizing effect in LDPE/L-LDPE increased with the relative content of L-LDPE [140]. Differences in the rate of the oxygen uptake were found between **28**, R $=$ H doped PP and PE oxidized at 90 °C or irradiated with light having $\lambda > 290$ nm [141].

Conventionally, HAS are blended with PO during processing. 2-(Diethylamino)-4,6-bis[butyl(1,2,2,6,6-pentamethyl-4-piperidyl) amino]-1,3,5-triazine may be fed with an olefin directly into the low pressure polymerization process catalyzed with a modified MgCl$_2$ supported Ziegler-Natta catalyst [142]. The catalytic activity was not impaired [143]. Tetramethylpiperidine was reported to be a useful component in MgCl$_2$-supported Ziegler-Natta catalysts as well. Very high stereospecificity of the synthesised PO was achieved. A complex of HAS with the alkyl aluminium activator was envisaged without interaction with the catalytically active alkyl titanium compound [144].

Data have been published dealing with successful applications of HAS in stabilization of other polymers than PO: elastomers, styrenic polymers, polyamides, polycarbonates, polyacetals, polyurethanes, linear polyesters, thermoplastic polyester elastomers, polyacrylates, epoxy resins, poly(phenylene oxide) or polysulfide [12]. In spite of their basicity, HAS may also be used for stabilization of PVC. This application includes less basic derivatives of piperidine and 1,4-dihydropyridine [12, 13, 145, 146].

HAS reduce the rate of weathering of coatings. Excellent results were obtained in topcoats or two-coat automotive finishes or protective coatings for photographic photoreceptors [147–156]. Using less basic N-substituted piperidines ($\geq$NOR, $\geq$NCOCH$_3$), efficient protection is also achieved in coatings containing acid catalysts [153, 154, 157, 158]. Mechanistic studies were performed with acrylic/urethane and acrylic/melamine coatings [153, 154, 159]. The efficiency of HAS was higher in the former type. In melamine coatings oxidizing by lower rates, the stabilizing effect increases with temperature [160], i.e. at higher oxidation rates of the polymer matrix.

HAS also have application in other sensitive substrates. This exploitation has been based almost exclusively on their radical scavenging capacity. They were recommended for stabilization of organic materials like vitamin D, organophosphorus insecticides and nematocides, cosmetic and dermatologic compositions or in anti-carcinogenic drugs as inhibitors of malignant tissue growth [161–163].

3.3.1 Activity Mechanisms of Hindered Amine Stabilizers

Excellent efficiency of HAS in commerical stabilization of PO and coatings has attracted an enormous theoretical interest in deciphering their activity mechanism. Radical scavenging by the parent HAS or their transformation products (nitroxides, O-alkylhydroxylamines, hydroxylamines, deactivation of peroxidic species by $\geq$NH or $\geq$NR, quenching of singlet oxygen and of polymer-oxygen CTC are the principal stabilizing pathways [13, 15, 125, 130, 134, 164, 165]. HAS function as LS and LTHS. Some pathways are common for the both stabilizing effects. The differences in the character of the polymer matrices and kinetic rules governing their oxidation are reflected in the actual HAS activity [134, 166]. Important information were obtained by comparing results measured with PP and PE [15, 16, 125]. For example, the performance in PP increases linearly with HAS concentration as long as it is not too high and limited by the compatibility of HAS with PP. The performance in PE is proportional to the square root of HAS concentration [15]. The effectiveness of HAS in particular polymers is a function of the oxidation initiating species, the kinetic chain length of oxidation and the lifetime of species formed from HAS in the protected polymer matrix. The excited complex $[PE, O_2]^*$ is important in the initiation phase of the PE photo-oxidation and the oxidation chains are very short. Photolysis of ROOH has a negligible role in this particular case. In PP, the photoinitiation is mainly due to photolysis of tertiary hydroperoxides and the oxidation chains are rather long [134, 166]. Two different species have therefore to be inactivated by HAS to reduce photoinitiation in PE and PP respectively. Because of mechanistic differences observed in HAS activity in PE and PP, the interpretation of mechanisms for other polymers than PO must be done very carefully. For example, HAS was significantly more effective and had a longer effective life in a urethane/acrylate coating than in a melamine/acrylate coating [167].

It has been shown [168, 169] that HAS change not only the kinetics but also the mechanism of the oxidation of the protected polymeric matrix. At a comparable level of the oxygen uptake, the changes in IR spectra of undoped PE reflected the expected oxidative transformation. On the contrary, only minor changes were observed in the presence of HAS.

3.3.1.1 Free-Radical Species Generated from Hindered Amines

Aminyls ($>N^\cdot$), aminium radicals ($>N^{\cdot+}H$) and nitroxides ($>NO^\cdot$) are the free radical species formed from HAS in oxidative processes. An intermediary formation of aminyls in HAS mechanism was proposed [170] as a result of the reactivity of $>NH$ with excited chromophores, acyl- and peracyl radicals, alkylperoxyls or alkylhydroperoxides (Scheme 12).

A light absorbing CTC is responsible for the aminyl formation according to [15]:

$$[>NH, O_2]^* \longrightarrow >N^\cdot + HOO^\cdot \tag{14}$$

Spectral evidence of the aminyl formation was obtained in photolysis of N-chloro-tetramethylpiperidine [171]. The aminyl was characterized by isolation of its N-benzyl derivative. It may hardly be expected as an intermediary species in $>NO^\cdot$ formation at ambient temperatures. The model elucidation was performed with $>N^\cdot$ generated by photolysis of tetramethylpiperidine [170]. It is stable in an oxygen-free atmosphere up to about $-100\,°C$. At higher temperatures, $>N^\cdot$ presumably abstracts hydrogen from organic molecules. Aminyl is oxidized with oxygen at temperatures as low as $-150\,°C$. ESR spectrum of the new intermediate fits to a peroxy radical [170]. At $-80\,°C$, the signal of the nitroxide was generated (Scheme 13).

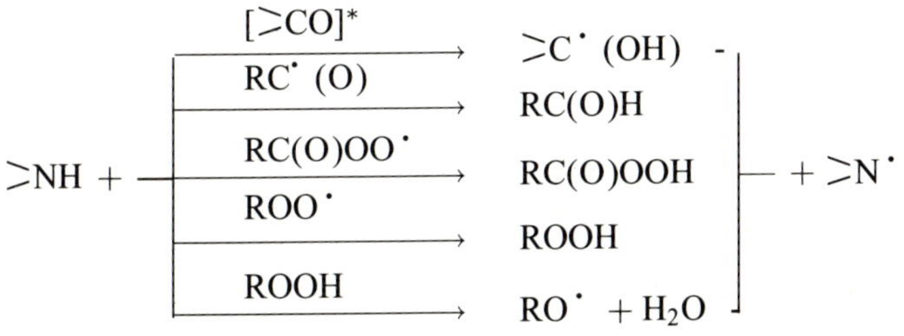

$$>NH + \begin{cases} \xrightarrow{[>CO]^*} & >C^\cdot(OH) \\ \xrightarrow{RC^\cdot(O)} & RC(O)H \\ \xrightarrow{RC(O)OO^\cdot} & RC(O)OOH \\ \xrightarrow{ROO^\cdot} & ROOH \\ \xrightarrow{ROOH} & RO^\cdot + H_2O \end{cases} + >N^\cdot$$

Scheme 12

$$>NH \xrightarrow[HX]{h\nu} >N^\cdot \xrightarrow{O_2} >NOO^\cdot \longrightarrow >NO^\cdot + 1/2\,O_2$$
$$\downarrow RH$$
$$>NOOH \xrightarrow{h\nu} >NO^\cdot + HO^\cdot$$

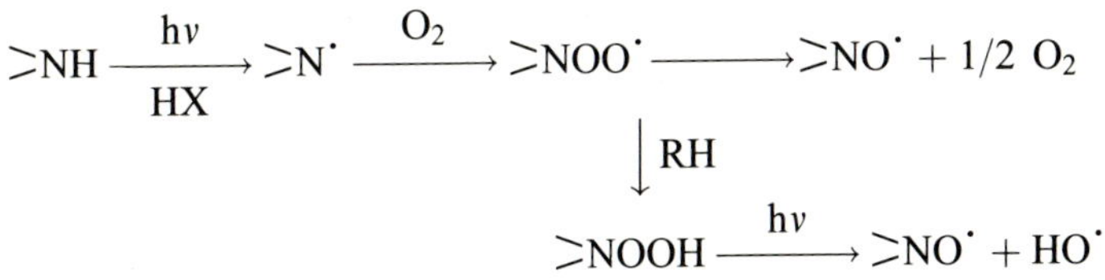

Scheme 13

Aminium radicals were not considered until now as intermediates in photosta-bilizing mechanisms of HAS. These unstable ion-radicals were reported to be formed transiently by photolysis of HAS [172].

In situ oxidation of secondary and tertiary HAS into the respective nitroxides $>NO^{\cdot}$ by various oxidants present in (photo)-oxidized polymers or in the envi-ronment (Eq.15) has been accepted as the first step in the HAS transformation mechanism in real polymer systems at ambient or increased temperatures.

$$>NH, >NCH_3 \quad \xrightarrow[RC(O)OO^{\cdot}, RC(O)OOH, O_3]{^3O_2, ^1O_2, ROO^{\cdot}, ROOH} \quad >NO^{\cdot} + X \tag{15}$$

It stipulates radical scavenging activity of HAS. Secondary HAS are converted directly into $>NO^{\cdot}$. Tertiary HAS exhibit a more complicated mechanism.

The photostabilizing efficiency of N-methyl HAS ($>NCH_3$, e.g. **28, 29, 31**, R = CH_3, **32**) is within experimental error comparable to that of secondary HAS $>NH$ [43, 111, 173, 174]. The simplest and most logical explanation includes a transformation of $>NCH_3$ into $>NH$. Some studies contributed to the mechanism of this conversion. Model N-substituted tetramethylpiperidines having an H-atom on the α-carbon in the N-substituent were found to be photo-oxidized more easily than the corresponding secondary HAS [173]. Besides, the tertiary HAS decom-posed *tert*-butylhydroperoxide more rapidly than did $>NH$. The reaction rates with *tert*-butylhydroperoxide at 132 °C were as follows:

$$
\begin{array}{cccc}
>N\Phi & >NCH_2CH_3 & >NH & >NCH_3 \\
0 & 17.6 & 21.7 & 33.4
\end{array}
$$

Salts of $>NH$ with carboxylic acids arising from the respective N-alkyl group, i.e. $[>N^+H_2^- OCOR]$ were identified. It was concluded that the N-alkyl is eas-ily oxidized via the α-CH moiety [26, 173]. The process accounts for oxidative dealkylation and formation of the mentioned salt. The latter does not hinder the stabilizing activity (see Sect. 3.3.2). Moreover, tertiary HAS offer more stabilizing steps than secondary HAS and may, therefore, impart a better stabilization effect. Similar mechanisms as proposed for transformation with ROOH [173] were re-ported for reactivity with free radicals, $^1O_2$ and excited complexes (Scheme 14) [111, 127, 130, 175].

$$>NCH_3 \quad \xrightarrow[{[>NCH_3, O_2]^*, ^1O_2}]{ROOH, RO^{\cdot}/ROO^{\cdot}} \quad >NC^{\cdot}H_2 + Products$$

$$O_2 \downarrow RH$$

$$>NCH_2OOH$$
121

$$[>N^+H_2 \ ^-OC(O)H] \quad \rightleftharpoons \quad >NH + HC(O)OH$$
122

Scheme 14

Radicals $>$NC$^{\cdot}$H$_2$ are formed intermediately and are oxidized via hydroperoxide **121** into the formate salt **122**. The ROOH and radical attacks are more prone to thermal processes taking part in heat stabilizing activity of HAS. Mechanism involving ^{1}O$_2$ and excited CTC fits participation of $>$NCH$_3$ in the photo-oxidation mechanism [43, 111].

Stable nitroxides, independent of the way of their formation, are traditionally considered as the prerequisite and the principal species of the free-radical scavenging activity of HAS [15, 125]. More pathways and oxidants certainly participate in $>$NO$^{\cdot}$ formation (Eq. 15). The specificity of $>$NO$^{\cdot}$ generation via oxidation with ROOH may serve for a rapid evaluation of the weatherability of HAS doped crosslinked melamine and urethane acrylic copolymers [152, 159].

The chemistry and physics of nitroxides derived from hindered piperidines attracted enormous attention for years [64, 108, 176, 177]. Information on their behaviour were available prior to the commercial boom of HAS. EPR spectral characteristics of $>$NO$^{\cdot}$ have been well defined. The structure was determined by mass spectroscopy. HAS derived nitroxides react in two canonical forms [178]. The contribution of the dipolar double bonded canonical structure **123b** (perhaps nearly 50%) makes the nitroxide **123a** a resonance hybrid and a species having a substantial dipole moment.

$$>\text{N} - \text{O}^{\cdot} \quad \longleftrightarrow \quad >\text{N}^{+\,\cdot} = \text{O}^{-}$$
$$\textbf{123a} \qquad\qquad\qquad \textbf{123b}$$

Hydrogen-bonding species stabilize the dipolar form **122b** and increase the spin density on nitrogen. This contributes to the formation of associates with molecules like ROOH or polar additives like phenols.

The conformation of nitroxides derived from polynuclear HAS has been studied recently [179].

HAS derived nitroxides are thermally stable, though reactive with various free radicals and yielding diamagnetic molecules. All theoretical information are of value for deciphering the stabilizing mechanism of HAS in synthetic polymers. The data are useful for biological macromolecular systems as well [180]. Hindered nitroxides are, therefore, interesting for elucidation of properties, transformations, organisation and mobility of macromolecular systems in vitro and in vivo.

Nitroxides were detected by ESR spectrometry in HAS doped photo-oxidized polymers [13, 15, 125]. Binuclear HAS, like **28** R = H are transformed into $>$NO$^{\cdot}$ stepwise [181].

The concentration of $>$NO$^{\cdot}$ in PO and acrylate/melamine coatings grows from the beginning of the photo-oxidation, reaches a maximum and then drops to a very low steady-state level [15, 125, 130, 147]. The equilibrium concentration of $>$NO$^{\cdot}$ varies markedly according to the nature of the polymer matrix, its thermal history and the parent HAS. A steady-state value of about 10^{-4} mol l^{-1} nitroxides was reported in photo-oxidized PO.

Besides the formation of $>$NO$^{\cdot}$ from the parent $>$NH or $>$NCH$_3$, the nitroxides arising in light-induced and dark regeneration steps from the diamagnetic product formed in the reaction of $>$NO$^{\cdot}$ with R$^{\cdot}$ participate at the creation of the equilibrium level [152, 159].

The rate of formation of nitroxides from $>$NH, $>$NCH$_3$ and $>$NCOCH$_3$ was a decisive factor of the measured photostabilizing efficiency of HAS in melamine-crosslinked acrylic coatings [159]. The rate of $>$NO$^{\cdot}$ formation from $>$NCOCH$_3$ and the oligomeric HAS **34** was roughly one order of magnitude slower than that from $>$NH or $>$NCH$_3$ and was considered as the reason for the lower efficiency in photostabilized coatings.

Nitroxides are also generated during oven ageing of HAS doped PO. A lower steady state concentration of $>$NO$^{\cdot}$ was observed in the presence of oligomeric HAS (used as HATS) than with LMW HAS [128]. For the LTHA of PP or PE doped with **35a** at 130 °C and 110 °C respectively, prior weathering substantially increased the level of $>$NO$^{\cdot}$ in the polymer matrix. As a consequence, the light stability of both PO rated by the embrittlement test was improved [182]. This demonstrates the importance of the presence of sufficiently high concentration of the nitroxides in the polymer matrix as a necessary condition for reaching high stabilization efficiency.

An important observation was published recently on solubility of the nitroxides in PO. Measurements in PP, HDPE, LDPE and L-LDPE revealed [183] that the nitroxides are always less soluble than the respective parent HAS and their diffusion coefficient is lower. Differences up to two orders of magnitude have been observed between PP and LDPE, the solubility being better in PE. The data show that $>$NO$^{\cdot}$ are less compatible with PO than the respective $>$NH or $>$NCH$_3$.

3.3.1.2 Free-Radical Scavenging by Hindered Amines

Scavenging mechanism has traditionally been considered as the principal contribution of HAS to polymer stabilization. It is more important in polymers where the photoinitiation is mainly due to ROOH photolysis, e.g. in PP [134]. Scavenging of C- and O-centred radicals participating in the initiation and propagation steps of the chain oxidation of polymers has been envisaged for the original structures of HAS and their active transformation products (nitroxides, O-alkylhydroxylamines, hydroxylamines).

3.3.1.2.1 Scavenging of Alkylradicals by Nitroxides

Because of the unequivocally confirmed formation of the nitroxides in HAS-doped photo- and thermo-oxidized PO and coatings, it was logical to suggest an involvement of $>$NO$^{\cdot}$ in R$^{\cdot}$ radicals trapping. The mechanism of this process has been well understood and intensively studied as a diagnostic tool [184, 185].

Deactivation of R$^{\cdot}$ is a necessary condition for efficient stabilization of polymers in the earliest step of the oxidative degradation [6] and was proposed in the very early stages of elucidation of the HAS mechanism. Formation of the respective O-alkylhydroxyakamines $>$NOR according to Eq. (16) was confirmed in model experiments [184]. The antioxidant effect of $>$NO$^\cdot$ observed in organic substrates fits this proposal. A more efficient formation of $>$NOR is expected in PE than in PP [15].

$$>NO^\cdot + R^\cdot \; \rightleftarrows \; [>NO^\cdot, R^\cdot] \; \longrightarrow \; >NOR \qquad (16)$$

Authors respectfully confirmed formation of $>$NOPP in radiation induced oxidation of PP doped with piperidinyloxyl [127]. About 75% of the added $>$NO$^\cdot$ derived from **28**, R = H was grafted onto PP. Another paper reported binding of 1–2% of the added HAS to PE or PP aged in mild conditions [183].

The self-reactivity of R$^\cdot$ and their reactivity with oxygen are serious competitive processes for the reaction (Eq. 16). Recombination of R$^\cdot$ is preferred in PE, disproportionation in PP [186]. Oxidation proceeds in both polymers. The low probability of the reaction (Eq. 16) in the air atmosphere has, unfortunately been, mostly not reported in discussions of the HAS mechanism. Serious doubts arise when analyzing the reaction possibilities of R$^\cdot$ and $>$NO$^\cdot$ [177, 178, 184, 185, 187]. Aliphatic $>$NO$^\cdot$ react rapidly with R$^\cdot$ at ambient and elevated temperatures [65]. The rate constants for the coupling with R$^\cdot$ are influenced by the resonance stability of alkyls [187]. The five membered $>$NO$^\cdot$ 1,1,3,3-tetramethylisoindoline-2-oxyl (TMIO) was found to be more reactive than 2,2,6,6,-tetramethylpiperidine-1-oxyl (TEMPO). *n*-Pentyl, *tert*-butyl and benzyl radicals were used at 20 ± 2 °C in deaerated isooctane (solvent) [178, 184, 185]. The rate constants for bimolecular reactions of R$^\cdot$ (disproportionation and recombination) were compared with those of R$^\cdot$ oxidation and TEMPO scavenging [178, 184, 185, 188] (Scheme 15). At room temperature, the scavenging of R$^\cdot$ by $>$NO$^\cdot$ is, by about one order of magnitude, slower than the self-termination of R$^\cdot$ [178].

Kinetic measurements envisage limits of the reaction (Eq. 16) in atmospheric oxygen [177, 184, 185, 188]. Oxygen is a better trap for R$^\cdot$ than $>$NO$^\cdot$ [187]. Scavenging of R$^\cdot$ according to Eq. (16) may be, therefore, the sole significant process under oxygen deficient conditions. This is rather unrealistic for polymer weathering.

It was calculated from analysis of γ-radiation induced oxidation of 2,4-dimethylpentane [188], using value 15×10^{-3} mol l^{-1} for the concentration of oxygen dissolved in the hydrocarbon, that only around 10% of R$^\cdot$ available in

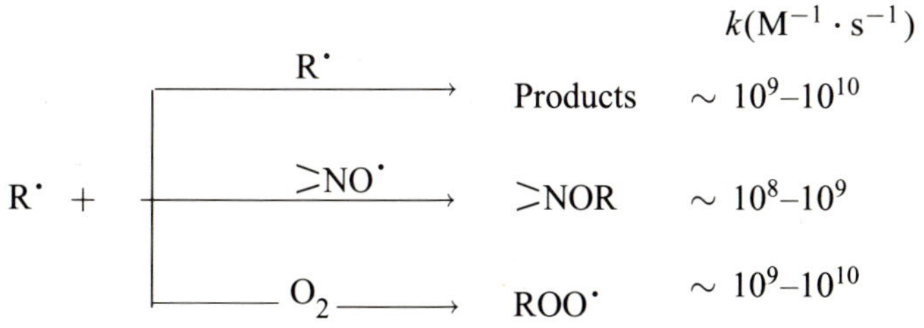

$$k(M^{-1} \cdot s^{-1})$$

$$R^\cdot + \begin{cases} \xrightarrow{\;R^\cdot\;} & \text{Products} & \sim 10^9 - 10^{10} \\ \xrightarrow{\;>NO^\cdot\;} & >NOR & \sim 10^8 - 10^9 \\ \xrightarrow{\;O_2\;} & ROO^\cdot & \sim 10^9 - 10^{10} \end{cases}$$

Scheme 15

the system can be trapped by $>NO^.$. This confirms the fact that the oxidation process propagates almost without restriction and its prevention by $>NO^./R^.$ recombination is negligible. A low efficiency of the reaction at Eq. (16) also results from other recent data. The value of the ratio of the rate constants $k_{>NO^.}/k_{O_2}$ for TEMPO, TMIO and other model $>NO^.$ in solution is approximately 0.15 [187]. This value is in good agreement with values between 0.1 and 0.25 reported from measurements in solid PO [15, 26].

A very low concentration of $>NO^.$ in (photo)-oxidized PO doped with HAS has already been mentioned. This concentration is about one order of magnitude lower than that of oxygen present in PP. No more than between 1 and 2.5% of $R^.$ available in the system can react with $>NO^.$ [189]. Hence, $R^.$ scavenging by $>NO^.$ at outdoor exposure of PO cannot be the only and principal process accounting for the light stabilizing activity of HAS. Most probably the possibility of $R^.$ scavenging is enhanced in areas of accumulated associates $[>NO^. ...$ HOOR], due to in situ formation of $R^.$ [125]:

$$[>NO^. ... HOOR] \xrightarrow{h\nu} [>NO^., HO^., RO^.] \xrightarrow{R'H} [>NO^., R''^.] \quad (17)$$

The probability of the formation of $>NOR$ according to Eq. (16) is still lower in thermal oxidation of PO. An equilibrium between both free radicals $R^.$ and $>NO^.$ and their caged associate has increasing importance at temperatures higher than $100\,^\circ C$ [15].

A stepwise formation and reactivity of nitroxides should be admitted in polynuclear HAS like **27, 28**, R = H or oligomeric HAS **35, 36** [179, 181]. Formation of a polymer-bound nitroxide **124** from **28**, R = H was reported [181].

124

Neighbouring nitroxide moieties in polynuclear HAS **35, 36** were proposed to react in photo-oxidized PP by abstraction of H and immediate scavenging of the in situ formed $R^.$ [179]:

$$(18)$$

3.3.1.2.2 Scavenging of O-Centred Radicals by Hindered Amines

Alkylperoxy radicals participate in the chain propagation step of oxidation. In photo-oxidized PP, most $ROO^.$ terminate after a few propagation steps [190]. The few $ROO^.$ that escape from this recombination propagate with a great rate. At the same time, they are easily scavenged by chain-breaking phenolic and aromatic aminic AO or HAS derived $>NOH$. Secondary and tertiary HAS associate in

$$[ROO^{\cdot}, >NH] \xrightarrow{h\nu} [ROO^{\cdot}, >NH]^* \begin{cases} \rightarrow >N^{\cdot} + ROOH \\ \rightarrow >N^{\cdot} + HOO^{\cdot} + R^{\cdot} \\ \rightarrow >NR + HOO^{\cdot} \\ \rightarrow >NO^{\cdot} + ROH \\ \rightarrow >NOH + RO^{\cdot} \end{cases}$$

125

Scheme 16

oxidized PP domains. It appears that they contribute to the stabilization by a rather inefficient scavenging of macroperoxyls:

$$>NR \xrightarrow{RO^{\cdot}/ROO^{\cdot}} >NH \xrightarrow{ROO^{\cdot}} >NO^{\cdot} + ROH \tag{19}$$

Formation of a CTC **125** and its photochemical transformations were proposed [15, 189] (Scheme 16). This CTC is expected to be more stable and a stronger absorber of UV light than CTC [ROO$^{\cdot}$, PO]. **125** may contribute to the light stability effect of HAS only at ambient temperatures [26].

HAS also react directly with acylperoxy radicals (Eq. 20). They play an important role only in the later stages of PO oxidation and react with $>NH$ faster than ROO$^{\cdot}$.

$$>NH + RC(O)OO^{\cdot} \longrightarrow >NO^{\cdot} + RC(O)OH \tag{20}$$

3.3.1.2.3 Regeneration of Nitroxides from O-Alkylhydroxylamines

O-Alkyl-hydroxylamines $>NOR$ formed according to Eqs. (16) or (24) (see Sect. 3.3.1.3) have been considered as a reservoir for new nitroxides [13, 15, 125, 127], due to participation in a regenerative inhibition cycle. Aspects of the nitroxide regeneration with respect to PO photo-oxidation [13] and heat stability [15] have been subjected to discussion for years. The role of $>NOR$ seems to be rather overrated in the light of the recent understanding of limits governing their formation under real aging conditions. The real concentration of $>NOR$ in aged PO certainly does not surpass 0.002% [26, 183].

It seems that $>NOR$ are more important for PE [15], where mainly O-sec-alkylhydroxylamines **126** are formed. Structurally analogous $>NOR$ may be formed in PP only as a minority product. The bulk will be constituted by less reactive O-tert-alkylhydroxylamines **127** [189].

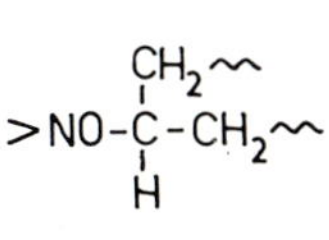

$$>NO-\underset{\underset{H}{|}}{\overset{\overset{CH_2\frown}{|}}{C}}-CH_2\frown \qquad\qquad >NO-\underset{\underset{CH_3}{|}}{\overset{\overset{CH_2\frown}{|}}{C}}-CH_2\frown$$

126 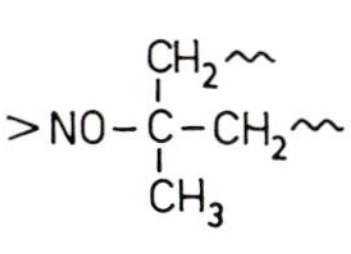   **127**

Two general pathways of regeneration of $>NO^{\cdot}$ from $>NOR$ – thermolysis and oxidation with ROO$^{\cdot}$– were proposed on the basis of model experiments [13, 15, 65, 68, 123, 125, 130, 134, 189, 191, 192]. Independently, on the way of the regeneration of the nitroxide, a repeated scavenging of R˙ via Eq. (16) may proceed in a cyclic process (with the already mentioned restrictions in the oxidizing environment).

Thermolysis of $>$NOR **128** yielding hydroxylamine $>$NOH and an olefinic fragment (Scheme 17, pathway *a*) was proposed as a two-step process [15, 65]. Pathway *a* is concerted with the reversible formation of $>$NO˙ and R˙ (pathway *b*), proceeding at temperatures below 100 °C [193]. In the absence of any other radical trap, the released $>$NO˙ and R˙ either recombine back into $>$NOR **128** or disproportionate into $>$NOH and olefin according to the pathway *c*. Hydroxylamine $>$NOH, the strong CB AO, may be therefore formed from $>$NOR either by thermolysis or disproportionation [65, 123]. The reversibility of the $>$NOR formation was evidenced by ESR.

The rate of the thermolysis of $>$NOR was higher in the presence of oxygen due to the scavenging of the released R˙ (Scheme 17, pathway *d*) preventing the formation of $>$NOR [193]. This process is more pronounced in the surface layers of the polymer, i.e. in direct contact with the air.

The nature of the bound (macro)alkyl influences the reactivity of $>$NOR [15, 134, 179, 194]. This problem was studied using isomeric $>$NOR generated from TEMPO and iso-octane. All three isolated isomeric alkylhydroxylamines were photochemically stable. They do not absorb light having wavelengths >300 nm [194]. Thermal properties depend on the R in $>$NOR. The isomer containing primary alkyl (R $= CH_2C_7H_{15}$) was thermostable up to 90 °C. *O-sec-* and *O-tert-*Alkylhydroxylamines thermolysed at 60 °C in nitrogen atmosphere into $>$NOH and olefins (according to pathway *a*, Scheme 17), in oxidizing atmosphere into $>$NO˙ and alcohols. The process is exemplified for the *sec*-alkyl isomer in Scheme 18.

Alcohols formed may participate in generation of esters with carboxylic acids arising among other products of oxidation of PO [68, 195, 196]. The mechanism

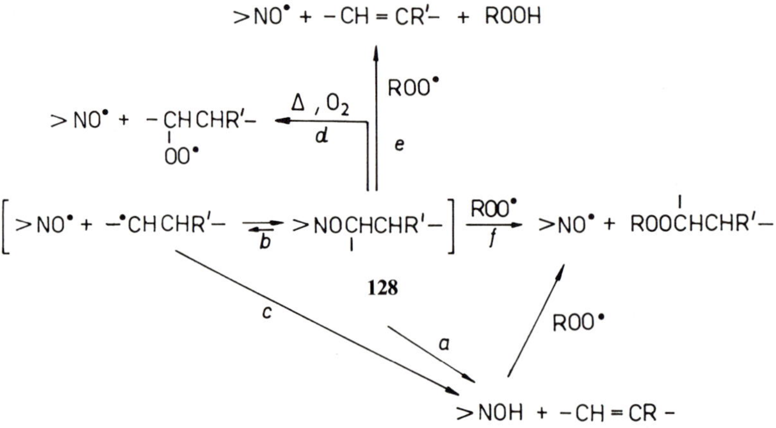

$$>\text{NO}^\bullet + -\text{CH}=\text{CR}'- + \text{ROOH}$$

Scheme 17

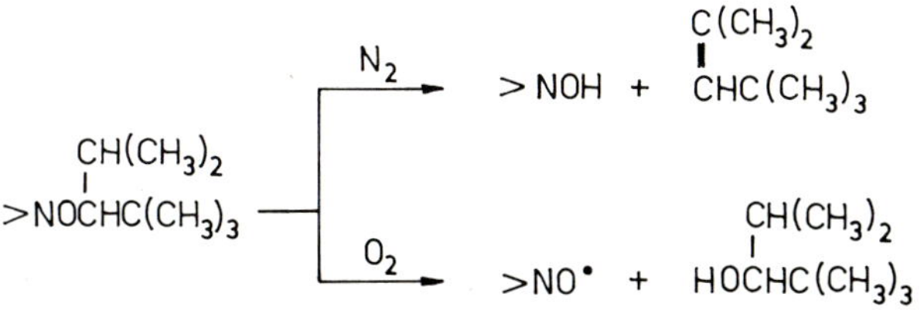
N2
O2
>NOH +
C(CH3)2
CHC(CH3)3
CH(CH3)2
>NOCHC(CH3)3
>NO• +
CH(CH3)2
HOCHC(CH3)3

Scheme 18

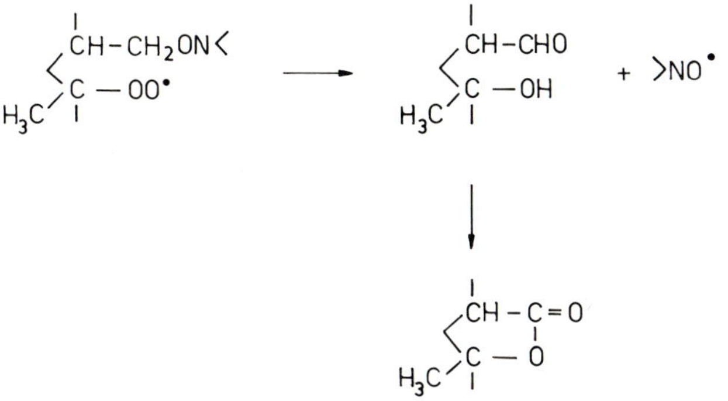
CH-CH2ON<
C — OO•
H3C
CH-CHO
C —OH
+ >NO•
CH - C=O
C — O

Scheme 19

of intramolecular formation of lactones in consecutive steps of the oxidative thermolysis of ⋝NOR was envisaged [179] (Scheme 19).

The regeneration of ⋝NO˙ via reactivity of ⋝NOR with ROO˙ was explained by either a *β*-hydrogen abstraction (Scheme 17, pathway *e*, for **128**, R′ = H) or by a somewhat slower radical substitution (Scheme 17, pathway *f*, for **128**, R′ = alkyl) [189]. The process explains the ROO˙ scavenging activity of HAS, which cannot be ascribed directly to ⋝NH or the derived ⋝NO˙.

Formation of an olefinic fragment from ⋝NOR was proposed as an explanation of the process. Its generation was, however, not confirmed in photo-oxidized PO. O-*sec*-Alkylhydroxylamines (**126**) characteristic of species formed in PE, were used in a recent mechanistic study [68, 191, 195–197]. A new mechanism was proposed for the regenerative process involving reactivity with O-centred radicals ROO˙ and RC(O)OO˙ (Scheme 20). The reactivity with the alkylperoxyl

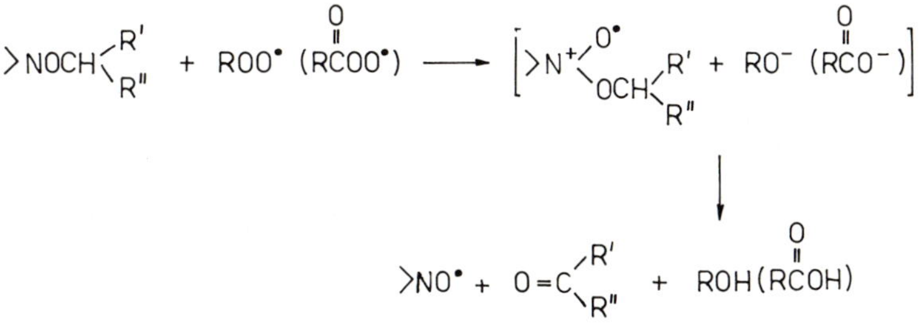
>NOCH
R'
R"
+ ROO• (RCOO•)
>N+
O•
OCH
R' + RO− (RCO−)
>NO• + O=C
+ ROH(RCOH)

Scheme 20

was estimated to be slower. This was confirmed by semiempirical calculations [198]. The reactivity with the acylperoxyl has a 36 kJ/mol lower energetic barrier than the reaction with ROO$^{\cdot}$.

The hydrogen atom on the α-carbon is particularly prone to abstraction by O-centered radicals [189]. According to the co-reactant, a carbonyl compound and either alcohol or carboxylic acid are formed besides the nitroxide. None of the co-reactants yielded peroxide proposed originally [199] for pathway f, Scheme 17. The mechanism exemplified in Scheme 20 was confirmed using labeled compounds [196, 197].

O-tert-Alkylhydroxylamines **127**, expected to be formed in PP, are of a rather low reactivity at ambient temperature and are accumulated in the photo-oxidized PP without any significant regeneration of $>$NO$^{\cdot}$ [15, 127]. However, their stabilizing reservoir potency is not lost and they can be activated by thermal treatment.

3.3.1.3 Deactivation of Organic Peroxides

Hydroperoxides formed in the chain propagation step of oxidation of polymers initiate new chains after thermal, photolytic or catalytic homolysis. Heterolytic deactivation of ROOH represents one of the principal modes of polymer stabilization [25]. HAS in their original form ($>$NH, $>$NR) reduce hydroperoxides effectively. The ROOH decomposing mechanism works in very early stages of PO oxidation, before radical scavenging HAS transformation products are formed [15, 26, 134, 136].

Differences in the importance of ROOH deactivation by HAS may be expected between PE and PP. This may be attributed to the differences in the role of ROOH in the two polymers [15, 134]. The secondary ROOH formed in PE are noninitiating species whereas the tertiary ROOH mainly arising in PP are more efficient in yielding free radicals playing a decisive role in the photo-initiation phase. HAS–ROOH reactivity is therefore of higher importance for the PP matrix.

Secondary HAS already decompose polypropylene hydroperoxide in the solid phase at ambient temperatures [200]. The process appears to be sufficiently rapid for the destruction of ROOH during the dark period, after the photo-oxidation phase. This enhances the photostability of PP in the repeated photo-oxidation.

Because of the lower role of ROOH in PE, the free radical scavenging activity of HAS is of greater importance in this polymer than ROOH deactivation [200].

The reactivity of aliphatic *sec*-amines with ROOH was studied in details. Both ionic and free radical intermediates may be involved [201] Eq. (21). This indicates the possibility of the dual role of amines as AO or pro-oxidants.

$$ROOH + >NH \rightarrow [>N^{\cdot+}H + RO^- + HO^{\cdot}] \rightarrow [>N^+H(OH)^-OR]$$

$$\downarrow \qquad\qquad (21)$$

$$>NO^{\cdot} \xleftarrow{\text{ROOH}} >NOH + ROH$$

The pro-oxidant effect is fortunately minimized in hindered secondary amines. Hydroxylamine $\geq$NOH is generated in the final phase of Eq.(21) and is readily oxidized in situ to $\geq$NO$^{\cdot}$ [130]. Under conditions of ROOH photolysis, the nitroxide may be formed directly [165]:

$$\geq NH \xrightarrow[h\nu]{ROOH} \geq NO^{\cdot} . \tag{22}$$

Tertiary HAS $\geq$NCH$_3$ are converted with ROOH into secondary HAS (Scheme 14) and, consecutively, into $\geq$NO$^{\cdot}$ [124]. Amine oxide **129** has been mentioned as an alternative product:

$$\geq NCH_3 + ROOH \longrightarrow \geq N^{+}(O^{-})CH_3 + ROH \tag{23}$$

129

Mechanistic differences exist between reactivity of HAS with randomly distributed (isolated) ROOH and hydroperoxidic sequences ("activated ROOH") accumulated in some areas of deeply oxidized PO [200]. The reactivity with isolated –OOH groups according to Eq. (21) is of minor importance in comparison with that with hydroperoxidic sequences. Associates between acid ROOH and basic $\geq$NH, $\geq$NCH$_3$ or $\geq$NO$^{\cdot}$ are formed in amorphous phases of PO [130, 202]. The association was proved using pre-oxidized PP films containing 0.06 mol OOH/kg PP. The HAS solubility increased about ten times in comparison with the identical unoxidized film [126, 130, 200]. This implies that the HAS species will concentrate in the oxidized domains of PP, provided they are able to diffuse through the amorphous phase of PP. Moreover, the association is favourable for scavenging of free radicals generated in situ in the oxidized areas. This significantly enhances the possibility of HAS inhibiting oxidation of PO. Associated ROOH could probably create $\geq$NOH and polymer bound $\geq$NOP without $\geq$NO$^{\cdot}$ formation [127, 130, 134, 202]:

$$
\begin{array}{l}
\geq N-H \\
\quad \vdots \quad \vdots \qquad \xrightarrow{h\nu} \\
H-O-OP
\end{array}
\left\{
\begin{array}{l}
\longrightarrow \geq NOH + POH \\
\\
\longrightarrow \geq NOP + H_2O \quad .
\end{array}
\right.
\tag{24}
$$

A similar association to that in Eq. (24) but creating POCH$_3$ and CH$_3$OH instead of POH and H$_2$O, respectively, was proposed for transformation of $\geq$NCH$_3$ with ROOH [202]. The association of $\geq$NO$^{\cdot}$ with ROOH increases the local concentration of $\geq$NO$^{\cdot}$ in regions where R$^{\cdot}$ are generated in situ. This promotes the possibility of R$^{\cdot}$ scavenging (Eq. 25). According to [203], ROOH also react with $\geq$NOR (Eq. 26).

$$[\geq NO^{\cdot} \cdot \cdot ROOH \xrightarrow{h\nu} [\geq NO^{\cdot}, RO^{\cdot}, HO^{\cdot}] \xrightarrow{RH} R^{\cdot} + \geq NO^{\cdot} \tag{25}$$

$$\geq NOR + ROOH \longrightarrow \geq NO^{\cdot} + RO^{\cdot} + ROH \tag{26}$$

Peroxyacids formed in photo-oxidized PP [104] transform $\geq$NH at ambient temperature to the corresponding $\geq$NOH and $\geq$NO$^{\cdot}$ more rapidly than hydroperoxides. A two-step process (Eqs. 27 and 28) or a one-step process (Eq. 29) were proposed [16, 64, 104, 204, 205]:

$$\text{>NH} + \text{RC(O)OOH} \quad \rightarrow [\text{>N}^+\text{H(OH)}^-\text{OC(O)R}]$$

$$\rightarrow \text{>NOH} + \text{RCOOH} \tag{27}$$

$$2\text{>NOH} + \text{RC(O)OOH} \rightarrow 2\text{>NO}^{\cdot} + \text{RC(O)OH} + \text{H}_2\text{O} \tag{28}$$

$$2\text{>NH} + 3\text{RC(O)OOH} \rightarrow 2\text{>NO}^{\cdot} + 3\text{RC(O)OH} + \text{H}_2\text{O} \tag{29}$$

A high reactivity of HAS with diacylperoxides was observed in low-temperature oxidation of PO and yields O-acylhydroxylamines [206]:

$$\text{>NH} + \text{PC(O)OO(O)CP} \longrightarrow \text{>NOC(O)P} + \text{PC(O)OH} \tag{30}$$

3.3.1.4 Hydrogen Abstraction Activity of Nitroxides

Nitroxides $\text{>NO}^{\cdot}$ at elevated temperatures are powerful abstractors of hydrogen atoms from organic molecules HX and are reduced into equimolar amounts of hydroxylamines [64] (Eq. 8). Liquid and polymeric hydrocarbons, particularly their C–H bonds in the neighbourhood of irregularities like $-\text{CH}_2\text{CO}-$ or $-\text{CH}_2\text{CH}=\text{CH}-$, secondary alcohols, phenols, mercaptans, primary amines or hydrazines are sources of hydrogen. Ketones, phenoxy radicals, benzoquinones, dithio compounds, Schiff's bases or azo compounds are formed. The activation energy between 125 and 145 kJ/mol was reported in thermal processes. The reaction with benzene thiol exemplifies concerted H-abstracting and free-radical scavenging activities of $\text{>NO}^{\cdot}$ [64]. The reaction mixture contains diphenyl disulfide, benzene sulfinic acid and 1-(benzene sulfinyl) derivative of HAS:

$$\text{>NO}^{\cdot} + \Phi\text{SH} \longrightarrow \Phi\text{SS}\Phi + \Phi\text{S(O)H} + \text{>NS(O)}\Phi. \tag{31}$$

Photochemically induced H-abstraction includes the electronically excited nitroxide $[\text{>NO}^{\cdot}]^*$ and occurs even at room temperature. The activation energy was reported to be 21 kJ/mol. The quantum yield is ca. 0.2 with light of wavelength <360 nm. Because of the air present during weathering of polymers, it is likely that the excited nitroxide is efficiently quenched by oxygen [189] and the photochemical formation of >NOH is only negligible (Eq. 32, pathway a):

$$[\text{>NO}^{\cdot}]^* + \text{HX} \quad \begin{array}{l} \longrightarrow \text{>NO}^{\cdot} + h\nu' \text{ or heat} \\ \xrightarrow{a} \text{>NOH} + \text{X}^{\cdot} \\ \xrightarrow{^3\text{O}_2} \text{>NO}^{\cdot} + {}^1\text{O}_2 \end{array} \tag{32}$$

More complicated interactions including competitive physical and chemical quenching of $[\text{>NO}^{\cdot}]^*$ arise in bi- and polynuclear HAS, like **28**, $\text{R}=\text{H}$ or **36** [179].

Abstraction of H from 2,4-dimethylpentane (a model of PP) was proposed [207] as an alternative stabilizing mechanism instead of R$^\bullet$ trapping in gamma-induced oxidation:

$$\mathord{>}NH \longrightarrow \mathord{>}NO^\bullet \xrightarrow{RH} \mathord{>}NOH \xrightarrow{ox.} \mathord{>}NO^\bullet \qquad (33)$$

3.3.1.5 Activity of Hydroxylamines

Hydroxylamines $\mathord{>}NOH$ arise from $\mathord{>}NO^\bullet$ according to Eqs. (8), (21), (24), (33) and Scheme 17. Independently on the way of formation, they function as CB AO scavenging radicals ROO$^\bullet$ (Scheme 21). The rate constant of the reaction is comparable with that of ROO$^\bullet$ scavenging by hindered phenols or aromatic amines [50]. Reactivity with ROO$^\bullet$ contributes to the integral scavenging effect of HAS. High efficiency of $\mathord{>}NOH$ was found in the stabilization of PP films against γ-initiated oxidation [130]. The scavenging efficiency of ROO$^\bullet$ was larger by a factor of ca. 10^5 compared with participation of ROO$^\bullet$ in the chain propagating step. $\mathord{>}NOH$ may be considered as another HAS transformation product contributing to the nitroxide regenerating mechanism [123, 126, 127].

Hydroxylamines are easily oxidized at ambient temperature with molecular oxygen and with ROOH (Scheme 21) [65, 134]. The last reaction is very fast [127]. In spite of formation of alkoxy radicals RO$^\bullet$, deactivation of ROOH contributes to the photostabilizing effect of $\mathord{>}NOH$ (formed RO$^\bullet$ are effectively scavenged by phenolic AO, mostly used in combination with HAS).

3.3.1.6 Interference of Hindered Amines with Singlet Oxygen, Excited Chromophores and Charge-Transfer-Complexes

Singlet molecular oxygen $^1O_2$ ($^1\Delta_g$) is active in aging of synthetic polymers and biopolymers [13, 14, 52]. Oxidation of allylic methylenes present in diene based elastomers or as defect structures in PO results in formation of ROOH acting as thermoinitiators and chromophores Ch. The role of $^1O_2$ in saturated polymers is not significant and may be enhanced in pigmented systems. Compounds acting as physical quenchers (Q) deactivate $^1O_2$:

$$Q + {}^1O_2 \longrightarrow Q^3 + {}^3O_2 \qquad (34)$$

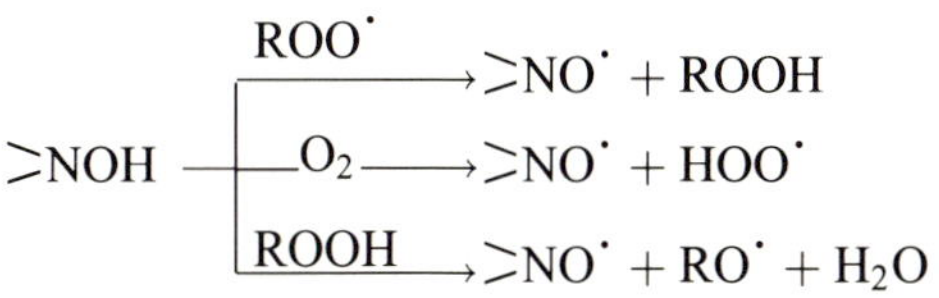

$$\mathord{>}NOH \begin{cases} \xrightarrow{ROO^\bullet} \mathord{>}NO^\bullet + ROOH \\ \xrightarrow{O_2} \mathord{>}NO^\bullet + HOO^\bullet \\ \xrightarrow{ROOH} \mathord{>}NO^\bullet + RO^\bullet + H_2O \end{cases}$$

Scheme 21

HAS are weaker quenchers of $^1O_2$ than aromatic amines [208]. Extensive quenching studies were performed with aliphatic and heterocyclic amines [209]. Data obtained in the inhibition of the self-sensitized photo-oxidation of rubrene revealed that the quenching efficiency of unhindered amines may be correlated with their ionization potentials. A low ionization potential means a better quenching activity. Substitution on the α-carbon atom to nitrogen by alkyls reduces the quenching rate predicted from the ionization potential [208, 209]. The sensitivity of $^1O_2$ quenching to steric effects may be exemplified with quenching rate constants k_q(in $dm^3 \ mol^{-1} \ s^{-1}$) [209]. 1,4-Diazabicyclo[2.2.2]octane (DABCO) was used as a standard.

DABCO	Et_2NH	N−CH$_3$	N−H	N−H	N−H
k_q	2×10^7	1.5×10^7	5.3×10^7	5.8×10^6	3.0×10^6	$< 2 \times 10^5$

Differences in experimental conditions may explain some differences in activity trials of the quenching by HAS. An analysis of the reactivity of HAS with $^1O_2$ revealed a strong quenching ability of tertiary HAS, surpassing that of DABCO. A flash photolytic study confirmed this fact [210]. The quenching increased in the order $>NO^\cdot < >NH < >NR$. The strong quenching effect of $>NCH_3$ should be attributed, to a large extent, to the reactive quenching. Tertiary HAS $>NCH_3$ is oxidized and the physical quenching is involved only in the last part of the process at Eq. (35) [175]. The quenching rate is a sum of the physical and reactive quenchings:

$$>NCH_3 \xrightarrow{^1O_2} >NH \xrightarrow{^1O_2} [>N^+(H)OO^\cdot] \longrightarrow >NO^\cdot + HO^\cdot$$
$$\downarrow {^1O_2}$$
$$>NO^\cdot + {^3O_2} \tag{35}$$

The nitroxide, the final product of HAS oxidation, acts as physical quencher of $^1O_2$. Cyclic nitrones can be formed among oxidation products of HAS and act as strong quenchers of $^1O_2$ as well [211]. It should be admitted that there is only a little contribution of $^1O_2$ quenching to the integral stabilizing effect of HAS in unpigmented PO [210]. It was shown in model measurements that no correlation exists between the quenching efficiency of HAS and their photostabilizing effect in PO.

An enhanced importance of $^1O_2$ quenching cannot be excluded in pigmented PO, polydienes and coatings. An efficient protection against $^1O_2$ induced photo-oxidation of cis-poly(1,4-butadiene) by HAS was confirmed [212]. Oligomeric HAS 34 and 35a imparted a good stability to the polydiene in the presence of the sensitizer Rhodamine 6G.

In photo-oxidized polymers, HAS may meet excited chromophores (Ch*) such as residues of polymerization catalysts or carbonyls arising in oxidized PO. Attempts were made to obtain information on the ability of HAS or derived $>$NO$^{\cdot}$ to quench Ch* [213, 214]. The phosphorescence emissions from sensitizers anthraquinone, benzophenone or benzhydrol were not affected by HAS. These data indicate that quenching ability is lacking in the HAS mechanism [214].

HAS derived $>$NO$^{\cdot}$ inhibits photo-oxidation of PP sensitized by anthraquinone or benzophenone [125]. The parent $>$NH was not effective. The authors consider involvement of the recombination of the nitroxide with semiquinone or ketyl radicals (HCh$^{\cdot}$) derived from the sensitizers. The latter were not protected from photodecomposition by the present HAS [214]. Some quenching activity of HAS was extrapolated from model reactions with excited carbonyls and polyconjugated systems. HAS derived $>$NO$^{\cdot}$ quenched singlet and triplet states of acetone (used as a model of ketonic chromophores in PO) and of anthracene [213]. Similar quenching experiments were not performed in a polymeric matrix. Formation of aminyls was proposed as a potential process resulting in quenching of excited carbonyls present in PO by HAS [215]. Energy transfer from excited carbonyls was mentioned in another paper [123] as a pathway to enhanced abstraction of hydrogen from the polymer by nitroxides:

$$[>CO]^* + >NO^{\cdot} \longrightarrow >CO + [>NO^{\cdot}]^* \xrightarrow{PH} >NOH + P^{\cdot} \qquad (36)$$

If HAS and Ch* are located in the polymer matrix in very close proximity and/or form a loose complex, HAS may compete with the polymer PH for the energy of Ch* [122]:

$$>NR + Ch^* \longrightarrow [>NR]^* + Ch \qquad (37)$$

$$PH + Ch^* \longrightarrow PH^* + Ch \longrightarrow P^{\cdot} + HCh^{\cdot} \qquad (38)$$

An excitation transfer (Eq. 39) was considered in PE as a process preventing formation of an activated exciplex via Eq. (40) and, consecutively, formation of *trans*-vinylene groups in oxidized PE. Reactions at Eqs. 37–40 were used for explanation of the involvement of HAS in quenching of chain initiation in oxidized PE [122, 134, 168, 189].

$$PH^* + [PH, >NR] \longrightarrow PH + [PH, >NR]^* \qquad (39)$$

$$PH^* + O_2 \longrightarrow [PH, O_2]^* \qquad (40)$$

3.3.1.7 Complexation of Metal Ions

HAS stabilize PO contaminated with ions of metals: Ti(III), Ti(IV), Fe(II), Co(II), Al(III) and Mg(II) [216]. Complexation of metallic impurities by the parent amine and the derived $>$NO$^{\cdot}$ has been envisaged [217] as a prevention of ROOH homolysis (Scheme 22).

$$\begin{array}{c} \xrightarrow{\text{ROOH}} \text{RO}^{\cdot} + \text{HO}^{\cdot} \\ \vert \\ \text{M}^{n+} + {>}\text{NH} \longrightarrow [\text{M}^{n+}, {>}\text{NH}] \xrightarrow{\text{ROOH}} [\text{M}^{n+}, {>}\text{NH}, \text{ROOH}] \\ \qquad\qquad\qquad\qquad\qquad\qquad\qquad\qquad\qquad \downarrow \\ \qquad\qquad\qquad\qquad\qquad\qquad\qquad \text{Nonradical Products} \end{array}$$

Scheme 22

3.3.1.8 Efficiency of Mixtures of HAS Transformation Products

Contribution of various individual transformation products of HAS ($>$NO$^{\cdot}$, $>$NOR, $>$NOH) to the integral effect of the parent amines was discussed in the preceding text. It should be admitted that the transformation products form in the polymer matrix a mixture of products differing in the relative contents of individual components. It was shown in a study [218] based on results of deactivation of ROOH, radical scavenging and quenching of $^{1}O_2$ or [$>$CO]* that the mixture and not the single product is responsible for the final effect. Even a synergism may be evoked in the product mixture due to some parallel and concerted processes accounting for the integral stabilizing effect.

3.3.1.9 Depletion of Hindered Amines

In spite of the suggested existence of the nitroxide regeneration cycle involving some HAS transformation products, the long-life activity of HAS – similar to that of other polymer stabilizers – does not go on forever. Finally, any active moiety of HAS is depleted, even though this may happen after many regenerative cycles. Concentration of HAS decreases during photo-oxidation of PE or PP according to the first order reaction, independent of the initial concentration [15]. Splitting-off of the ester group from the commercial HAS **28**, R = H results in a 4-oxoderivative, e.g. **132**, more prone to deeper destruction of the piperidine cycle. An intramolecular abstraction of hydrogen atom from the position 4 of **28**, R = H by exciplexes **130** or **131** formed in photo-oxidized PO [15] (Scheme 23, pathway a) or a hydrogen abstraction by any efficient free radical, including $>$NO$^{\cdot}$ (Scheme 23, pathway b) [134, 219] were proposed as pathways for the loss of the ester group. The pathway b, an analogue of the reaction at Eq. (8), proceeds at a significant rate at 50 °C in a nitrogen atmosphere. In the presence of air, 2,2,6,6-tetramethyl-4-oxopiperidinyl-1-oxyl **133** and a peracid are formed.

The loss of the ester group and formation of **134** was also observed with *N*-acylpiperidine in hydrochloric acid environment [220].

A photolytic fragmentation of the oligomer **34** resulting in formation of traces of gaseous acetaldehyde and methane proceeds at 50 °C with light having $\lambda > 300$ nm [221]. Intermediary formation of $>$N$^{\cdot}$ and $>$NO$^{\cdot}$ was suggested.

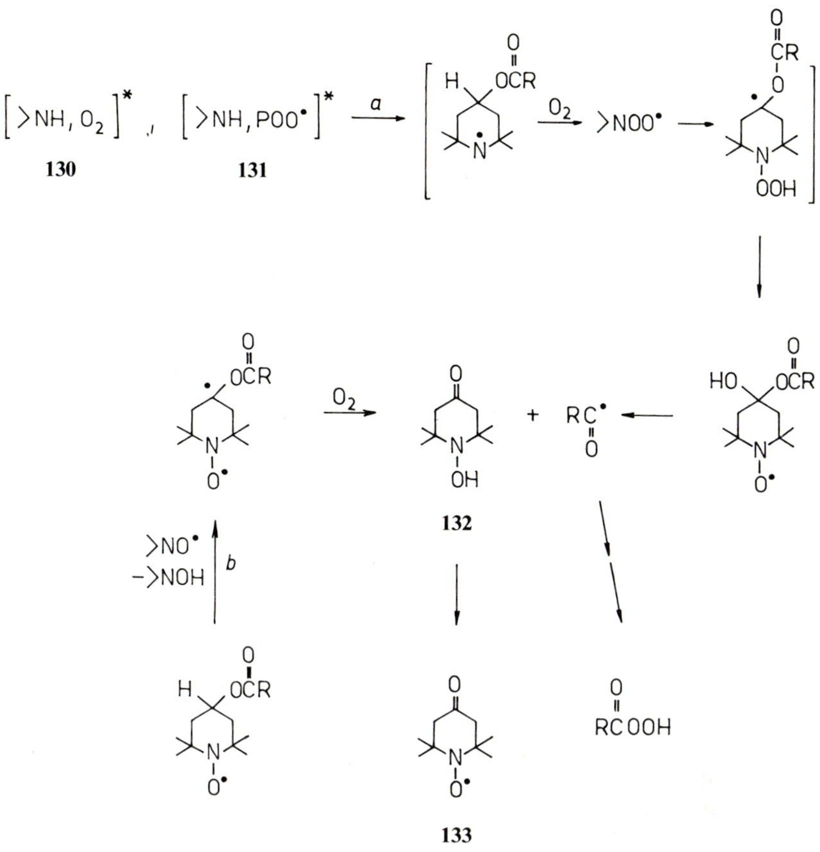

Scheme 23

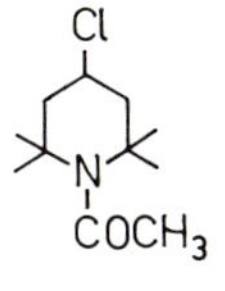

134

Two main processes participate in the ultimate phase of HAS depletion as a result of its sacrificial activity: (a) formation of stabilizing inactive derivatives of piperidine, and (b) destruction of the nitrogen heterocycle.

Inactive *N*-acylhydroxylamines **135** result in free-radical scavenging of acyl radicals by nitroxides (Scheme 24, pathway *a*) [126, 173, 174, 191]. Alternative pathways *b* and *c* starting with $>$NOH were proposed as well [26] and include reactivity with aldehydes or carboxylic acids formed in oxidized PO. Aldehydes generate hemiacetals **136** in the first step and are transformed into **135** after scavenging ROO·.

$$\geq NO^{\cdot} + RC^{\cdot}(O) \xrightarrow{\quad a \quad}$$

$$\geq NOH + RC(O)H \xrightarrow{\quad b \quad} \geq NOCH(OH)R \xrightarrow{\quad ROO^{\cdot} \quad} \geq NOC(O)R$$

$$\phantom{\geq NOH + RC(O)H \xrightarrow{\quad b \quad}} \mathbf{136} \mathbf{135}$$

$$\geq NOH + RC(O)OH \xrightarrow{\quad c \quad}$$

Scheme 24

According to [220], *O*-acylhydroxylamines **135** are destroyed during PO processing by heat and shear. This process may be involved in recycling of the waste PO.

The ultimate loss of the HAS activity occurs by destruction of the heterocycle initiated thermally, photochemically, chemically or by high-energy radiation. An intramolecular H-abstraction from the β-carbon atom in thermolysis of 2,2,6,6-tetramethyl-4-oxo-piperidinyl-1-oxyl **133** via a general reaction (Eq. 8) was proposed as a pathway of thermal selfdestruction of the piperidine cycle [25] (Scheme 25). The respective hydroxylamine was isolated in the yield of 66.5%. The biradical intermediate **137** either dimerizes to nitroxide **138** or thermolyses via **139** to a nitrogen-free fragment **140** (phorone) and nitric oxide.

Elimination of the nitrogen moiety from **139** may also proceed during photolysis [64]. A similar photolytic destruction was described for 2,2,5,5-tetramethylphenyl-3-imidazoline-1-oxyl [222]. A reversible photo-α-cleavage/thermal spin trapping sequence (Eq. 41) was observed with 4-hydroxy-2,2,6,6-tetramethylpiperidyl-1-oxyl **141** as an energy-wasting process [211]. The consequence of this process for PO stabilization in alternating light and dark periods was not declared.

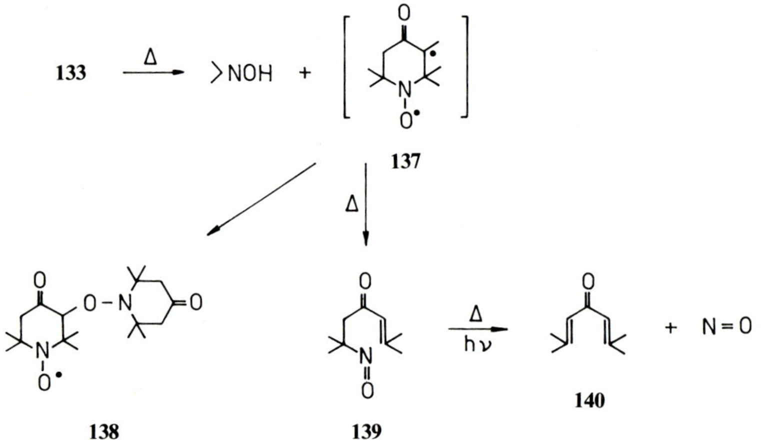

Scheme 25

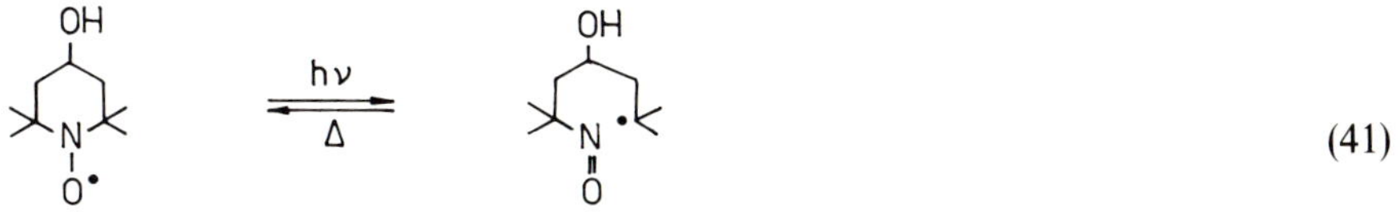

$$(41)$$

141

Ozone is a dangerous deteriogen for unsaturated polymers. Its role in oxidation of PO is rather underrated. Little information is available on the influence of O_3 on HAS. Photo-ozonization of tetramethylpiperidine yielded in the first step the respective nitroxide [223]. 2,6-Dimethyl-2-hydroxy-6-nitroheptane **142** was isolated in 91–98% yield:

$$\geq NH \xrightarrow{O_3} \geq NO^{\cdot} \xrightarrow{O_3} HOC(CH_3)_2(CH_2)_3C(CH_3)_2NO_2 \qquad (42)$$

142

A chemically induced opening of the piperidine cycle was described [173]. 1-Acyl-2,2,6,6,-tetramethylpiperidine was converted in a hydrogen chloride environment into **143**.

Open-ring nitroso- [224] **144** and nitrocompounds [126] **145** were reported as the ultimate products of ROO$^{\cdot}$ reaction with HAS derived nitroxides. N-Oxide **146** was mentioned among the ultimate products of HAS as well [225]. Some alkyl radical trapping and $^1O_2$ quenching activities should be preserved in **146**.

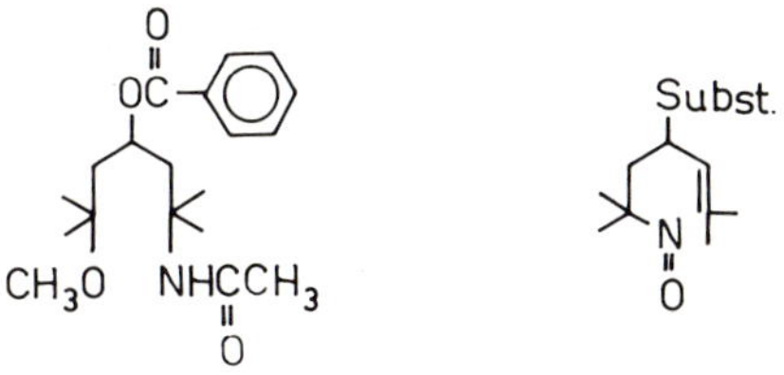

143   **144**

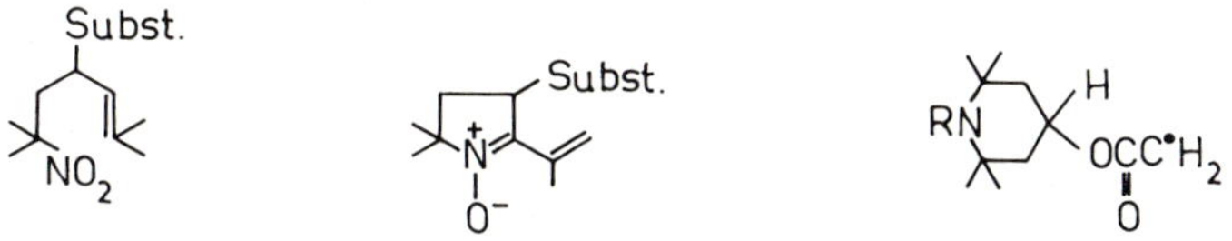

145   **146**   **147**

A detailed mechanistic elucidation was performed to decipher HAS depletion during gamma-irradiation of PP films at a sterilization dose of 2.5 Mrad and 35 °C [226, 227]. It was found that **28**, R = H, **28**, R = CH$_3$, and **34** lose the ester group. The piperidine nucleus was not degraded in this phase [226]. The primary attack is probably on the methylene group in the α-position to the carbonyl. Formation of a free-radical fragment **147** is followed by further scission during the irradiation in air. The radiolysis of the ester group in HAS was reduced in the presence of a phenolic AO.

Some radiolysis of the piperidine ring of **28**, R = H was observed during the post-γ-irradiation storage of PP under accelerated test condition [227]. Formation of a nitrone **146** and open-chain nitroso- **144** and nitrocompounds **145** was reported (Subst. = an ester group as in **28**).

An amine-oxide intermediate may be formed from $>$NCH$_3$ [227]. In this case, the ring-opened nitrosocompound **144** may be formed without detectable amounts of $>$NO$^{\cdot}$.

3.3.2 Hindered Amines in Contact with Acid Species

Acid pollutants formed in terrestrial environment as a result of anthropogenic activities (oxides of sulfur and nitrogen, acid rain), acid species arising from halogen containing pesticides, flame retardants (FR), of polymers like PVC, acid curing catalysts in coatings, carboxylic acids formed in oxidized polymers or acid transformation products of secondary AO (activated sulfides in particular) have been considered as potential reactants interfering with the activity mechanism of HAS. Deactivation of HAS has been explained by formation of a salt **148** (Eq. 43, HX = protonic acid) preventing generation of $>$NO$^{\cdot}$ as the principal intermediate in HAS mechanism.

$$>\text{N-Subst.} + \text{HX} \longrightarrow [>\text{N}^+(\text{H})\text{Subst.}\,\text{X}^-] \tag{43}$$
$$\textbf{148}$$

A reduced formation of $>$NO$^{\cdot}$ was confirmed in HAS doped preoxidized PP films exposed to hydrogen chloride [228], although the salt **148** was reported to decompose ROOH. Salts of hydroxylamine **149** and oxoammonium salt **150** are formed from $>$NO$^{\cdot}$ and a protonic acid (Eq. 44) and block the free-radical scavenging activity of $>$NO$^{\cdot}$ [64, 229]. Protonation of $>$NO$^{\cdot}$ takes place in the first step of Eq. (44).

$$2>\text{NO}^{\cdot} + 2\,\text{HX} \longrightarrow 2[>\text{N}^{\cdot+}\text{OH}\ X^-]$$
$$\downarrow \qquad\qquad\qquad\qquad\qquad\qquad (44)$$
$$>\text{NOH} + \text{HX} \longrightarrow\ >\text{N}^+(\text{OH})\text{H}\ X^- +\ >\text{N}^+ = \text{OX}^-$$
$$\textbf{149}\qquad\qquad\quad\textbf{150}$$

The strength of the acid HX plays a decisive role in the fate of HAS. Salts of weak acids, like carboxylic acids formed in oxidized PO or used as processing aids, benzoic acid or carbonic acid are evidently not dangerous from the point of view of HAS depletion [228]. This was also envisaged in mechanistic studies dealing with the conversion of N-methylated HAS into the salt of formic acid and secondary HAS **122**, not depleting the photo-antioxidant effect.

A detailed elucidation confirmed [220, 228] that the extent of deactivation of sec- and tert-HAS in PP films by volatile acids correlates with their increasing strength as measured by pK$_a$ values.

The extent of the reduction of the efficiency of HAS due to the salt formation according to Eq. (43) depends evidently also on the basicity of the piperidyl species, i.e. the substitution effects in position 1 [228]. The pK_a values of various categories of HAS were determined for sake of comparison [153, 155, 230, 231]: $>$NH (8.0–9.7), $>$NCH$_3$ (8.5–9.2), $>$NO$^{\cdot}$ (7.4–9.6), $>$N(CH$_2$)$_2$ substituent (6.5), $>$NOH (4.3–6.1), $>$NOR ($\sim$4.2), $>$NCOCH$_3$ ($\sim$2.0). The respective less basic species should resist better acid conditions.

Aliphatic hydroxylamines like N,N-didodecylhydroxylamine or their functionalized derivatives are efficient photo-antioxidants in PS, PO, diene based polymers and lubricants [232]. Good light stabilizing performance of the hydroxylamine derived from **28**, R = OH was claimed in coatings and PO [233]. However, $>$NOH is still too basic and inclined to interfere with acid catalysts in coatings. Moreover, **28**, R = OH is very sensitive to oxidation and may be prematurely converted into $>$NO$^{\cdot}$ during baking of coatings [155].

Other derivatives of piperidine matching better application conditions in acid environment were therefore developed and commercialized. These are O-alkylhydroxylamines $>$NOR, e.g. **28**, R = O-isoC$_7$H$_{17}$, O–C$_4$H$_9$ or analogous mono- and polynuclear piperidines substituted in position 1 with alkoxy groups [153, 155, 230, 231, 233] and N-acylpiperidines $>$NCOCH$_3$, e.g. **25**, R = COCH$_3$ or **151** [173]. Both $>$NOR and $>$NCOCH$_3$ survive processing of PO [220].

$$CH_3\overset{O}{\underset{\parallel}{C}}N\left\langle\;\right\rangle\!-NH\overset{O}{\underset{\parallel}{C}}\overset{O}{\underset{\parallel}{C}}NHC_{12}H_{25}$$

151

Introduction of O-alkylhydroxylamines $>$NOR represents a successful commercial exploitation of the mechanistic knowledge of HAS [155, 230, 231, 233–236]. It should be anticipated that this kind of HAS reacts as a scavenger of O-centred radicals, independent of the way of regeneration of $>$NO$^{\cdot}$. O-Alkylhydroxylamines are resistant to acid species. This was exemplified by excellent photoprotection of greenhouse films in contact with vapours of some acid generating pesticides having deleterious effect on conventional HAS [237]. HAS **28**, R = O-isoC$_8$H$_{17}$ is a very efficient free-radical scavenger in clearcoats based on styrene-alkyl acrylate-hydroxyalkyl acrylate copolymers or in acrylate-melamine coatings and does not interact with acid catalysts, pigments or transition metals driers [231, 234, 235]. It has an unparalleled performance in PO containing halogenated FR [230, 231]. Various $>$NOR may also be used for stabilization of PVC.

N-Acylderivatives of piperidine, e.g. **25**, R = COCH$_3$ or **151** have the lowest basicity of all HAS commercialized at present [173]. They were designed particularly for stabilization of coatings containing acid catalysts, e.g. for high-solid thermosetting acrylics or stoving lacquers based on alkyl resins [124, 153, 173]. In tests performed in photo-oxidized PP with various derivatives of 2,2,6,6-tetramethyl-4-stearoylpiperidine, the 1-acyl derivative was less efficient than the parent amine $>$NH, the derived $>$NO$^{\cdot}$ or $>$NOH [238]. In an analogous series of 4-benzoyloxypiperidines, the $>$NCOCH$_3$ was a weaker stabilizer than $>$NO$^{\cdot}$ as well [173]. $>$NOR **28** R = O-isoC$_8$H$_{17}$ was rated as a better stabilizer than **25**, R = COCH$_3$ [174, 238]. A similar conclusion was obtained in rating various 1-substituted 4-benzoyloxy- and 4-stearoyloxytetramethylpiperidines in photo-oxidized PP. The efficiency decreased in the order of N-substituents $>$NOH $>$ $>$NO$^{\cdot}$ $>$ $>$NOBu, $>$NBu $>$ $>$NH $>$ $>$NCOCH$_3$ $>$ $>$NOC(O)CH$_3$. The difference between the conversion of $>$NOR and $>$NCOCH$_3$ into $>$NO$^{\cdot}$ or $>$NH is envisaged. The process proceeds under photo-oxidation conditions probably faster with $>$NOR than with $>$NOCH$_3$ [153].

A difference in the behaviour between various $>$NCOCH$_3$ was observed in photo-oxidized coatings [153]. The mode of the substitution of the position 4 influences the splitting-off of the acyl group. A quantitative monitoring of the photochemical deacylation during irradiation at >300 nm either under model conditions or in HAS doped coatings revealed that **25**, R = COCH$_3$ was more stable than compound **151**. Secondary HAS was generated in both air and nitrogen atmosphere (Eq. 45). The concentration of $>$NH passes through a maximum and then diminishes. The differences in the rate of deacylation reveal that the stabilizer **25**, R = COCH$_3$ needs more time to be converted into a chain-breaking species $>$NH than the compound **151**. As a consequence, formation of cracks and loss of gloss were more efficient in coatings doped with **151**.

$$>\text{NCOCH}_3 \xrightarrow{\text{h}\nu} >\text{NH} \tag{45}$$

A mechanistic elucidation of the conversion according to Eq. (45) was performed with 1-acyl-4-benzoyloxy-2,2,6,6-tetramethylpiperidine [173]. O-Alkylhydroxylamine $>$NOR and acetate **152** were formed among products of interaction with *tert*-butylhydroperoxide in pentane solution at 132 °C. The salt **152** may be transformed into $>$NH and $>$NO$^{\cdot}$ in consecutive steps [124]

$$>\text{NCOCH}_3 \xrightarrow{\text{ROOH}} >\text{NOR} + >\text{N}^+\text{H}_2^-\text{OC(O)CH}_3 \rightarrow >\text{NH} \tag{46}$$

The transformation pathway at Eq. (46) explains the CB activity of $>$NCOCH$_3$.

A tendency to open the heterocycle of $>$NCOCH$_3$ under acid conditions and formation of **143** [173] or splitting-off of the 4-octadecyloxy group and formation of **134** [220] were reported.

A very unfavourable interaction resulting in depletion of HAS activity and categorized as antagonism was observed between HAS and sulfidic hydroperoxide decomposing (HD) AO. Antagonism was experimentally proved with thioethers and phenolic sulfides or HAS substituted xanthates in photo-oxidized PO or *cis* poly(1,4-butadiene) [5, 13, 64, 239, 240]. The knowledge of the chemistry of transformation products of thiosynergists and phenolic sulfides formed during polymer stabilization may be used for explanation of the deactivation of HAS [5]. Sulfoxides are formed in the first step of transformation of sulfur containing AO as a consequence of deactivation of ROOH. Both free radicals and

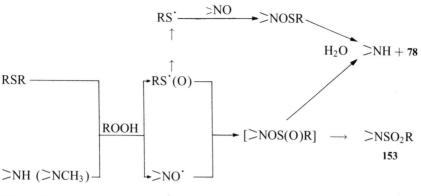

Scheme 26

acid species arise via thermolysis or photolysis of sulfoxides and participate in the HAS depleting process. Sulfinyl radicals $RS\dot{}O$ react with HAS derived $>NO\dot{}$ in an irreversible process [64, 241] resulting in inactive sulfonamide **153** (Scheme 26). The primary products of recombination of $>NO\dot{}$ with S-centred radicals are hydrolyzed similarly as in the aromatic series (Eq. 11) into HAS and protonic S-acids **78**, $n = 1$–3. The latter are also formed among the ultimate products of transformation of thiosynergists. Salts of S-acids **78** formed according to Eqs. (43) and (44), $X = RS(O)_n^-$, have been considered as another species responsible for the antagonism [239] in sulfide and HAS doped photo-oxidized PO.

Environmental temperature is a factor governing transformation of sulfoxides into sulfinyl radicals and sulfinic acid [5]. Therefore the antagonism does not appear at temperatures lower than those of thermolysis of sulfoxides. Examination of the LTHS of PP multifilaments stabilized with a combination of 0.05% phenolic AO, 0.05% phosphite and 0.1% oligomeric HAS **34** revealed at testing temperatures of 100 and 120 °C no antagonism when at the same time 0.1% of didodecyl 3,3-thiodipropionate was added [26]. On the contrary, the long-term heat stability was enhanced.

4 Antiradiant Activity of Amines

Deterioration of polymers, PO in particular, resulting from γ- or other high-energy irradiation, takes place in diverse areas of their application. This includes controlled modification of polymers for industrial purposes, use of polymers in some radiation exposed parts of nuclear reactors and radiation sterilization of food packaging materials or of equipment and materials for medical uses. Radiation-induced primary free radicals result from either the scission of the main chain

or scission of bonds between the main chain and pending groups [11]. The radicals are localized very inhomogeneously and have a long lifetime due to their
reduced mobility in the polymer matrix. The degradation of polymers continues
in the post-irradiation period during storage or application in air. The primarily
formed macroradicals, "frozen" in the crystalline domains or PO, migrate slowly
into amorphous areas where they are trapped by oxygen. Radical-chain radiation-
oxidation is triggered by this mechanism and proceeds by a slow decomposition
of ROOH according to mechanisms valid for auto-oxidation and photo-oxidation.
Cross-linking, chain scission, gel formation, changes in mechanical properties or
discoloration result in the irradiated polymer.

Due to the free-radical character of both the γ-radiation induced transformation and oxidation in the post-irradiation phase, free-radical scavenging stabilizers
– phenols and amines – effectively protect PO [11, 226, 227]. This was confirmed
by excellent antioxidant/antiradiant effects of DPA, PNA and PD in particular in
the post-γ-irradiation phase of PO and elastomers [11]. Oligomeric DHQ **21** or
its combination with **9b** efficiently stabilized EPDM or X-LPE used as insulation
materials for electric cables exposed to irradiation doses up to 2 MGy at a dose
rate 300 Gy h^{-1} [242].

Application of aromatic and partially hindered heterocyclic amines as AO/AR
in irradiated polymers results in analogous chemical transformations as reported in
Sects. 3.1 and 3.2. Discoloration due to QI is the principal disadvantage. Dealkylation of **9a** was observed during radiation vulcanization of SBR. HMW and
oligomeric HAS are best at concentrations ca. 0.1% as AR for PO and are suitable for applications in biomedical products needing γ-sterilization [226, 227].
No discoloration of the polymer matrix takes place and the stabilization is
effective during both sterilization and post-irradiation oxidation phases. **28**, R = H
for example very efficiently reduced formation of ROOH and >CO functions in
poly(propylene-ran-2.3% ethylene) films irradiated up to 20 Mrad in air at room
temperature [243]. Substitution of the piperidine in position 1 modifies to some
extent the protective effect. HAS **28**, R = Cl was more active than **28**, R = H under
the same irradiation conditions in reduction of recrystallization, density increase
and the main-chain scission of the copolymer. **28**, R = H or CH$_3$ imparted an
excellent post-irradiation stability to PP films, >NCH$_3$ was more efficient in sup-
pressing oxidation during the irradiation step [227]. The respective >NO$^\cdot$ derived
from **28** was a weaker AR than the parent HAS. **28**, R = H also protected very
efficiently the polybutadiene phase of LDPE/SBS blends against radio-oxidation
[244].

The similarity between the CB AO mechanism of phenols and HAS was
reported at 25 °C in radiation oxidation of 2,4-dimethylpentane performed in the
dark [207]. It was concluded that the respective >NOH formed from **28**, R = H
during the stabilization process was the species responsible for the antioxidant
effect (Scheme 27). The >NOH is considered to be formed via trapping H$^\cdot$ by
>NO$^\cdot$. The authors declared that >NOR are not involved in the stabilization
of the radiation-oxidation of hydrocarbons and have doubts about the importance
of the involvement of >NOR in the regenerative cycle in an oxygen containing
environment.

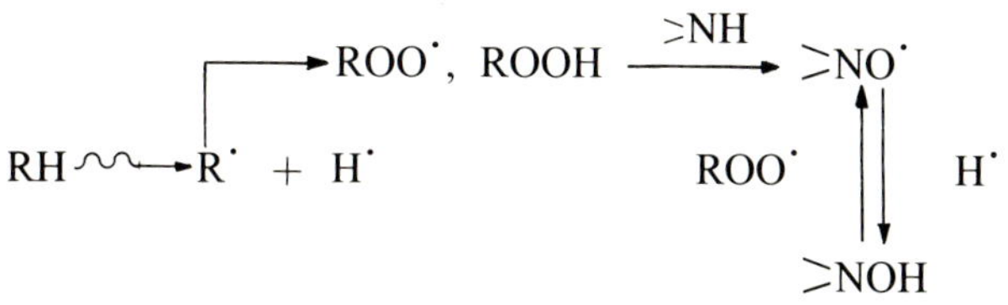

$$RH \rightsquigarrow R^{\bullet} + H^{\bullet} \qquad \begin{array}{c} \rightarrow ROO^{\bullet},\ ROOH \xrightarrow{\ >NH\ } >NO^{\bullet} \\ \\ ROO^{\bullet} \quad \updownarrow \quad H^{\bullet} \\ \\ >NOH \end{array}$$

Scheme 27

Both the polynuclear phenolic AO (Irganox 1010) and HAS (**28**, R = H or CH$_3$) did not affect the initial formation rate of oxidation products during the γ-irradiation of PP in air [227]. Free-radical scavengers, however, suppressed formation of ROOH during the post-irradiation oxidation at 60 °C due to scavenging of ROO$^{\bullet}$. At equivalent brittle failure values of PP, the stabilized samples showed only little total ROOH concentration as compared to the unstabilized PP. Nevertheless, ketonic products levels were similar in stabilized and unstabilized samples, and surprisingly, high levels of carboxylic acids end groups were found in the long-lived stabilized samples.

Phenols are stronger AO/AR than HAS. Contrary to the transformation of phenols into quinone methides, discoloration-free products are formed from HAS. Combinations HAS/phenol are, therefore, used to obtain protection at minimum discoloration. The efficient protection of PP or of a block copolymer of propylene with 1.5% ethylene irradiated with a dose of 5 Mrad was reached with 0.1% **28**, R = H and 0.05% Irganox 1076. Moreover, it was found that hindered phenolic AO prevent radiolysis of the ester type HAS (e.g. **28**) [226]. Combinations of HAS with phenolic AO may, therefore, be recommended for commercial stabilization of γ-irradiated hydrocarbon polymers [227].

5 Antiozonant Activity of Amines

Characteristic cracks at stressed rubber surfaces, formation of new chemical functions and scission of macromolecular chains are the consequences of the ozone attack on unsaturated rubbers. Their inherent resistance against O$_3$ is insufficient and may be enhanced by blending with 20–50% of saturated elastomers (EPM or EPDM) or with CR. Suitable antidegradant must be used to obtain full protection. Microcrystalline paraffin wax provides static protection by forming an inert surface barrier and improving efficiency of chemical AOZ [1]. The latter must be very reactive with O$_3$ to be able to compete with rubber ozonation. Besides, AOZ must be resistant to direct oxidation with oxygen and to interactions with cure ingredients, have antioxidant properties and good solubility in vulcanizates [1,3,21,31]. To achieve antiozonant protection, amines must be present in rubbers in much higher concentrations (1–3.5 phr) than when used as AO. N,N'-Disubstituted PD **11** are the only class of aromatic amines having properties of chemical AOZ [1, 3]. 6-Ethoxy-2,2,4-trimethyl-1,2-dihydroquinoline **11b** is also used as AOZ. It is mostly inferior to PD; it may, however, replace a part of the more expensive PD in the rubber [32].

PD possess unique multiple properties [32]. Depending on N, N'-substitution and concentration, PD are AO, AF, AOZ or mild MD. Rating by the inherent chemical efficiency, the antiozonant activity increases in N, N'-disubstituted PD in the series diaryl-PD < *sec*-alkyl-aryl-PD < di-*sec*-alkyl-PD [3]. An optimum balance with other technically important properties, like a low effect on the scorch time, low depletion with O_2, high physical persistency, good compatibility and migration in the vulcanizate, low toxicity, must, however, be fulfilled. A careful rating of all these factors was performed and relations between structure and activity were interpreted to optimize structure of PD. N, N'-Di-*sec*-alkyl-1,4-PD **11a** are excellent AOZ increasing the critical stress required for cracking to occur, providing protection in static and dynamic tests and having extreme reactivity with O_3. They are unfortunately rather scorchy due to their basicity, volatile and easily depleted by oxygen [18]. N-*sec*-Alkyl-N'-phenyl-PD **11b,c** are commercially dominant AOZ. They react a little slower with O_3 than **11a** and do not increase the critical stress required for cracking to occur. Excellent protection at dynamic conditions, a smaller scorch effect, synergism with waxes and lower reactivity with O_2 than **11a** are their advantages. Currently, the best commercial protection has been obtained with N-(1,3-dimethylbutyl)-N'-phenyl-1,4-PD **11c**. Heterocyclic 1,3,5-triazine moiety may serve as one of the N-substituents, as in **12**: A is either an amino (**12a**) or QI (**12b**) moiety [245]. The antidegradant **12** contains 12% tris-QI, 60% bis-QI and 31% mono-QI structures.

Mixed N, N'-diaryl-1,4-PD **11d** are of increasing importance. They are excellent AO, give a minimum scorch, depletion by O_2 and physical losses. However, they are less reactive with O_3 [246]. The limited solubility in vulcanizates enables us to use them only on levels which are in the concentration range of AO.

To achieve a balanced optimum protection of stressed rubber vulcanizates under a combined attack of O_2 and O_3, blends of **11a**–**11d** are used. Practical experience indicates that combinations of two or three different PD are more beneficial than the use of individual ingredients [246]. Strong protection is achieved for an extended period of time. The observed final effect cannot be explained as a synergism: it is a benefit of additivity of the optimal physical and chemical properties of the individual components. For example, the use of diaryl-PD in mixtures with alkylated PD is beneficial, since it reduces the integral pro-oxidative effect.

To attain an optimum integral protection, combination of PD with DHQ, derivatives of DPA or PNA have also been successfully used.

The mechanism of the antiozonant protection has been the matter of studies and discussions for a long time [3, 4, 18, 21, 31] and it is still an area of active research. Due to the diversity of the deteriogens and reactive products of rubber degradation involved in ozone ageing, more concerted protective mechanisms are certainly included [247]. The antiozonant activity accounts for a combination of a direct scavenging of ozone by the chemical AOZ and formation of a protective film (an inert barrier) on the rubber surface due to HMW products of AOZ ozonation and reactivity of ozonized rubbers with AOZ accounting for partial reparation of rubber macromolecules by relinking of severed chains and formation of physically persistent rubber bound nitrogen compounds [3, 4]. Ozone scavenging, i.e. a direct reactivity of the AOZ with O_3, is certainly the true primary step and prevents the unsaturated rubber from ozonation during the initial phases of ozone ageing, in the absence of any protective surface layer [19, 247] (wax layers provide insufficient protection under dynamic conditions). The critical stress of rubber surfaces is reduced by ozonation [3, 21]. The other mechanisms may be triggered only after formation of some ozonation products of the AOZ and rubber. The protective layer can be built-up providing the fresh AOZ migrates to the rubber surface and replenishes the consumed amine. HMW transformation products of the AOZ remain in the surface layer and constitute an inert and relaxed film barrier.

The competitive reactivity for O_3 between an AOZ and rubber has been reported to be the key primary mechanism [3, 4, 18, 21, 247]. The principal mechanistic studies with aliphatic amines disclosed the primary formation of an ozone adduct $\geq N^+OOO^-$ and its conversion into amine oxides $\geq N^+O^-$ and nitroxides $\geq NO^{\cdot}$. The amine oxide pathway was demonstrated to be more probable for secondary aromatic amines, although nitroxides should be considered as free-radical intermediates for some products isolated from PD. The ozonation pathways of phenylenediamine AOZ were studied in detail [3, 247, 248]. The direct ozonation of PD gives rise to a complex mixture of products including derivatives of QI, nitrones and aromatic nitro and nitroso compounds. Products having lower and higher molecular weights than the parent PD are formed. The HMW products were suggested to be more important for the high antiozonant effect due to their contribution to the protective film formation [18].

N, N'-bis(1-Methylheptyl)-1,4-PD **11a** reacts mainly via amine oxide (route a) and oxidation of the aliphatic side chains pathways (route c, Scheme 28,

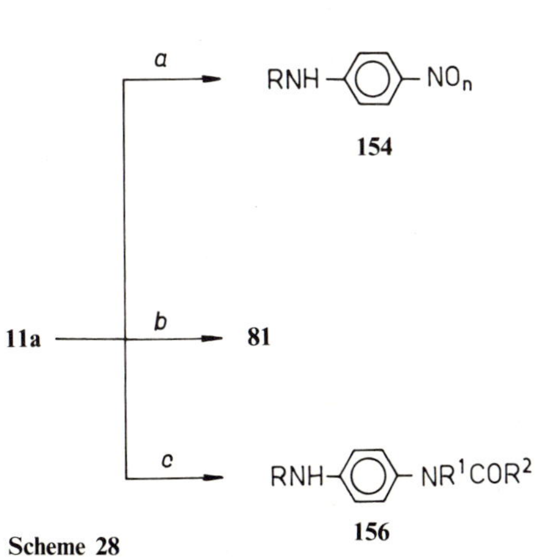

Scheme 28

R = 1-methylheptyl) [18, 247]. Pathway *a* results in nitroso- **154** (n = 1), or nitrocompounds **154** (n = 2) and, via hydroxylamine intermediates, ultimately to Bandrowski's base **155** (Y = NR–C_6H_4NHR). The pathway *c* results in various amides, e.g. **156** (R^1, R^2 = alkyls), via oxidation of the 1-methylheptyl group. Pathway *b* is of minor importance for **11a** and results in bisnitrone **81** ($R^1 = R^2$ = 1-methylheptyl).

Wurster's cation-radical **49** may arise in the primary step of ozonation of *N,N'*-dialkyl-PD **11a** as well and form a CTC **157** with O_3 [18]. CTC **157** disproportionates into BQDI **53** (R^1,R^2 = 1-methylheptyl) and **11a** (Eq. 47), and is transformed in situ by the released O_3 into compounds derived from the respective BQDI, e.g. various BQMI, more complicated structures similar to that exemplified earlier, e.g. **84**, **86**, **87** or a base **158** (R = 1-methylheptyl). Simple **53** are formed only transitionally.

$$\left[>\overset{\cdot\cdot}{N}-\hspace{-2pt}\left\langle\bigcirc\right\rangle\hspace{-2pt}-\overset{+\cdot}{N}< \quad \bar{:}O\ 00^{\bullet} \right] \longrightarrow \left[53, 11a, O_3 \right] \longrightarrow \text{Products} \tag{47}$$

157

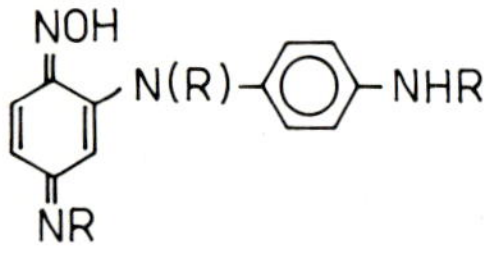

158

Earlier studies revealed a better antiozonant efficiency of PD substituted with *sec*-alkyls than with *n*-alkyls or *tert*-alkyls [3, 18, 247]. A recent elucidation shows that the behaviour of *N,N'*-di(*n*-octyl)-1,4-PD closely parallels ozonation mechanism of **11a** (Scheme 28), except that the formation of the base **155** was reduced in importance [18, 247]. *N,N'*-bis(1,1-Dimethylethyl)-1,4-PD bearing *tert*-alkyls was rather unstable and was oxidized with oxygen even in the absence of O_3.

N-(1,3-Dimethylbutyl)-*N'*-phenyl-1,4-PD **11c** is ozonized according to the all three pathways *a* to *c* exemplified in Scheme 28. Moreover, the aminyl pathway *d* contributes due to the presence of one *N*-phenyl group (Scheme 29,

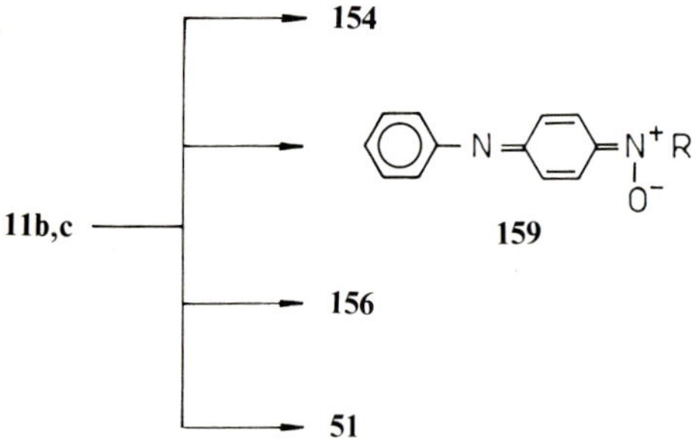

159

Scheme 29

R = 1,3-dimethylbutyl) [18, 247]. Pathway *a* results in **154** (no Bandrowski's base was formed). Pathway *b* yields nitrone **159**, pathway *c* amide **156**, pathway *d* hydrazine **51**, R^1 = 1,3-dimethylbutyl, R^2 = phenyl) as the N–N coupling product. The compound **51** is further oxidized via aminyl and nitroxide into polynuclear nitrones. The aminyl pathway *d* was observed only with **11b** and **11c** containing one N-aryl substituent and is lacking in **11a** due to the extremely low stability of the respective aminyl. The reactivity according to Scheme 29 may be attributed to all *N-sec*-alkyl-*N'*-phenyl-1,4-PD [3, 18].

N,N'-Diphenyl-1,4-PD **11d** is transformed due to the exclusive aromatic substitution on both amino groups via nitroxide (*b*) and aminyl (*d*) pathways, analogically to the transformations in antioxidant processes shown in Schemes 4 and 8.

Formation of the protective film from ozonized rubber and AOZ creates a barrier against penetration of O_3 into rubber [3, 4]. Reactions of AOZ with ozonides **3**, zwitterions **4** and aldehydic fragments **8** are envisaged. The respective reactivities are lower than the direct ozonation and the relevant contribution to the antiozonant mechanism is therefore inferior to the ozone scavenging. The reactivity of the ozonized rubber with AOZ accounts for rubber chain repairing, classified also as chain relinking or selfhealing mechanisms [4, 21, 247–249]. This contributes to formation of relaxed surface films.

Ozonides oxidize *N,N'*-disubstituted PD **11** in weak acid catalyzed reactions into the respective BQDI **53**, Scheme 8 [3, 4]. The intermediary formed Wurster's cation-radical **49** disproportionates. Formation of α-hydroperoxides of the type **160** was proposed for the reaction of **11** with **4**. A relinking of the broken rubber chains may arise via the reaction of **160** with the second zwitterion.

$$>C<^{OOH}_{N(R^1)} \!\!-\!\!\bigcirc\!\!-\! NHR^2$$

160

Aldehydic fragments **8** formed in ozonized rubbers or by PD/acid assisted ring-opening of ozonides react with **11** under formation of bound-in species **161** or various relinked systems, e.g. **162** able to form crosslinks [3, 4] (Scheme 30).

Polymer-bound species **160–162** contribute, together with species like **54, 55, 68, 67, 82, 91–94** (R = a polymeric residue), to the formation of various rubber-bound nitrogen-containing moieties called "nonextractable" nitrogen. Other processes taking place in PD doped rubbers, like vulcanization, may also contribute to the "nonextractable" nitrogen. It should be kept in mind that the antioxidant and antifatigue activities of amines have been regarded as parallel with that of the antiozonant activity in rubber protection. Moreover, other rubber processing chemicals are certainly involved in the complicated antidegradant chemistry. Whichever antiozonant mechanism prevails, transformation products of aromatic amines are always formed and are involved in the antidegradant process.

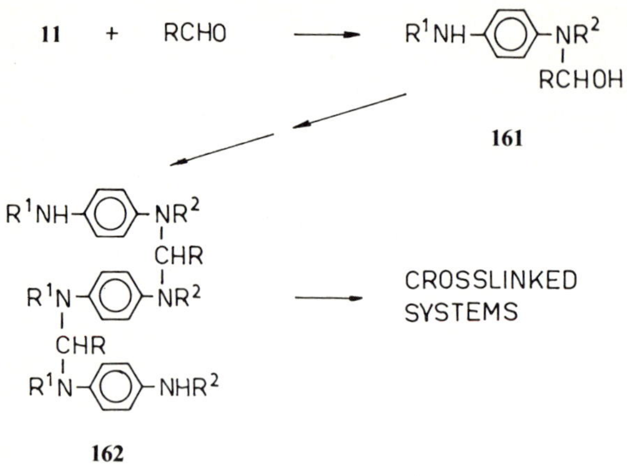

$$11 \ + \ RCHO \ \longrightarrow \ R^1NH-\!\!\left\langle\bigcirc\right\rangle\!\!-NR^2$$

$$\underset{\displaystyle 161}{\overset{\displaystyle RCHOH}{}}$$

$$
\begin{aligned}
&R^1NH-\!\!\left\langle\bigcirc\right\rangle\!\!-NR^2 \\
&\qquad\qquad\quad CHR \\
&R^1N-\!\!\left\langle\bigcirc\right\rangle\!\!-NR^2 \\
&\;CHR \\
&R^1N-\!\!\left\langle\bigcirc\right\rangle\!\!-NHR^2
\end{aligned}
\qquad \longrightarrow \qquad
\begin{aligned}
&\text{CROSSLINKED}\\
&\text{SYSTEMS}
\end{aligned}
$$

162

Scheme 30

6 Blends of Amines with Other Polymer Stabilizers

Effective protection of polymers against degradation by different deteriogens and several distinguished processes requires application of combinations of stabilizers differing in their activity modes [250]. Most stabilizers act only by one major mechanism. Involvement of other mechanisms currently has a supporting or complementary character. In this respect, amines possess a specific position. Some aromatic amines, like PD **11** are AO, AF and AOZ, DHQ **16** are AO with more or less expressed anti-flex-crack activity, HAS are LS/photo-antioxidants contributing to long-term heat stability. This multifunctionality of amines does not accomplish all requirements for a practically efficient integral protection of a particular polymer. Blends of stabilizers are therefore mandatory. Each component should contribute in a specific way, reflecting its inherent chemical efficiency, resistance to depletion and physical persistence.

Polymer stabilizers are very reactive systems. This is expressed not only in scavenging chain-carrying radicals or deactivation of ROOH but also in interactions between various stabilizers or their transformation products [250]. This chemical cooperation may have a positive effect accounting for additivity or even synergism. Unfortunatelly, some interactions are negative, i.e. antagonistic, and should be avoided.

Stabilizing activity is improved in mixtures of aromatic amines varying in their structures. This is due to different contributions to antioxidant, antifatigue or antiozonant effects. Combinations of DPA **9a** with PD **11c, d**, PNA with PD, ternary mixtures PD/DPA/PNA, PD **11b** with DHQ **16b** or **16c**, oligomeric **21** with **11** or **16** ($R = NHC_6H_5$), condensate **22** with **11b**, PT **20** with **9** or **11b** are examples of combinations effective in rubber protection. Mixtures containing

PT are also active in lubricating oils and vinylic monomers. Multicomponent mixtures comprising HAS **28** (R = H), PT **20** and DPA **9a** or a phenolic AO were proposed as antiwear additives for lubricating oils. PNA or dinaphthyl-PD provide excellent thermostability and suppress discoloration of PUR or PE doped with halogenated FR. Any negative influence of potentially released hydrogen halogenides from FR on amines was observed in these combinations.

Cooperation between aromatic amines and hindered phenolic AO or phenolic sulfides was used for explanation of homosynergism [3, 49, 250]. Due to the higher rate constants of ROO˙ with amines in comparison with that of phenols, amines react in the sacrificial process faster than phenols. The latter serve as hydrogen donors for the aminyl radical and a part of the parent amine is cyclically regenerated. The rate constant of the reaction (Eq. 48, ArOH = phenolic AO) is of the order 10^7 dm^3 mol^{-1} s^{-1} and is equal or higher than that for Eq. (1) [50].

$$>N˙ + ArOH \longrightarrow >NH + ArO˙ \tag{48}$$

Formed phenoxyls ArO˙ are irreversibly consumed in coupling and ROO˙ scavenging [251]. The aminic antidegradant starts to be consumed according to processes reported in Sect. 3.1 only after the disappearance of all of the phenolic AO. QI are the principal products of transformation of PD (Sect. 3.1.1.3). A reductive regeneration of PD from QI proceeds in mixtures of PD and phenols on account of the oxidation of the phenol [3]. This reactivity explains very effective cooperation between **11d** and 6-hydroxy-DHQ **16** (R = OH). QI **53** (R^1, R^2 = phenyl) arising from **11d** was reduced and the derivative of DHQ was oxidized into **105** [252].

Chemical interaction between transformation products of both the amine and phenol is involved. Derivatives of cyclohexadienone like **77**, deeply coloured compounds with structure of indophenols or BQMI **57** are expected to be formed as intermediates [251]. The compound **77** is formally analogous to quinol ethers of 1,4-benzoquinone dioxime **163** (R = *tert*-butyl, methoxy, phenyl) thermally releasing the corresponding phenoxyl and therefore imparting the antioxidant effect [253].

163   **164**

The homosynergism according to Eq. (48) was exploited in melt and colour stabilization of PP with a combination of 0.05% **9b**, 0.05% phenol Irganox 1010 and 0.1% of tris(2,4-di-*tert*-butyl) phosphite [47] or a mixture of DNQ with hindered phenolic AO.

Other combinations of aromatic amines with phenols or phenolic sulfides having homosynergistic character were proposed for stabilization of radiation resistant poly(ethylene-*co*-ethyl acrylate), PE cable jackets in contact with copper, poly(ethylene-*co*-carbon monoxide), PUR, polysiloxanes or polydienes. A combination of dibenzylhydroxylamine or alkylated PT with phenolic AO was proposed for PP and PUR foams, respectively [254, 255]. Mixtures of aromatic amines with phenolic AO are also efficient in stabilization of other organic substrates like gasoline, synthetic ester lubricating oils and lipids and were proposed as inhibitors of polymerization of styrene [256–258].

Aromatic amines are not able to deactivate ROOH. The effect may be reached in PO, PUR or diene based elastomers with heterosynergistic combinations of amines with secondary AO like organic phosphites or sulfides.

Aromatic and heterocyclic amines are commonly used in rubbers or PO together with carbon black. Surface catalysis or adsorption by the pigment and its acidity may negatively influence the efficiency of basic amines. Amines were found to be less effective AO in PE containing carbon black than in clean PE [259]. A similar observation was made with rubbers and categorized as antagonism. The effect seems to be higher on light exposed surfaces than on the dark side where thermal oxidation prevails. Depletion of aromatic amines is enhanced by transition metal impurities. Disubstituted PD were transformed by metal contaminated carbon black into BQDI [260].

Although imparting outstanding light stability and some heat stability to PO, HAS fail as processing stabilizers. Blends of HAS with hindered phenols and hydrolysis resistant phosphites are therefore mandatory for efficient melt and heat stabilization of PO [3, 135]. Systems phenol/HAS also enhance the outdoor stability of EPDM, acrylic resins, epoxy resins, PUR or polyurea/acrylate coatings and antiwear properties of lubricating oils. Mechanistic studies dealing with cooperation phenol/HAS were performed only in PO.

Phenomenological elucidation revealed some features of antagonism during PP photo-oxidation [125]. Thermal history of melt processing certainly affects the interactions HAS/phenol in subsequent outdoor exposure. A detailed analysis of results obtained with PE and PP revealed that excellent long-term heat stability at temperatures below 120 °C was obtained with a combination of oligomeric HAS **34** or **35a** with efficient polynuclear phenolic AO Irganox 1010 [129]. The principal contribution to the heat stability is unequivocally due to the hindered phenol [12, 13]. Application of the stabilizer blends improved additively the integral effect.

There is a principal difference between the mechanistic and phenomenological contribution of HAS and phenols to the stabilization process and a difference in the response on the testing temperature as well [26, 136]. Phenols break the autoxidation chains predominantly by scavenging ROO˙ [251]. Photo-antioxidant and heat stabilizing action of HAS has been based principally on peroxide deactivation and R˙/ROO˙/RC(O)OO˙ scavenging. Phenolic AO protect PO effectively up to 150 °C. The mechanical properties of PO remain for a long period (induction period, IP) without change and then decline rapidly. Practically no oxidation products can be detected during IP [26, 136]. It is only at the end of the IP that a rapid formation of oxidation products starts. The process is often so fast ("catastrophic failure") that it is difficult to monitor it by means of IR spectroscopy or by mechanical tests. Phenol doped PP disintegrates into powder similarly to unstabilized PP.

The effectiveness of HAS as LTHS is more pronounced at temperatures below 100 °C, drops with increasing temperature and is almost negligible at 150 °C. The drop of mechanical properties and growth of oxidation products ($>$CO,$-$OH) in HAS doped PP develop gradually from the very beginning of oven aging. The test specimens became brittle after a long period of time, with no external signs of degradation or discoloration [26, 136].

It is evident that the differences in the oxidation mechanisms at temperatures below and above 100 °C are operative with respect to the response of phenols or HAS. At the lower temperature level, the oxidation of some oxidation products of PP (alcohols, aldehydes) is important and may be reduced or prevented by HAS. At temperatures above 120 °C, the free-radical oxidation of PP proceeds statistically by means of ROO·, cannot be effectively stopped by HAS or its transformation products and is formally expressed by the gradual loss of properties of PP. Contrary to this process, phenols are able to scavenge effectively the ROO· up to 150 °C and until the oxidation level of PP is too high and the residual concentration of phenol cannot interfere effectively any more with ROO· [261].

The interference with degradation products has an additive effect if HAS is used as LTHS below 120 °C in combination with phenolic AO and an efficient phosphite that represents the state-of-the-art processing base stabilization [26, 122, 129]. The same IP is observed using this combination as in systems without HAS. That means HAS did not contribute to the length of the IP. There is, however, not a "catastrophic failure" of the tensile strength as characteristic for the basic stabilization without HAS. After the end of the IP, the tensile strength decreases and the concentration of $>$CO increases more or less rapidly, according to the concentration and structure of the added HAS. The secondary oligomeric HAS **35a** contributes more effectively to the integral heat stability than tertiary HAS **34**. At testing temperatures unfavourable for contribution of HAS as HS, i.e. at 150 °C [26, 122, 219], the positive effect of the combination HAS/phenol was lost.

The thermo- and photobehaviour of all components arising in particular degradation phases of the HAS/phenol doped polymer plays a role and mechanistic studies performed in model systems and polymers are very useful [262]. Chemical interactions between phenols, HAS and their transformation products were elucidated and key products were identified [5]. Phenols function as H-donors according to Eq. (8) and are oxidized with HAS derived $>$NO· via the respective phenoxyls (Eq. 49) [64, 262]. Coloured C–C coupling phenolic products and substituted benzoquinones (BQ), diphenoquinone, stilbenequinone **164** or more complicated quinone methides (QM) are currently formed [251]. BQ and QM contribute slightly to the stability of hydocarbons and polymers. Their effect is strongly inferior to that of the original phenolic AO and formation of quinoide compounds via reactivity of phenols with $>$NO· is, therefore, a depleting process [251].

$$\underset{\substack{\text{CH}_2\text{Subst.}}}{\overset{\text{OH}}{\bigcirc}} \;\; + \;\; >\text{NO}^{\bullet} \;\; \longrightarrow \;\; \underset{\substack{\text{CH}_2\text{Subst.}}}{\overset{\text{O}^{\bullet}}{\bigcirc}} \;\; + \;\; >\text{NOH} \tag{49}$$

Potentially, nitroxides can convert into phenoxyls other stabilizers containing phenolic moieties, e.g. phenolic derivatives of benzotriazole or 2-hydroxybenzophenones.

Fortunately, substituted BQ or QM do not diminish the activity of HAS **28** (R = H, CH$_3$) in thermal or sensitized photo-oxidation of heptane [263, 264]. On the contrary, 3,5,3'5'-tetra-*tert*-butyl-4,4'-stilbenquinone **164**, the most common QM derived from hindered phenols, enhanced the photoprotective effect of **28**. The light screening ability of **164** is favourably cooperating in the model system.

Substituted alkylperoxy- **165** (R = alkyl) and hydroperoxy-cyclohexa-2, 5-diene-1-ones **165** (R = H) having properties of thermo- and photoinitiators and formed from phenolic AO in sites with high concentration of ROO$^{\bullet}$ or with excited sensitizers do not reduce the effect of **28** even when used in concentrations exceeding those in the real system [263, 264]. An enhanced formation of $>$NO$^{\bullet}$ from **28** (R = H) in heptane in the presence of **165** R = H, *tert*-butyl) in comparison with heptane doped with 2,6-di-*tert*-butyl-4-methylphenol reveals that **28** deactivates the peroxidic species **165**. A similar mechanism as that taking place with alkylhydroperoxides was expected.

Interference of HAS with quinones is certainly very affected by the reaction microenvironment. According to [265], HAS derived $>$NO$^{\bullet}$ inhibits thermal reduction of BQ in PP resulting in hydroquinone (BQH$_2$, Eq. 50) by reoxidation into semiquinone (BQH$^{\bullet}$, Eq. 51) and BQ (Eq. 52).

$$\text{BQ} \xrightarrow{\text{PH}} \text{BQH}^{\bullet} + \text{P}^{\bullet}$$
$$\raisebox{1ex}{$\llcorner$}\!\!\!\longrightarrow \text{BQ} + \text{BQH}_2 \tag{50}$$

$$\text{BQH}_2 + >\text{NO}^{\bullet} \longrightarrow \text{BQH}^{\bullet} + >\text{NOH} \tag{51}$$

$$\text{BQH}^{\bullet} + > \text{NO}^{\bullet} \longrightarrow \text{BQ} + >\text{NOH} \tag{52}$$

Under actinic irradiation, the excited [BQ]* abstracts hydrogen from $>$NH and generates $>$N$^{\bullet}$ [266]. At concentration of BQ exceeding that of HAS (fortunately, this does not happen in phenol doped PO), both the thermal- and photoprocess diminish the resulting stabilizing action of HAS. Contrary to these results, the action of anthraquinone used in PP as photosensitizer under similar conditions as in the above mentioned reaction with BQ, was reported to be diminished by nitroxide derived from **28** (R = H) [125].

Substituted cyclohexadienonyls **166b** and 4-hydroxybenzyls **167**, the free-radical intermediates of phenol oxidation existing in C-centred radical forms [251], recombine with $>$NO$^{\bullet}$ [5, 6, 64, 267] (Scheme 31; for the sake of simplification, bulky substituents in positions 2,6- are not drawn in the structures of phenol and the derived products). Relatively complicated and rather unstable products arise from free-radical species derived from polynuclear phenolic AO and phenolic C–C coupling products [251]. Their phenolic nuclei are transformed stepwise and randomly. One and the same parent phenol may generate systems differing in the degree of transformation. Compounds **168** and **169** formed from the mononuclear hindered phenol are examples of both possible types of recombination products. Various nitroxides derived from piperidine based HAS and DHI react in this way with benzyl- and propionate-type phenolic AO [64, 262, 267]. **168** or analogues arise by coupling of the mesomeric form of a phenoxyl **166b** with $>$NO·. **168** decomposes either by a retrorecombination to the starting phenoxyl **166a** and $>$NO· (Scheme 31) (the stabilizing activity of the system is not

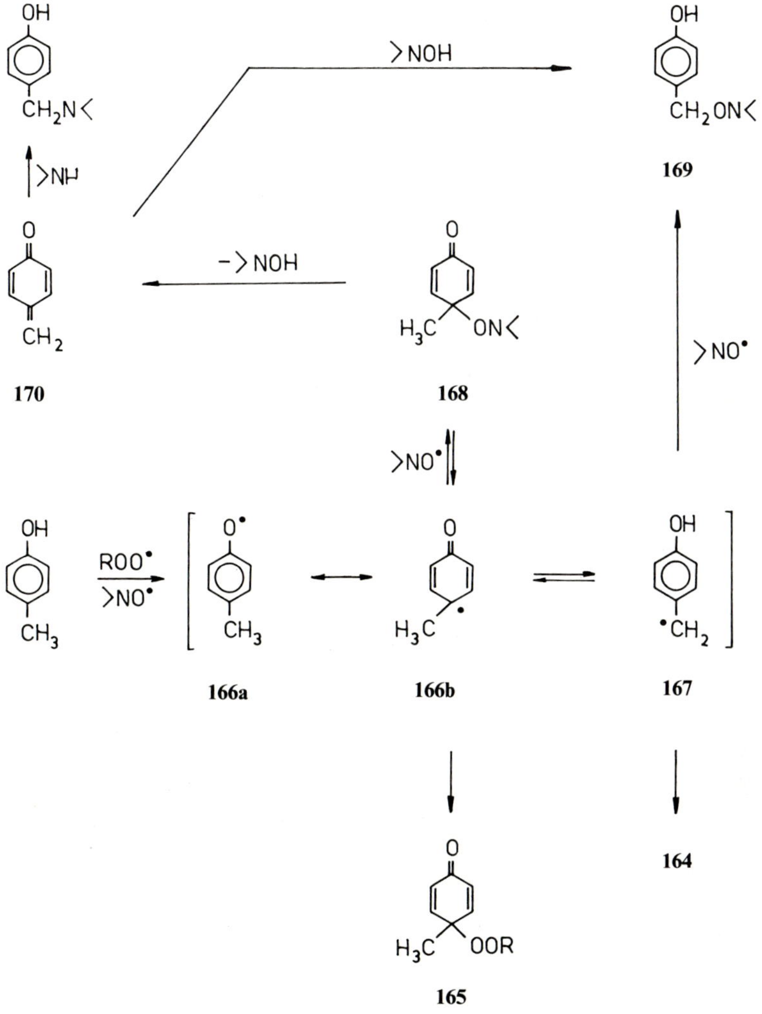

Scheme 31

diminished, **168** acts as a reservoir of two active components) or by elimination of $\gt$NOH and formation of QM **170** (activity of the system is reduced by formation of QM). The compound **169** may be regenerated either by recombination of the nitroxide with the 4-hydroxybenzyl radical **167** or by 1,6-addition of $\gt$NOH on QM **170**. In accordance with the reported mechanisms of *O*-substituted hydroxylamines (Scheme 17), **169** and analogues arising from HAS or partially hindered heterocyclic amines (e.g. **15** to **18**) should be considered as transformation species preserving their activity in polymer stabilization.

To obtain an integral protection of polymers, HAS is also combined with other stabilizers, aromatic phosphites in particular [129]. Blends HAS/phosphite impart excellent light and heat stability to PO, polyamides, PC or PVC [13, 129]. Multicomponent blends HAS/phenol/phosphite, combined moreover with UVA or FR were proposed for protection of PUR, PO or PC [268, 269]. A complementary mechanism should be accepted for the combinations.

Application of thiosynergists in PO processing is rather infrequent on account of phosphites. In the presence of HAS, depleting process may arise due to their reactivity with transformation products of thiosynergists (Sect. 3.3.2).

Reduction of HAS activity in applications in flame and weather resistant polymers and coatings containing halogenated FR is solved by means of HAS having reduced basicity (Sect. 3.3.2). The danger of antagonism in blends HAS/FR is reduced by addition of hydrotalcite [270]. *tert*-HAS **34** may be used for stabilization of PVC in combination with the conventional HS (salts of carboxylic acids, organotin compounds) [271].

HAS are excellent LS/photo-antioxidants, but they fail as UVA. This disadvantage may be eliminated by application of blends HAS/UVA/phenol in PO and other plastics; blends without phenols are suitable for coatings. Combinations HAS/UVA are mostly superior to HAS used alone in unpigmented and pigmented PO, PS, PUR, ABS and coatings, where the failure is determined by polymer-related failure due to insufficient light fastness of pigments and /or the loss of colour [13]. HAS/UVA or HAS/quencher blends were proved as excellent systems in stabilization of greenhouse foils exposed to some sulfur containing pesticides. The depleting effect of volatile acid components was reduced due to the lower amount of HAS necessary to achieve a protection. Features of synergism were observed between HAS and UVA in PO, PS, ABS, POP, *cis*-1,4-polybutadiene, PVC, clear and high-solid coatings [13, 155, 156, 272]. HAS protect UVA against photodegradation by scavenging ROO˙ and in this way preventing the depletion of the phenolic moieties in UVA. Inhibition of photolysis of UVA has been mentioned as well.

The presence of UVA is not necessary in HAS doped polymers containing pigments screening UV radiation, like carbon black, titanium dioxide or phthalocyanine pigments [273].

7 Bifunctional Stabilizers Containing Amine Moiety

Different stabilizing moieties bound in one molecule have been considered to contribute, by an intramolecular cooperative effect, to the integral stabilizing activity. This suggests a well balanced molecular architecture of the bifunctional systems allowing assertion of individual mechanistic contributions of each particular function [250]. It should be anticipated that each function acts by a mechanism characteristic of that known from the respective monofunctional stabilizer. However, it must be expected that the action of both centres is mostly not entirely concerted. This depends on chronological succession or cyclical character of the attack of deteriogens and their intensity. Many bifunctional systems have been synthesised. However, the expectations of advantageous properties have not always been fulfilled.

An attempt was made to crosslink NR or IR by bisperoxycarbamate **171** and bind the released free-radical fragment $\cdot NH–C_6H_4–CH_2–C_6H_4–NH\cdot$ as antioxidant function into the rubber [274].

171   172

An oxamide/DPA combination in **172** imparts antioxidant and metal deactivating effects. Activated thiosynergists combined with DPA or PD moieties in 4-(β-alkylthioethoxy)-DPA, N-(4-dodecylthiomethylphenyl)-N'-sec-alkyl-PD or analogous derivative of DPA are designed for synthetic rubbers and impart hydroperoxide decomposing effect in CB AO [275]. N,N-bis(2-Ethylhexyloxycarbonylmethylthiomethyl)hydroxylamine improves antiwear properties in lubricating oils. An intramolecular cooperation between sulfur and amino moieties may also take place in ring substituted phenothiazines.

Substituted aminophenols and phenolic or anilino moieties bearing N-heterocycles contain two distinguished CB functional centres. Substituted BQMI **57** are formed via 4-aminophenoxyl after ROO· trapping by aminophenols. For example, the aminophenolic moiety creates the active part of 2,4-bis(octylthio)-6-(3,5-di-tert-butyl-4-hydroxyanilino)-1,3,5-triazine, a very efficient AO for diene based polymers. The respective QI **173** is formed transiently and 2,6-di-tert-butyl-1,4-benzoquinone is released by hydrolysis of **173** in the ultimate phase of its lifetime [251]. The respective QI are also formed in sacrificial transformations of 6-hydroxy- or 6-anilino-2,2,4-trimethyl-1,2-DHQ, 6- or 8-hydroxy-2,2,4-trimethyl-1,2,3,4-tetrahydroquinolines or 6-hydroxycarbazole [37]. All these bifunctional heterocyclic amines are strong AO.

Synthetical skill was exploited to obtain systems containing HAS together with other stabilizing moieties. Various combinations with hindered phenols, e.g.

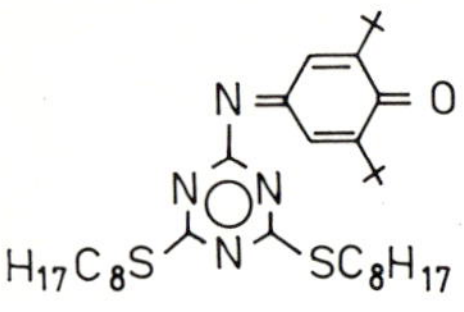

173

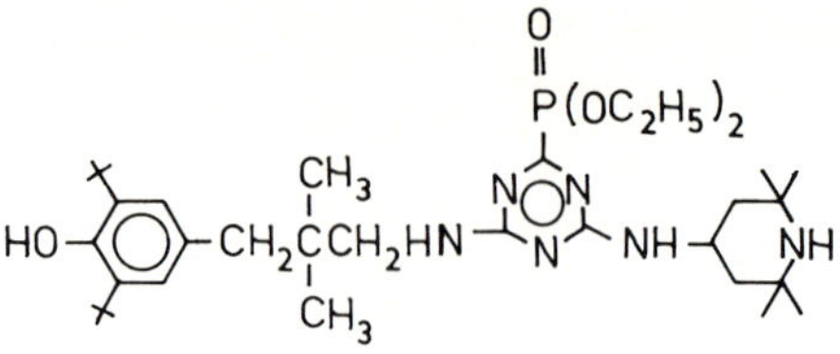

174

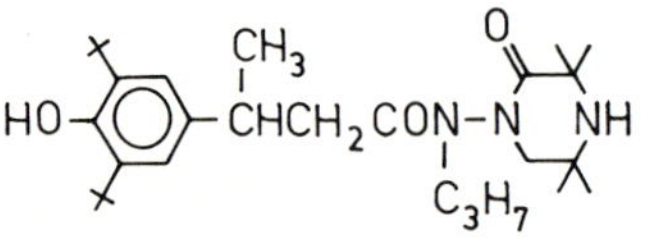

175

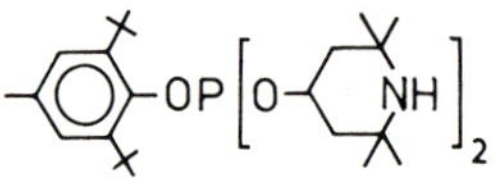

176

37, 38, 174 or **175** were synthesized [13, 14, 40]. Compounds **37** and **38** are examples of efficient commercialized structures. Vacuum photolysis of **37** yields the respective phenoxyl, formation of aminyl was not observed. Regenerative QM/phenol mechanism is involved in **38** bearing a propionate phenolic moiety. As a result, **38** imparts at comparable molar basis the same processing and heat stability to PP as hindered phenolic AO and about 70% of photostability as compared with **28** (R = H) [40]. An intramolecular cooperation has been suggested for phenolic HAS.

Heat and light stability has been attributed to *sec-* and *tert*-piperidines, derived *O*-alkylhydroxylamines or hydroxylamines containing hydrolysis resistant phosphite moieties, e.g. [bis(2,2,6,6-tetramethylpiperidin-4-yl)2,6-di-*tert*-butyl-4-methylphenyl)]phosphite **176** [276]. The light/heat stabilizing effect was confirmed in PP films [277]. The bifunctional piperidine stabilizers were superior to mixtures of structurally analogous aromatic phosphites and HAS in the both thermooxidized (110 °C) and photo-oxidized (30 °C) PP. Features of synergism were observed.

Strong heat and light stabilizing effect was attributed to unconventional HAS, 1,4-dihydropyridine substituted with a phosphonate moiety and to piperidine containing a nitrone moiety [278].

UV light absorbing radical scavengers containing oxamide moiety, e.g. **39** (R = H, CH₃CO) [279], 2-hydroxybenzophenone moiety e.g. 2-hydroxy-4-[(4-amino-2,2,6,6-tetramethylpiperidyl)butoxy]benzophenone **177** [280], 2-hydroxyphe-

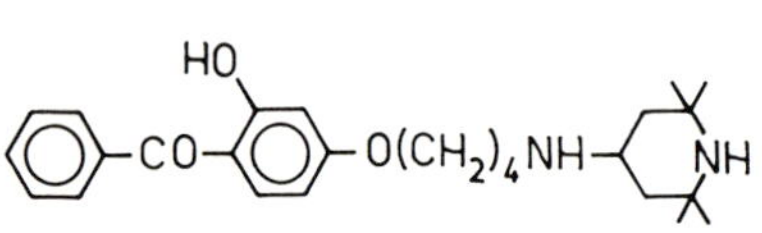

177

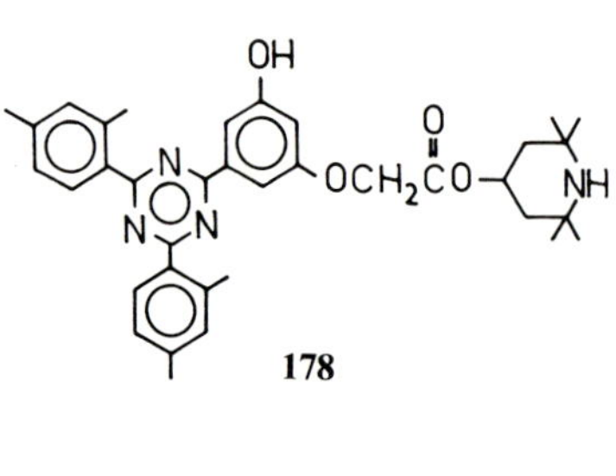

178

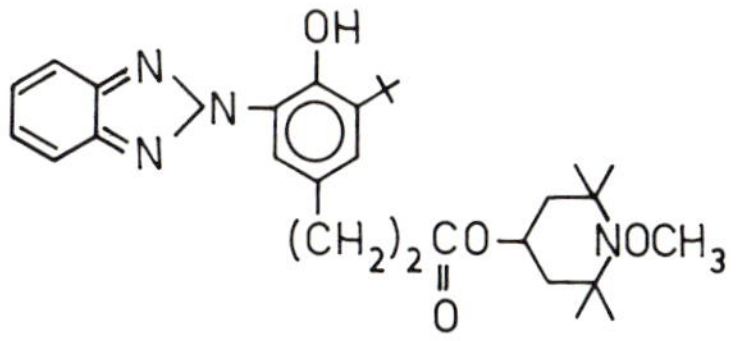

179

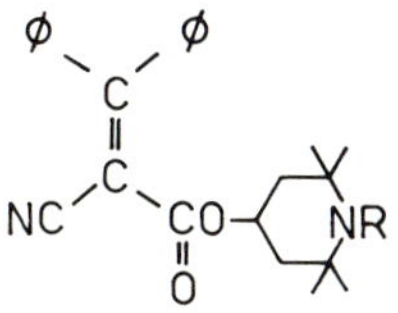

180

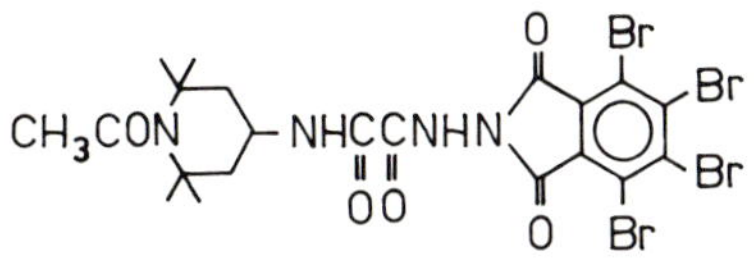

181

nyltriazine moiety, e.g. **178** [281], 2-(2-hydroxyphenyl)benzotriazole moiety, e.g. **179** [282] or α-cyanocrylate moiety **180** are designed for PO and coatings.

A mechanistic study performed with **177** and analogues showed thermal (at 110 °C and 130 °C) and photostabilizing (at 50 °C) effects in PP. The authors suggest involvement of the ortho-hydroxy group in the benzophenone moiety in the CB mechanism [280].

Compound **181** was proposed to impart concerted light stabilizing/fireproofing protection to polymers [283].

8 Physical Factors Affecting Stabilization of Polymers

Plastics, rubbers and coatings have been used in various demanding household, engineering, agricultural and medical applications under aggressive environmental conditions. Stabilizers used to improve their performance should maintain their efficiency as long as possible. They are, however, stepwise chemically depleted due to their sacrificial role. Moreover, they are physically lost as a consequence of volatilization due to high environmental temperature, leaching into extractive media and rubbing-off after stabilizer exudation to the polymer surface as a precipitate. The physical losses may dramatically reduce the efficiency of the stabilization process. They are influenced by molecular parameters of the stabilizer, i.e. its molecular weight and polarity and by physical relations between the stabilizer and the polymer matrix. Solubility of stabilizers, their compatibility with the host polymer and migration in the polymer matrix are the decisive factors. Physical phenomena affecting the integral efficiency of stabilizers have been elucidated theoretically in detail [7].

The assessment of physical factors can give a qualified explanation of some phenomena observed in antidegradant activity of various amines. It is rather difficult to define the borderline in the molecular weight of stabilizers assuring the physical persistence. The volatility problem is mostly solved using conventional stabilizers with mol. wt. > 500. Extractability is a more serious problem. It has been accepted that stabilizers having mol. wt. approximately 3000 and higher assure full resistance to leaching [8]. The requirements on the resistance differ according to the application mode. Extraction tests by means of food-stimulating liquids have been used for packaging materials to assess the danger of health risks due to additives. Extractability by water having different p_H is important for tyres, extraction by hot oils and aggressive fluids for materials used in the automotive industry, by solvents and detergents for plastics in laundry and washing machines [8]. Underground water leaching of waste materials in landfills emerges as a new problem.

The common application of aromatic amines is in thick-walled goods made from rubber vulcanizates. The physical loss has been considered as the main factor in the ultimate failure of goods like wire and cable coatings or automotive components. Tests report lower volatility of **11d** and **11b, c** in comparison with **11a**. Excellent resistance against fugitivity at 150 °C was reported for **9b** (1.5% loss after 2 weeks) in comparison with 7.8% loss of a condensate like **22** and 17.5% loss of **9a** [1]. Extraction tests reveal good resistance of **9b, 11d** and **22** [8].

Solubility in the polymer matrix is a decisive factor determining the mechanism of the physical loss of additives [7]. Data dealing with aromatic amines reveal mostly good solubility in PO and vulcanized rubbers [8]. They are more soluble in more polar rubbers, like NBR. A low solubility of 4,4'-disubstituted diphenylhydroxylamines or **11d** in vulcanized NR or SBR accounts for limits in their application as AO and AOZ, respectively. The surface blooming of amines is a characteristic phenomenon for amine-oversaturated rubbers. Differences in solubility of amines in various polymers result in partitioning of stabilizers in individual phases of polymer blends and multiphase rubber modified plastics [1]. The partitioning is influenced by the microheterogeneity of the system, differences in polarities of phases and the stabilizer and differences in compatibility of stabilizers with individual phases. For example, highly alkylated aromatic amines or DHQ have a very high solubility in the rubber phase of impact-PS.

Diffusion of aromatic amines decreases with polymer crystallinity, with increased solubility in the polymer and the mol. wt. of the amine [8]. A high migration rate of aromatic amines to PO and rubber surfaces is considered as a reason for their enhanced physical losses [34]. Fillers having large surface areas, like carbon black, reduce the migration rate. A limited migration arising from polymeric amines strongly reduces the stabilizing activity in applications where a high surface concentration of stabilizers is mandatory, i.e. in the antiozonant protection [34].

Diffusion and migration of HAS were elucidated in connection with their application in PO [284] or coatings [148, 149, 157]. An increase in mol. wt. of HAS improves the observed efficiency only to a certain range of molecular weights (3000–5000). A further increase may be connected with reduced solubility in the PO matrix [7] accounting for inhomogeneities in distribution of HAS, leaving a great part of the PO bulk unprotected. The random distribution of HAS is one of the reasons for reduction of the protection effect. A HMW non-migrating HAS can protect PO provided that it is homogeneously distributed. The high solubility and compatibility of HAS in PO is therefore mandatory.

A microtome 2 μm-section technique applied in two-coat automotive finishes revealed [148, 149, 157] the distribution profiles through the entire coating due to migration. HAS migrate during weathering between layers and are envisaged as physical losses from the surface layer. HMW and oligomeric HAS are, therefore, of top importance in coatings and PO thin section products, i.e. multifilaments and foils.

8.1 Physically Persistent Amines

Because of the importance of physical persistence, a great effort has been paid to the synthesis of stabilizers having both high inherent chemical efficiency and advanced physical properties. The optimum commercial solution is usually a compromise between these two requirements [8]. Attention has been paid for a long time to the synthesis and evaluation of new generations of macromolecular stabilizers. Numerous synthetical approaches exploit polyreactions (polymerization, polyaddition, polycondensation) and polymer-analogous reactions ("reactions on polymers", like polymer grafting and functionalization with reactive compounds bearing stabilizing moieties). Stabilizer-functionalized monomers, oligomeric/polymeric stabilizers and their properties were reviewed recently [8]. Some new approaches dealing with amines will be mentioned here. The material has been organised according to the individual synthetical methods.

Continuing attention has been paid to the synthesis of amine-bearing monomers, e.g. N-(4-anilinophenyl)methacrylamide, used for the synthesis of the terpolymer 13, and other monomers containing PD or PT moieties [8]. Copolymers of the acryloyl-PT with conventional acrylates were proposed for optical lenses and hydrogels [118]. Formation of a polymer-bound amine via copolymerization of butadiene with acrylonitrile in the presence of N-phenyl-N'-alkoxy-BQDI was mentioned.

Piperidines (>NH, >NR, >NOH, >NOR) bearing polymerizable vinylic or glycidyl moieties, e.g. 182, 183 were proposed for free-radical homo- and copolymerization or grafting of conventional polymers. HAS-functionalized copolymers have a potentially important application in stabilization of hydrogels [285]. Monomer 184 is a crosslinking agent in acrylate hydrogels [286].

Piperidine-functionalized radical initiators, like O,O-tert-amyl O-(1,2,2,6,6-pentamethyl-4-piperidinyl)monoperoxycarbonate 185 [287] or bis-(1-octyloxy-2,2,

182   **183**

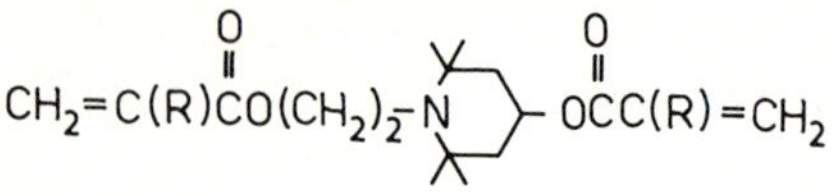

184

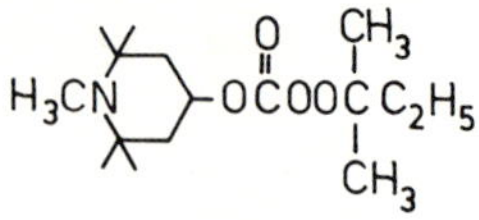

185

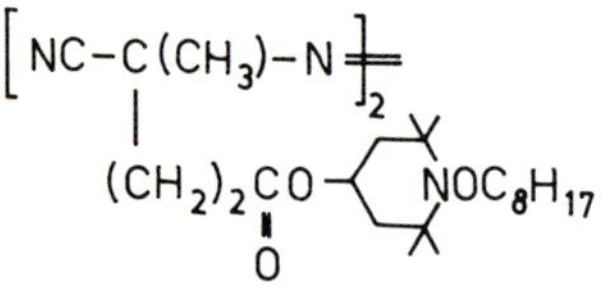

186

6,6-tetramethylpiperidin-4-yl)-4,4′-azobis (4-cyanovalerate) **186** [288] represent another approach to binding the HAS moiety into the polymer chain of impact modified PS or acrylates during polymerization.

4-(Hex-5-enyl)-2,2,6,6-tetramethylpiperidine was successfully copolymerized with propylene over supported Ziegler-Natta fourth-generation MgCl$_2$-supported catalyst [289].

Friedel-Crafts alkylation of **11b** with 1,3-di-isopropenylbenzene was used to prepare an oligomeric stabilizer for diene based polymers. SBR may serve as a co-reactant in the alkylation of the pendant phenyl group with DHQ and yields **187** [8].

Aniline polycondensate **188** was reported to impart ozone and heat resistances to NR. Condensates of BQ with primary aromatic diamines are of potential interest as stabilizers [290].

Various piperidine-functionalized polymers were obtained by polyaddition or polycondensation and contain ether, ester, siloxane, carbonate, amine, amide, urea or urethane construction units [8]. Oligomeric siloxane **36** having n = 4 and analogous HAS-functionalized siloxanes [291] or polymers like **189** obtained by

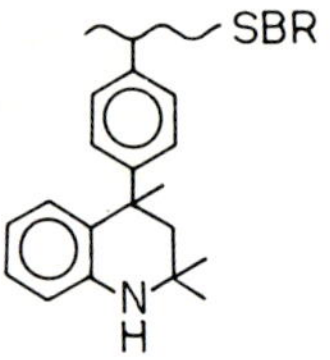

187

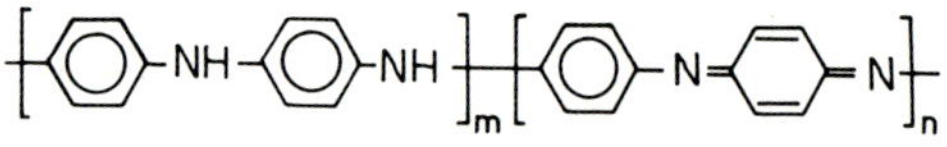

188

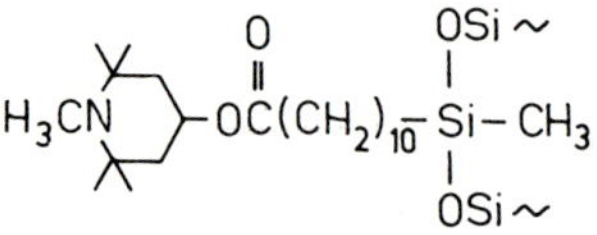

189

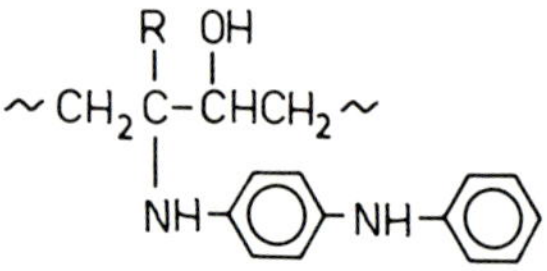

190

hydrosilylation [292] are designed for PO. HAS-containing alkoxysilane was used for functionalization of titanium dioxide.

Transesterification yields oligomeric esters analogous to **34**. Reactive HAS containing hydroxy, oxirane or carboxy groups in position 4 can be chemically attached to acrylic PUR by condensation. The products obtained are intended as persistent stabilizers for coatings, transformed into oligomeric condensates with cyanuric chloride or used for endcapping of terminally functionalized macromonomers [293].

In spite of the relatively low performance/cost ratio of the respective products, application of reactions on polymers are of continuous interest for synthesis of amine functionalized polymers. The technique of modification of solid polymers by exploitation of their reactive moieties under heterogeneous conditions allows properties to be tailored and optimized according to specific needs and requirements. Telechelic polymers are used for the synthesis of materials with predictable and controlled properties. Application of carefully designed macromolecules provides systems with functional groups dispersed along the polymer chain or bond at the end of macromolecules.

Various systems containing aromatic amine moieties were described [8]. Functionalization of epoxidized rubber is an example of exploitation of a reactive commercially available polymeric matrix. Polymer-bound moiety **190** is formed with 4-amino-DPA. Vulcanization with **171** enhances the stability of NR by means of introduction of bound-in 4,4'-diaminodiphenylmethane bridges. Allylchloride moieties in CR were exploited for functionalization with **11b, c** during high-shear mixing or vulcanization [294]. Terminal PD groups can be introduced by end-capping of hydroxy- or carboxy group-terminated macromonomers with 4-amino-DPA [8].

Functionalized mercaptans and disulfides like **191**, 2-mercapto-N-[(4-phenylamino)phenyl]acetamide, 2,2'-dithiobis[N-(4-phenylamino)phenyl]acetamide or 2-mercapto-4,6-bis(4-anilinophenyl)amino-1,3,5-triazine have been used to bind PD moieties into PO, EPDM or diene based rubbers via recombination of the respective thiyl and macroalkyl radicals generated mechanochemically [8, 295]. The process was followed using ^{35}S-labelled compounds. High level of binding of PD-moieties was found at 180 °C with EPDM. The mode of the attachment of PD via sulfur atoms was confirmed using model compounds [295].

$$\langle\bigcirc\rangle - NH - \langle\bigcirc\rangle - \underset{\underset{CH_3}{|}}{N}HCHCH_2\overset{\overset{O}{\|}}{C}O(CH_2)_n O\overset{\overset{O}{\|}}{C}CH_2SH$$

191

Interesting stabilizers were obtained by functionalization of poly(maleic anhydride-co-α-methylstyrene) or epoxidized PP with 4-amino-2,2,6,6-tetramethylpiperidine [296]. Maleic anhydride grafted PP, EPDM and SAN or polyacrylates with pending anhydride or epoxy groups were functionalized with N-(2,2,6,6-tetramethyl-4-piperidyl)-N-aminooxamide **192** [297]. Polymer of the type **193** containing HAS moiety is an example.

192   193

HDPE, SBR or NBR were grafted with 2-(4-anilinophenylamino)-4,6-diallyl-oxy-1,3,5-triazine or analogous functionalized systems. Methacryloyloxyethyl-decahydro-2,2,4-trimethylquinoline is suitable for grafting NR [8].

A targetted binding of amines into PP by reactive processing at 180 °C in the presence of organic peroxides was studied in detail [298]. Excellent results and gel-free polymer grafts were obtained with N-[4-anilinophenyl]maleic anhydride and 1-acryloyl-4-acryloyloxy-2,2,6,6-tetramethylpiperidine. Functionalized monoacrylates and acrylamides have an increased tendency to homopolymerization, a competitive process to grafting. Functionalized PP was blended as masterbatch with unstabilized PP. A HAS functionalized acrylate was used together with styrene and acrylonitrile for grafting of diene based rubber and blended with SAN to obtain photostable ABS, for grafting of fluoropolymers functionalized with diisocyanate or for photografting of PP films.

Oligomeric stabilizers with molecular weights not exceeding 5000 have been exploited commercially and belong to the most successful developments in polymer stabilization. Oligomeric DHQ **21** (n = 2–3) belong among the oldest stabilizers. It is commercialized under various trade names by the principal producers of rubber chemicals. Polycondensates containing 9,10-dihydroacridine **22** (Naugard A, Uniroyal) or acridine/carbazole moieties **23** (Naugard BG, Uniroyal) are applied in rubbers.

Structures **31** (R = H or CH$_3$, Mark LA-62 and 67, respectively, Asahi-Denka), **32** (Chimassorb 119, Ciba-Geigy), **33** (R = H or CH$_3$ Mark LA-68 and 63, respectively, Asahi-Denka), **34** (Tinuvin 622, Ciba-Geigy), **35a** (Chimassorb 944, Ciba-Geigy), **35b** (Cyasorb UV 3346, American Cyanamid), **36** (Uvasil 299, Great Lakes) are examples of commercialized HMW and oligomeric HAS.

A general purpose application of polymeric stabilizers cannot be expected at the present time. They have been commercialized only exceptionally and applied for rather special purposes. Terpolymer **13** (Chemigum-HR, Goodyear) contains 22–45% acrylonitrile and has been available as custom-made masterbatch. The degree of NBR protection against aging in hot oil may be controlled by the level of the masterbatch in the blend. Blends **13**/NBR are compatible with PVC and ABS [299] and enhance the application possibilities of **13**. A commercialized polymeric HAS of the type **193**, prepared from maleic anhydride grafted PP [296] was introduced recently (Luchem HA-B18, Lucidol/Pennwalt). Synthetical approaches like reactive processing, improving both the cost/performance ratio and physical relations with the host polymeric matrix may potentially broaden application of polymeric stabilizers.

8.2 Properties of Macromolecular Amines

Modern technology is making ever-increasing demands on durability of polymers. Highly demanding processing operations and more discriminating consumers account for enhanced requirements on extremely high persistence of stabilizers in aggressive environment. Only some recent HMW amines, like **31** or **32** approach this standard. Development programmes of producers of stabilizers have been aimed at amine functionalized oligomers and polymers. It has been generally accepted that macromolecular stabilizers protect polymers by the same chemical and physical mechanism as HMW stabilizers containing a comparable functional moiety [8]. Differences arise in physical properties due to the macromolecular character of the stabilizer. The compatibility with the host matrix mostly drops with increasing mol. wt. of the stabilizer. Incompatible blends may arise in semicrystalline polymers. Separation of phases diminishes the expected stabilization effect. The exact rules enabling generalization in the choice of a proper macromolecular aminic stabilizer for a particular host polymer have, until now, been lacking. Experiences obtained in blending polymers are exploited. A lot of data exist on the efficiency of amine functionalized macromolecules before and after solvent extraction of the host polymer. The physical state of these "blends" after long-term ageing was not reported.

Macromolecular stabilizers are immobile in the polymer matrix [284]. This is unfavourable for applications where surface concentration of stabilizers in thick-walled products should remain high. It was reported [34] that rubber-bound derivatives of PD provide only very poor antiozonant protection. Their antioxidant efficiency was only comparable with that of conventional HMW PD. This indicates that application of polymer-bound amines in rubbers has the prospect of exclusive long-term use in extracting media.

Most papers analysing properties of functionalized macromolecular stabilizers deal with HAS. The rating conditions and the host material may cause some discrepancies in results.

Efficient protection in coatings exposed to increased temperatures and leaching environment was imparted with copolymers end-capped with *O*-alkylhydroxylamines [288] or with siloxane based HAS **36** [150]. Gloss of the acrylic or polyester topcoat is conserved. The ability of **36** to form bonds with aluminium powder or mica increases its performance in paints.

Compatibility with the host polymer plays a principal role in application of oligomeric HAS [13, 284]. This phenomenon was studied in PP doped with poly(2,2,6,6-tetramethyl-4-piperidyl acrylate-*co*-*n*-octadecyl acrylate) having $\overline{Mn}$ between 5000 and 116 000. The light stabilizing effect diminished with increasing mol. wt. of the stabilizer [300].

To achieve an optimum efficiency of a copolymeric HAS in a particular polymer, the stabilizer has to have an optimum mol. wt. The principle of photodegradable polymers was exploited for a new class of degradable polymeric HAS. A rubbery terpolymer **194** containing 50–60% of 2,2,6,6-tetramethylpiperidyl acrylate moiety as a stabilizing component, *n*-octadecyl

acrylate as a unit assuring compatibility with the host polymer and 2.5–6%
1-phenyl-2-propenone as a photosensitizing unit was prepared [301] and 0.2%
of **194** was blended with PP. Under solar irradiation, **194** undergoes photolysis
and the released fragments migrate throughout the stabilized PP. As a consequence, the photodegradable **194** having a mol. wt. of 420 000 provided a better light stability for PP than a comparable copolymer having molecular weight
450 000 but without the photosensitizing unit.

194

Terpolymer **194** having a higher content of the sensitizing moiety but a comparable level of the HAS units imparted a higher stability to PP, due to the
formation of smaller photolytic fragments migrating through the polymer matrix.
Photolysis of **194** occurs before the start of PP photo-oxidation [301].

The comonomer used for copolymerization with tetramethyl-4-piperidyl
methacrylate influences the observed light stabilizing effect in blends with PP
[302]. The best efficiency was imparted by copolymers with styrene and alkyl
acrylates. The most efficient copolymer with styrene was also tested in photo-
oxidized *cis*-1,4-polybutadiene and was found to be less efficient than **35a**.

Commercial oligomers **34** and **35a** are well compatible with PO [137], do
not bloom from L-LDPE even at concentration 1.2% after a period of 2.5 years
and provide excellent protection to products having high specific surface area, e.g.
PP fibres or PE foils [13, 134]. They protect thick products against weathering
and LTHA as well. It was reported that **34** or **35a** perform better in PE than in
PP. The action of oligomeric HAS is also excellent in the presence of titanium
dioxide and carbon black [132, 273].

In LTHA and photoaging of PP and LDPE, **35a** is a more efficient LTHS
than **34** [138]. In combinations of the two HAS, the increase of the relative
concentration of **35a** improved the observed long-term heat stability effect. In
applications, where oligomeric HAS are the principal stabilizers of choice, e.g.
in PP fine multifilaments, 1:2 combinations of the two HAS **34** and **35a** provide
excellent synergism [138]. This combination represents the state-of-the-art in the
light stabilization of LDPE/L-LDPE/EVA films.

Oligomeric **35a** and HMW HAS **32** were reported [273] to have comparable
activity in TiO$_2$ pigmented or unpigmented PP fibres. Both **32** and **35a** have
excellent physical persistence and protect PP against gas fading and loss of tensile
strength.

Tests of the light stabilizing activity of monomeric HAS and the corresponding homo- and copolymers reveal mostly better properties of the monomers if physical persistence is not the decisive testing factor [8]. This was found e.g. in comparison of the functionalized urethane **182** and its copolymers with styrene or methyl methacrylate [303]. The macromolecular architecture is expressed very distinctly. For example, a PP photografted HAS-functionalized acrylate was more efficient than the respective monomer or homopolymer. Another observation performed with N-(2,2,6,6-tetramethyl-4-piperidyl)methacrylamide, piperidyl acrylate and methacrylate, their homopolymers and copolymers with dodecyl methacrylate and octadecyl acrylate revealed that the stabilizing effect in PP was in favour of copolymers [304]. Similar HAS-functionalized monomers were copolymerized with styrene. In this case, the copolymers were substantially less efficient in PS than the monomers. Masterbatches of PP-bound HAS prepared by reactive processing imparted a comparable effectivity as conventional HAS when tested at an equimolar basis [298].

Differences between activities of commercialized LMW HAS, e.g. **28** (R = H) and oligomers like **34** and **35a** were reported. The compound **28** does not contribute practically to the basic long-term heat stabilization formulations of PP. An appreciable increase in heat stability was observed after addition of 0.2% **34** [138]. The rating criteria and shape of the polymer influence the decision on the superiority of the individual type of HAS. When surface photoprotection of PP exposed in Florida was rated using development of $>$CO, both classes of HAS imparted a comparable protection. When mechanical properties are a criterion, then oligomeric HAS provided a better protection than LMW HAS [122, 273]. The experimental results show the advantage of LMW HAS over oligomeric HAS, mainly in outdoor weathering of PO [189]. This fact was explained by the contribution of the mobility of LMW HAS and their diffusion to the surface layers of thicker samples. However, a behaviour like this is a disadvantage in systems having a high specific surface, like PP filaments, where physical persistency is a condition for high light stability [7].

The most recent development in the stabilization of PO consists of introduction of combination of LMW and HMW or oligomeric HAS. Interpretation of the stability of HAS doped plastics in natural and accelerated weathering and oven tests performed below 130 °C revealed that the combinations provide an excellent synergistic protection in both light and heat induced processes [3, 122, 133, 138]. Individual specific contributions of components differing in their molecular weight result in the excellent integral effect in thin and thick sections of PO. Physical factors like compatibility and migration most probably coooperate with the inherent chemical efficiency [6]. Generally, the combination of HAS differing in molecular weight represents the state-of-the-art for most PE and PP applications [122]. Experimental data indicate that such combinations exceed activity of LMW or oligomeric HAS when applied alone at the same concentration level. Moreover, the simultaneous application of the oligomeric HAS allows an increase in the concentration of the active piperidine moieties in the system: the solubility of **28** (R = H) in PP is limited by 0.3–0.4%. At higher concentrations, a surface blooming of **28** occurs. This problem is eliminated by simultaneous application of more soluble **35a**.

A structural analysis revealed that several types of oligomers are present in the commercial HAS **36**. The most representative species contain from three to eight repeating units. The siloxane backbone adopts the most stable either open or cyclic conformation. The tetrameric cyclic species **195** is the principal component (reaching up to 90%). A theoretical study on the molecular mechanism and dynamics was performed with defined model compounds [305]. The most stable acyclic component **196** is folded-up in a way to resemble a cycle. The adjacent Si–O bonds prefer the alternating gauche$^+$/gauche$^-$ states. The shape and molecular volume of **195** and **196** are therefore almost identical. The simulation of the molecular dynamics evidences a high degree of flexibility in the cyclic tetramer. The high miscibility of **36** with PP may be due to an intimate interaction between the cyclic tetramer **195** and segments of PP.

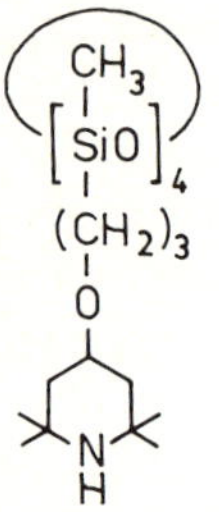

195

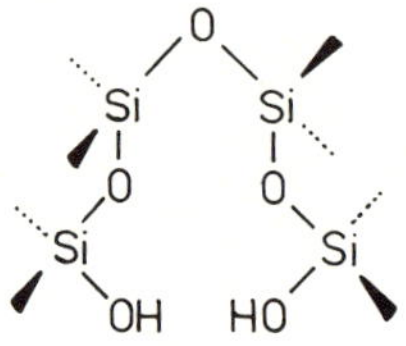

196

Polymer-bound amines are of specific importance in multiphase systems where partitioning of migratable stabilizers may diminish the stability of more sensitive phases. System **193** representing a matrix and rubber-phase bound amine in acrylonitrile/EPDM/styrene terpolymer is an example [297]. The superior performance was obtained when the migratable monomer **192** was melt-blended into the HAS-functionalized multiphase system **193**.

9 Ecological Problems Encountering Application of Aminic Stabilizers

The public concern is aimed at the danger of any kind of intoxication arising from the contact of human beings and living nature with chemicals in our environment. The scientifically based quantitative toxic risk assessment methods have been advanced to extrapolate the animal studies to human exposure levels. In the stabilizer business, the health risk must be assessed for all opportunities where human beings are expected to come into contact with stabilizers, i.e. during the production of additives, processing of polymers, use of final products and disposal of the polymer waste. The laws of industrial hygiene govern aspects of direct contact with stabilizers in the production sphere. Health and handling advice for industrial stabilizers are given in the respective safety data sheets issued by stabilizer producers. Legislation rules determine indirect contact limits during application of stabilized plastics and rubbers where migration of additives into food or leaching into biological systems is of potential danger [306, 307]. This involves extractability of packaging materials with defined oily, alcoholic and aqueous food simulants and rules for use of stabilized polymers in toys, sanitary goods or medicinal aids. The environmental protection legislation of individual countries issues regionally valid Positive Lists defining maximum levels and modes of use of stabilizers approved for contact with food. An uncertainty arises from a potential danger of long-term leaching of additives from the plastic waste in landfills and agriculture or from the rubber dust arising from abrasion of tyres on highways.

Toxic and physiological impacts of polymer stabilizers have been carefully elucidated. Pharmaco-kinetic models for determining the chemical dose delivery to the target tissue were examined and doses responsible for allergies, acute and chronic toxicities were defined [306]. Information on potential irritant effects, dermatitis, mutagenicity or carcinogenicity due to amines is of prime importance. It has been estimated that about 75–80% of all human cancers are environmentally induced (30–40% of them by contaminated diet [308]).

Historically, the greatest attention has been paid to rubber chemicals, many of them being aromatic or nonhindered heterocyclic amines. The aim has been to avoid the human urine bladder cancer due to amines. Another interest includes reduction of formation of N-nitrosamines from rubber amine sources. Fortunately, conventional amine stabilizers are not precursors of N-nitrosamines. The extended application of plastics in food packaging materials and foils for agriculture enhances interest in toxic properties of HAS. Companies producing pharmaceuticals and having experiences with health risk assessment, like Ciba-Geigy, Hoechst or Sankyo are fortunately among the principal suppliers of HAS. Application of HMW and oligomeric stabilizers, more resistant to environmental extracting media, is receiving attention as a tool to prevent environmental contamination [8].

As with many other chemicals, some amines can irritate the skin of supersensitive individuals. This cannot be the reason for elimination of application of amines when the rules of the industrial hygiene are respected. Data indicating chronic or acute toxicity and arising from amines are rigorously respected by producers and users of stabilizers. The human community is protected against misuse of stabilizers much more than in other areas of the use of chemicals in the human environment.

Toxicity screening revealed cancerogenity of some primary aromatic and heterocyclic amines [309]. According to data obtained with rodents, the mutagenicity cannot be correlated with amine carcinogenity. Health damage analysis of workers in the rubber industry has been done very carefully [310]. Primary amines may not be used as stabilizers. However, some commercial stabilizers may be contaminated by traces of primary amines arising during synthesis by side-reactions. This makes antidegradants like N-phenyl-1 (or 2)-naphthylamine or N,N′-bis(2-naphthyl)-1,4-phenylenediamine candidates for chemical cancerogenity. Early analyses of commercial samples confirmed the presence of 15–50 p.p.m. of free 2-naphthylamine (NA) in phenyl-2-naphthylamine and 40–90 p.p.m. of NA in naphthyl derivatives of phenylenediamine. Modern technologies can deliver products with contamination lower than 10 p.p.m. NA. The acceptable top level of free NA, safe for workers in the rubber industry, was reported to be as high as 50 p.p.m. NA [3, 24]. Bladder tumours were, however, typical of men working in rubber compounding and anxiety about possible chronic intoxication led to complete abandonment of any commercial application of phenylnaphthylamines in the rubber industry.

Fortunately, no such conflict of interests arises with diphenylamines **9** and phenylenediamines **11a–d**. Various DPA were tested and approved as AO not only in rubbers, their traditional application field: the physically persistent **9b** and **9a** have also been approved for stabilization of plastics in contact with non-fatty food [307]. Derivatives of PD are the most versatile nowadays as rubber antidegradants. Oral administration to experimental animals indicated some danger of necrosis of skeletal and/or cardiac muscles [311]. This has been explained by in vivo oxidation of PD **11d** into quinone imine by muscle mitochondria and establishing an alternative pathway for electron transport in the physiological respiratory chain.

Some human skin sensitizing activity of **11b** (IPPD) has been reported for many years. Although this activity does not limit the commercial application of **11b** at all, safer derivatives, like the respective 1,3-dimethylbutyl analogue **11c** have been preferred and have been delivered by all principal producers. Moreover, new environmentally safer technologies were introduced for PD production [312].

Interest in the carcinogenic response or allergic activity of DHQ **16b** was evoked mainly due to non-polymer applications. A weak chronic nephrotoxicity of rats was reported as a consequence of the dietary administration [313]. The response was, however, very dependent on age and sex of the experimental animals. No cancerogenicity or mutagenicity was revealed in tests with phenothiazine and its derivatives [116].

Great attention has been paid to HAS and their safety application in plastics and coatings. The 4-unsubstituted 2,2,6,6-tetramethylpiperidine is considered as relatively toxic, the acute oral toxicity being about 1 g/kg. The substitution in position 4 (i.e. the general mode in the synthesis of HAS for polymer purposes) dramatically improves the situation. Therefore, commercial HAS like **28** (R = H), **34**, **35a** or **35b** were approved for stabilization of packaging materials in contact with food [307]. Some data are available on properties of TEMPO (2,2,6,6-tetramethylpiperidinyl-1-oxyl) and its 4-amino or 4-hydroxy derivatives. They were found to act as weak intrinsic direct mutagens in *Salmonella typhimurium*. TEMPO increases intracellular hydroperoxide concentration. This may indicate its pro-oxidative effect which does not result, however, in cellular toxicity [314].

In spite of these data, TEMPO has been widely used for diagnostic applications in biology.

The environmentally positive role of amines having the same or very close structures to that of polymer stabilizers should also be mentioned. All secondary and tertiary amines are active radical scavengers. Therefore, they interfere with free-radical metabolites of chemical carcinogens and inhibit these processes in vivo [315]. Amines have been declared as biological anticarcinogens or antimutagens, antioxidants in disease defense mechanisms and biological antioxidants in pathophysiology, physiological aging, cardiovascular diseases and oxidative stress in animal tissues. Various amines have been admitted and recognized as medicaments against chemically induced hepatotoxicity, as antihypertonia, analgesica, spasmolytica, antiarythmica, antihistaminica or tranquilizers. These properties were reported for various derivatives of 2,2,6,6-tetramethylpiperidine, indole, carbazole, dihydroquinoline, decahydroquinoline, 1,4-dihydropyridine or phenothiazine, i.e. amines having structures analogous to those adopted for commercial polymer stabilizers. This review does not aim at enumeration of structures and medical applications of amines. The public should be informed, however, that aminic polymer additives have been selected really very carefully and with full respect to environmental safety.

10 Conclusions

Practical experience and research data constitute a satisfactory base for the postulation of antidegradant mechanisms of three typical groups of stabilizers, i.e. aromatic, partially hindered and hindered heterocyclic amines. Principal differences in the chemistry of nonhindered and hindered amines govern the inherent efficiency. Model product studies, proper interpretations of mechanistic pathways, the possibility of the regeneration of some active moieties in sacrificial processes in particular, knowledge of intermolecular and intramolecular cooperations between different stabilizing functions, reduction of the depleting effects of the acid microenvironment – these are the principal theoretical topics enhancing effective exploitation of amines in the commercial use in polymers. Calls for an improved integral protection of polymers strengthen the importance of the research data. The plastics, rubbers and coatings markets become more competitive. An economical protection of polymers in all uses and without health hazard is the general aim. Commercial aminic stabilizers are exemplified in the appendix. All mechanistic data applicable in achieving improved structures at acceptable cost/performance relations have been appreciated and elaborated to perfection by producers of rubber chemicals and plastics additives.

11 Appendix

Trade Names and Assumed Structures of Principal Commercialized Aminic Stabilizers

Aromatic Amines

Trade Name	Structure	Producer
Agerite Antozite	**11a**	Vanderbilt
Agerite DPPD	**11d**	Vanderbilt
Agerite Gel	**9a**	Vanderbilt
Agerite HP	**10**	Vanderbilt
Agerite White	**11d**	Vanderbilt
Amoco 532	**11**, $R^1 = R^2 = sec$-butyl	Amoco
Antage 3C	**11b**	Kawaguchi
Antage DP	**11d**	Kawaguchi
Antigene 3C	**11b**	Sumitomo
Antigene 6C	**11c**	Sumitomo
Antigene A, D	**10**	Sumitomo
Antigene DPT, P	**11d**	Sumitomo
Antioxidant CD	**11b**	Istrochim
Antozite 1	**11**, $R^1 = $ phenyl, $R^2 = $ 1-methylheptyl	Vanderbilt
Antozite 2	**11**, $R^1 = R^2 = $ 1-ethyl-3-methylpentyl	Vanderbilt
Antozite 67	**11c**	Vanderbilt
ASM CD	**11b**	Bayer
Chemigum HR	**13**	Goodyear
Cyanox 8	**9a**	Cyanamid
Cyzone DH	**11**, $R^1 = R^2 = $ 1,4-dimethylbutyl	Cyanamid
Eastozone 30	**11a**	Eastman
Eastozone 33	**11**, $R^1 = R^2 = $ 1,4-dimethylbutyl	Eastman
Eastozone 34	**11b**	Eastman
Flectol ODP	**9a**	Monsanto
Flexzone 3C	**11b**	Uniroyal
Flexzone 4L, 7P	**11c**	Uniroyal
Flexzone 6H	**11**, $R^1 = $ phenyl, $R^2 = $ cyclohexyl	Uniroyal
Flexzone 8L	**11**, $R^1 = R^2 = $ 1-ethyl-3-methylpentyl	Uniroyal
Goodrite Amine	**9a**	Goodrich
Goodrite R 59	**10**	Goodrich
Irganox LO 1	**9a**	Ciba
Naugard 438	**9a**	Uniroyal
Naugard 445	**9b**	Uniroyal
Naugard J	**11d**	Uniroyal
Naugard PBN	**10**	Uniroyal
Neozone D	**10**	DuPont
Nocrack HP	**10**	Ouchi Shinko
Nonflex F, H, TP	**11d**	Seiko
Nonflex OD-3	**9a**	Seiko
Nonox 2C	**11c**	ICI

Trade Name	Structure	Producer
Nonox CI, HP	**11d**	ICI
Nonox CN	**11b**	ICI
Nonox D, E	**10**	ICI
Novazone AS	**11d**	Uniroyal
Octamine	**9a**	Uniroyal
Ozonone 3C	**11b**	Seiko
Ozonone 6C	**11c**	Seiko
Ozonone 35	**11**, R^1 = phenyl, R^2 = 1-methylheptyl	Seiko
Perflectol X	**11d**	Monsanto
Permanax 6PPD	**11c**	Akzo, Vulnax
Permanax 49	**9**, R = cumyl	Vulnax
Permanax DPPD	**11d**	Akzo, Vulnax
Permanax HD	**9**, R = heptyl	Akzo
Permanax IPPD	**11b**	Akzo, Vulnax
Permanax OD	**9a**	Akzo
Santoflex 13	**11c**	Monsanto
Santoflex 36	**11b**	Monsanto
Santoflex 77	**11**, $R^1 = R^2$ = 1,4-dimethylpentyl	Monsanto
Santoflex 217	**11a**	Monsanto
Santoflex 9010	**11d**	Monsanto
Santoflex IPPD	**11b**	Monsanto
Santoflex HP	**10**	Monsanto
Sumilizer 9A	**9**, proprietary	Sumitomo
Sumilizer BPA	**11**, $R^1 = R^2$ = *sec*-butyl	Sumitomo
Tenamene 30	**11a**	Eastman
UOP 26	**11**, $R^1 = R^2$ = cyclohexyl	UOP
UOP 62	**11a**	UOP
UOP 88	**11**, $R^1 = R^2$ = 1-ethyl-3-methylpentyl	UOP
UOP 288	**11**, R^1 = phenyl, R^2 = 1-methylheptyl	UOP
UOP 588	**11c**	UOP
UOP 788	**11**, $R^1 = R^2$ = 1,4-dimethylpentyl	UOP
Vanlube 81	**9a**	Vanderbilt
Vanox 12	**9a**	Vanderbilt
Vulkanox 3100	**11d**	Bayer
Vulkanox 4010 NA	**11b**	Bayer
Vulkanox 4020	**11c**	Bayer
Vulkanox 4030	**11**, $R^1 = R^2$ = 1,4-dimethylbutyl	Bayer
Vulkanox DDA	**9**, R = α-methylbenzyl	Bayer
Vulkanox OCD	**9a**	Bayer
Vulkanox PBN	**10**	Bayer
Wingstay 29	**9**, R = α-methylbenzyl	Goodyear
Wingstay 100, 200	**11d**	Goodyear
Wingstay 250, 300	**11c**	Goodyear
Wingstay 400X	**11b**	Goodyear

Nonhindered Heterocyclic Amines

Trade Name	Structure	Producer
Aceto POD	21	Aceto
Agerite AK, MA	21	Vanderbilt
Agerite Resin D	21	Vanderbilt
Agerite Resin PE	16a	Vanderbilt
Agerite Superflex	22	Vanderbilt
Aminox	22	Uniroyal
Anchor TMQ	21	Anchor
Antage AW	16b	Kawaguchi
Antage RD	21	Sumitomo
Antigene AW	16b	Sumitomo
Antigene RD-G	21	Sumitomo
Cyanox 12	21	Cyanamid
Flectol A, B	21	Monsanto
Flectol H	16a	Monsanto
Flectol Flakes	21	Monsanto
Goodrite 3140	16a	Goodrich
Naugard A, BLE	22	Uniroyal
Naugard BG	23	Uniroyal
Naugard Q	21	Uniroyal
Nocrack AW	16b	Ouchi Shinko
Nonflex AW	16b	Seiko
Nonflex BA	22	Seiko
Nonflex RD	21	Seiko
Nonox TQ	21	ICI
Nonoxol PT	20	ICI
Ozonone AW	16b	Seiko
Permanax B, BL, BLN, BLW	22	Akzo, Vulnax
Permanax ETMQ	16b	Vulnax
Permanax TQ	21	Akzo, Vulnax
Polyflex	16b	Uniroyal
Santoflex AW	16b	Monsanto
Santoflex DD	16, $R = C_{12}H_{25}$	Monsanto
Santoflex R	21	Monsanto
Santoquin	16b	Monsanto
Vanox AT	23	Vanderbilt
Vulkanox EC	16b	Bayer
Vulkanox HS	21	Bayer

Hindered Heterocyclic Amines

Trade Name	Structure	Producer
Chimassorb 119	**32**	Ciba
Chimassorb 944	**35a**	Ciba
Cyasorb UV 3346	**35b**	Cyanamid
Goodrite UV 3034	**27**	Goodrich
Hostavin N-24	Proprietary	Hoechst
Hostavin N-30	Proprietary	Hoechst
Hostavin TM N-20	**24**	Hoechst
Luchem HA-B18	**113**	Pennwalt
Mark LA-57	**31**, R = H	Asahi Denka
Mark LA-61	**31**, R = CH_3	Asahi Denka
Mark LA-63	**33**, R = CH_3	Asahi Denka
Mark LA-68	**33**, R = H	Asahi Denka
Mark LA-77	**28**, R = H	Asahi Denka
Sanduvor 3055	**25**, R = H	Sandoz
Sanduvor 3056	**25**, R = CH_3	Sandoz
Sanduvor 3058	**25**, R = $COCH_3$	Sandoz
Sanduvor 3212	**151**	Sandoz
Sanol LS 765	**28**, R = CH_3	Sankyo
Sanol LS 770	**28**, R = H	Sankyo
Sanol LS 1114	**25**, R = benzyl	Sankyo
Sanol LS 2626	**38**	Sankyo
Sumisorb MM 006	**30**	Sumitomo
T-163	Proprietary	Asahi Denka
Tinuvin 123	**28**, R = O-isoC_8H_{17}	Ciba
Tinuvin 144	**37**	Ciba
Tinuvin 292	**28**, R = CH_3	Ciba
Tinuvin 622	**34**	Ciba
Tinuvin 765	**28**, R = CH_3	Ciba
Tinuvin 770	**28**, R = H	Ciba
Topanex 500 H	**29**, R = H	ICI
Topanex 516 H	**29**, R = CH_3	ICI
Uvaseb 770	**28**, R = H	Enichem
Uvasil 299	**36**	Enichem

Producers of Aminic Stabilizers

Aceto Chemical Co., Ltd
    162-02 Northern Blvd, Flushing, NY 11363, USA
Akzo Chemicals bv
    Stationstraat 48, 3800 AE Amersford, The Netherlands
American Cyanamid Co.
    P.O. Box 6885, Bridgewater, NJ 08807-6885, USA
Amoco Chemicals Corp.
    200 East Randolph Drive, Chicago, IL 60601, USA

Anchor Chemical Co., Ltd
    Clayton Lane, Clayton, Manchester M11 4SR, UK
Asahi Denka Chemical Co., Ltd
    5-2-13 Shirahata, Urawa, Saitama Prefecture, 336 Japan
Bayer AG, Sparte Kautschuk
    Bayerwerk, D-5090 Leverkusen, FRG
BF Goodrich Co., Specialty Polymers & Chemicals Division
    6100 Oak Tree Boulevard, Cleveland, OH 44131, USA
Ciba-Geigy Ltd, Additives Division
    CH-4002 Basle, Switzerland
Eastman Company, Chemical Products Inc.
    P.O. Box 431, Kingsport, TN 37662, USA
E.I. DuPont de Nemours, Co.
    Nemours Building, Wilmington, DE 19898, USA
Enichem Synthesis SpA, Plastics Additives
    Via Medici des Vascello 40, 20138 Milano, Italy
Goodyear Chemicals Co.
    1485 E Archwood Ave, Akron, OH 44316, USA
Hoechst AG, Marketing Wachse und Kunststoffadditive
    Postfach 101567, D-8900, Augsburg 1, FRG
ICI Petrochemicals and Plastics Division
    Wilton, Cleveland, UK
Istrochim Chemical Co.
    83603 Bratislava, Slovak Republic
Kawaguchi Chemical Co., Ltd
    3-8 Nihonbashi, Hon-cho, Chuo-ku, Tokyo, 103 Japan
Mobay Chemical Co.
    Penn-Lincoln Parkway, W Pittsburgh, PA 15205, USA
Monsanto Chemical Co.
    800 N Lindbergh Blvd, St Louis, Mo 63166, USA
Ouchi Shinko Chemical Industries Co., Ltd
    1-3-7 Nihonbashi, Kobune-cho, Chuo-ku, Tokyo, 103 Japan
Pennwalt Corp.
    Pennwalt Bldg, Three Parkway, Philadelphia, PA 19102, USA
RT Vanderbilt Co., Ltd
    30 Winfield Street, Norwalk, CT 06855, USA
Sandoz Huningue S.A.
    Avenue de Bale, BP 29, F-68330, Huningue, France
Sankyo Co., Ltd
    7-17 Ginza, 2-chome, Chuo-ku, Tokyo, 104 Japan
Seiko Chemical Co., Ltd
    Hiranuma Bldg, 6-2-chome, Kanda Tsukasa-cho, Chiyoda-ku,
    Tokyo, 101 Japan
Sumitomo Chemical Co., Ltd
    5-33 Kitahama, 4-chome, Chuo-ku, Osaka, 541 Japan

Uniroyal Chemical Co., Ltd, Specialty Chemicals
Middlesbury, CT 06749, USA
UOP Chemical Co., Process Division
20 UOP Plaza, Des Plaines, IL 60016, USA
Vulnax International Ltd
Delaunays Road, Blackley, Manchester, M60 1EP, UK

12 References

1. Hoffman W (1989) Rubber technology handbook. Hanser, Munich
2. Dominighaus H (1976) Die Kunststoffe und ihre Eigenschaften. VDI Verlag, Düsseldorf
3. Pospíšil J (1984) In: Scott G (ed) Developments in polymer stabilization, vol 7. Elsevier, London, p 1
4. Pospíšil J (1989) In: Patsis AV (ed) 11th Intern Confer on stabilization and degradation of polymers, 24–26 May 1989. Luzern, Proc p 163
5. Pospíšil J (1991) Polym Degrad Stab 34: 85
6. Pospíšil J (1994) Angew Makromol Chem 216: 135
7. Billingham NC (1990) In: Pospíšil J, Klemchuk PP (eds) Oxidation inhibition in organic materials, vol 2. CRC Press, Boca Raton, p 249
8. Pospíšil J (1991) Advan Polym Sci 101: 65
9. Johnson MD, Korcek S (1991) Lub Sci 3(2): 95
10. Sohma J (1989) Progr Polym Sci 14: 451
11. Clough RC, Gillen KT (1990) In: Pospíšil J, Klemchuk PP (eds) Oxidation inhibition in organic materials, vol 2. CRC Press, Boca Raton, p 191
12. Gugumus F (1990) In: Pospíšil J, Klemchuk PP (eds) Oxidation inhibition in organic materials, vol 1. CRC Press, Boca Raton, p 61
13. Gugumus F (1990) In: Pospíšil J, Klemchuk PP (eds) Oxidation inhibition in organic materials, vol 2. CRC Press, Boca Raton, p 29
14. Pospíšil J (1991) Chem Listy 85: 904
15. Gugumus F (1993) Polym Degrad Stab 40: 167
16. Felder BN (1985) In: Klemchuk PP (ed) Polymer stabilization and degradation. ACS Symp Ser 280: 69
17. Trainer M, Perrish DD, Buhr MP (1993) J Geophys Res 98 (D2): 2918
18. Layer RW, Lattimer RP (1990) Rubber Chem Technol 63: 426
19. Brück D (1989) Kaut Gummi Kunstst 42: 760
20. Lake GY, Mente PG (1992) J Nat Rubber Res 7(1): 1
21. Kuczkowski J (1990) In: Pospíšil J, Klemchuk PP (eds) Oxidation inhibition in organic materials, vol 1. CRC Press, Boca Raton, p 247
22. Pickett JE (1994) Polym Degrad Stab 43: 453
23. Kerr JB, McElroy CT (1993) Science 262: 1032
24. Pospíšil J (1985) In: Klemchuk PP (ed) Polymer stabilization and degradation. ACS Symp Ser 280: 157
25. Pospíšil J (1990) In: Pospíšil J, Klemchuk PP (eds) Oxidation inhibition in organic materials, vol 1. CRC Press, Boca Raton, p 33
26. Gugumus F (1993) In: 18th Confer. of the polymer degradation discussion group, 15–17 Sept 1993. Bolton
27. Upadlyay NB, Warrach W (1990) Rubber World 203(1): 38
28. Kashiwaraki S, Takahata N, Kashimura H (1993) Jpn Kokai Tokkyo Koho 05 81, 936
29. Doe LA (1987) US Pat 4. 704, 426
30. Adorisio PA, Chasan DE, Pastor SD (1992) US Pat 5. 160, 647
31. Brück D, Engels HW (1991) Kaut Gummi Kunstst 44: 1014
32. Lévy M (1989) Kaut Gummi Kunstst 42: 129
33. Kimura T, Ko M (1990) Jpn Kokai Tokkyo Koho 02 55, 366

34. Engels W, Hammer H, Brück D, Redetzky W (1989) Rubber Chem Technol 62: 609
35. Kuczkowski JA, Cottman ES, Hoppstock FH (1993) Polym Prepr 34(2): 162
36. Alberti A, Carloni P, Greci L, Stipa P, Neri C (1993) Polym Degrad Stab 39: 215
37. Hagasaki H, Takemoto Y, Yoshimura M (1986) Jpn Kokai Tokkyo Koho 61: 31,444
38. Son PN (1980) Polym Degrad Stab 2: 295
39. Layer RW, Lai JT, Lattimer RP, Westfahl JC (1985) In: Klemchuk PP (ed) Polymer stabilization and degradation. ACS Symp Ser 280: 99
40. Toda T, Kurumada T (1983) Ann Rep Sankyo Res Lab 35: 1
41. Tanimoto S, Toshimitsu A, Inoue Y (1991) Bull Inst Chem Res, Kyoto Univ 69: 234
42. Layer RW, Lai JT, Son PN (1986) US Pat 4. 629, 752
43. Gugumus F (1991) Angew Makromol Chem 190: 111
44. Toda T, Kurumada T, Murayama K (1985) In: Klemchuk PP (ed) Polymer stabilization and degradation. ACS Symp Ser 280: 37
45. Müller H (1985) In: Klemchuk PP (ed) Polymer stabilization and degradation. ACS Symp Ser 280: 55
46. Kurumada T (1989) J Japan Soc Colour Mat 61: 215
47. Chucta T (1988) Soc Plast Eng, Tech Pap 34: 1451
48. Tagawa K (1990) Jpn Kokai Tokkyo Koho 02 219, 841
49. Pospíšil J (1979) In: Scott G (ed) Developments in polymer stabilization, vol 1. Elsevier, Barking, p 1
50. Denisov ET, Khudyakov IV (1987) Chem Rev 87: 1313
51. Adamic K, Bowman DF, Ingold KU (1970) J Am Oil Chemist's Soc 47: 109
52. Carlsson DJ, Wiles DM (1974) Rubber Chem Technol 47: 991
53. Zeman A, von Roenne V, Trebert Y (1987) J Synth Lubr 4: 179
54. Varlamov VT (1989) Kinet Katal 30: 786
55. Hunter M, Klaus EE, Duda JL (1993) Lubr Eng 49: 492
56. Zeman A, Trebert Y, von Roenne V, Fuchs HJ (1990) Tribologie + Schmierungstechn 37: 158
57. Lorenz O, Haulena F, Braun B (1985) Kaut Gummi Kunstst 38: 255
58. Pospíšil J (1990) Angew Makromol Chem 176/177: 347
59. Tsurugi J, Murakani S, Goda K (1971) Rubber Chem Technol 44: 857
60. Adamic K, Ingold KU (1969) Canad J Chem 47: 295
61. Adamic K, Bowman DF, Gillan T, Ingold KU (1971) J Am Chem Soc 93: 902
62. Kuzminskij AS (1981) In: Scott G (ed) Developments in polymer stabilization, vol 4. Elsevier, London, p 71
63. Bowman DF, Brokenshire JL, Gillan T, Ingold KU (1971) J Am Chem Soc 93: 6551
64. Murayama K (1971) J Synthet Org Chem Japan 29: 366
65. Berger H, Bolsman TABM, Brower DM (1983) In: Scott G (ed) Developments in polymer stabilization, vol 6. Applied Science Publishers, London, p 1
66. Thomas JR, Tolman C (1962) J Am Chem Soc 84: 2930
67. Berti C (1983) Synthesis –: 793
68. Klemchuk PP, Gande ME (1988) Polym Degrad Stab 22: 241
69. Ravichandran R, Snead TE (1988) US Pat 4. 720, 517
70. Cowley DJ, Waters WA (1970) J Chem Soc B –: 96
71. Ravichandran R (1987) US Pat 4. 696, 964
72. Müller E, Mayer R (1961) Ann 645: 1
73. Aurich HG, Weiss W (1975) Topics in current chemistry, vol 59, Organic syntheses. Springer, Berlin, p 65
74. Dahmiwal NR, Grewal D (1991) Indian J Chem 30A: 660
75. Ducháček V, Kuta A, Sošková L, Taimr L, Rotschová J, Pospíšil J (1990) Polym Degrad Stab 29: 217
76. DeKoninck D, Aarts RJ, Burkin H, Orband A (1988) Gummi Asbest Kunstst 41: 440
77. Grünanger P (1979) In: Grundmann Ch (ed) Methoden der organischen Chemie, 4th edn, vol 7, part 3b. Thieme, Stuttgart, p 233
78. Padva A (1977) Chem Rev 77: 37
79. Pedersen CJ (1957) J Am Chem Soc 79: 2295, 5014
80. Lattimer RP, Gianelos H, Diem HE, Layer RW, Rhee CK (1986) Rubber Chem Technol 59: 263
81. Billingham NC, Calvert PD (1989) Advan Polym Sci 90: 1
82. Neoh KG, Kang ET, Tan LL (1994) Polym Degrad Stab 43: 141

83. Sidhu KS, Bansal WR, Malhotra K (1988) Indian J Chem 27A: 747
84. Bremer E, Aurich HG, Nielsen A (1989) In: Patai S, Rappoport Z (eds) The chemistry of functional groups: nitrones, nitronates and nitroxides. Wiley, New York, p 231
85. Mazaletskaya LI, Karpukhina GV (1983) Izv Akad Nauk, ser khim –: 279
86. Scott G (1984) Polym Eng Sci 24: 1007
87. Pospíšil J (1980) Advan Polym Sci 36: 69
88. Downs BW, Stott PE, Barry LB, Richardsson MC (1992) PCT Int Appl WU 92 06, 133
89. Gatechair LR, Seltzer R, Hyun JL (1992) Eur Pat Appl 467, 850
90. Friedman HS, Stott PE (1992) PCT Int Appl WU 92 06, 939
91. Winkler RE, Naumann EB (1988) J Polym Sci 26A: 2853
92. Thinius K (1969) Stabilisierung und Alterung von Plastwerkstoffen, vol 1. Akademie Verlag, Berlin
93. Tabor MW, Coats E, Sainsbury M, Shertzer HG (1990) In: Witner CM (ed) Biological reactive intermediates. Plenum Press, New York
94. Greci L (1983) Tetrahedron 39: 677
95. Carloni P, Greci L, Stipa P, Eberson L (1991) J Org Chem 56: 4733
96. Alberti A, Greci L, Stipa P, Sgarabotto P, Uguzzoli F (1987) Tetrahedron 43: 3031
97. Kememitsuya K, Tachibara T, Okruda T (1993) Jpn Kokai Tokkyo Koho 04 314, 752
98. Thorisson S, Gunstone F, Hardy R (1992) J Am Oil Chemist's Soc 69: 806
99. Chen PM, Varga DM, Mielke EA, Facteau TJ, Drake SR (1990) J Food Sci 55: 167, 171
100. Shaare JV, Hentiksen T (1975) J Sci Food Agric 26: 1647
101. Taimr L, Prusíková M, Pospíšil J (1991) Angew Makromol Chem 190: 53
102. Malkin JA, Piroga NU, Ivanov JA, Pokrovskaya IE, Kuzmin VA (1981) Izv Akad Nauk, ser khim –: 2008
103. Fentsov DV, Lobanova TV, Kasaikina OT (1990) Neftekhimiya 30(1): 103
104. Zahradníčková A, Sedlář J, Dastych D (1991) Polym Degrad Stab 32: 155
105. Thorisson S, Gunstone FD, Hardy R (1992) Chem Phys Lipids 60: 263
106. Taimr L (1994) Angew Makromol Chem 217: 119
107. Taimr L, Šmelhausová M, Prusíková M (1993) Angew Makromol Chem 206: 199
108. Dagonneau M, Ivanov VB, Rozantsev EG, Sholle VD, Kagan ES (1982/1983) J Macromol Sci, Rev Macromol Chem Phys C22: 161
109. Meier HR, Evans S (1988) Eur Pat Appl 273, 868
110. Levin PP, Chudjakov IV, Kuzmin VA, Ivanov JA (1980) Izv Akad Nauk, ser khim –: 421
111. Gugumus F (1991) Polym Degrad Stab 34: 205
112. Bowman DF, Gillan T, Ingold KU (1971) J Am Chem Soc 93: 6555
113. Lai JT, Son PT, Jennings A (1985) In: Klemchuk PP (ed) Polymer stabilization and degradation. ACS Symp Ser 280: 91
114. Carloni P, Greci L, Marin A, Stipa P (1994) Polym Degrad Stab 44: 201
115. Evans S (1992) Eur Pat Appl 475, 904
116. Okafor CO (1986) Dyes Pigm 7: 249
117. Levy LB (1992) J Polym Sci, Part A, Polym Chem 30: 569
118. Chandra R, Saini R (1992) Polym Degrad Stab 37: 131
119. Hiatt R (1972) In: Swern D (ed) Organic peroxides, vol 3. Wiley, New York, p 1
120. Wassermann D, Jones RE (1964) US Pat 3. 143, 560
121. Matisová-Rychlá L, Velikov AA, Rychlý J, Schulz M (1992) Polym Degrad Stab 37: 77
122. Gugumus F (1993) In: 18th Confer. of the polymer degradation discussion group, 15–17 Sept 1993. Bolton
123. Faucitano A, Buttafawa A, Martinotti F, Greci G (1992) Polym Degrad Stab 35: 211
124. Kurumada T, Ohsawa H, Oda O, Fujita T, Toda T, Yoshioka T (1985) J Polym Sci, Polym Chem Ed 23: 1477
125. Allen NS (1986) Chem Soc Rev 15: 373
126. Carlsson DJ, Wiles DM (1984) Polym Degrad Stab 6: 1
127. Wiles DM, Jensen JPT, Carlsson DJ (1983) Pure Appl Chem 55: 1651
128. Allen NS, Katecha JL, Gardette J-L, Lemaire J (1985) Polym Degrad Stab 11: 181
129. Drake WO (1992) In: Patsis AV (ed) 14th Intern Confer on stabilization and degradation of polymers, June 25–27 1992, Luzern, Proc p 57
130. Carlsson DJ, Tovborg-Jensen JP, Wiles DM (1984) Makromol Chem, Suppl 8: 79
131. Gugumus F (1987) In: Scott G (ed) Developments in polymer stabilization, vol 8. Elsevier, London, p 239

132. Gray RL (1991) In: Vigo TL, Turbak AF (eds) Composites, biomedical materials, protective clothings, geotextiles. ACS Symp Ser 457: 320
133. Gugumus F (1989) Polym Degrad Stab 24: 289
134. Gugumus F (1993) Polym Degrad Stab 39: 117
135. Drake WO, Pauquet JR, Zingg J, Zweifel H (1993) Polym Prepr 34(2): 174
136. Gijsman P (1994) Polym Degrad Stab 43: 171
137. Meyer FK, Pedrazetti E (1987) Plast Rubber Process Appl 8: 29
138. Gugumus F (1988) SPE Tech Pap 34: 1447
139. Al-Malaika S, Omokorede EO, Scott G (1987) J Appl Polym Sci 33: 703
140. La Mantia F, Gratani F (1990) Polym Degrad Stab 30: 257
141. Vink P, Wisse JDM (1982) Polym Degrad Stab 4: 51
142. Mülhaupt R, Rody J, Slongo M (1990) Eur Pat Appl 350, 444
143. Mülhaupt R, Dubs P (1990) Eur Pat Appl 351, 360
144. Mülhaupt R (1990) In: Patsis AV (ed) 12th Intern Confer on stabilization and degradation of polymers, 25–27 May 1990. Luzern, Proc p 182
145. Cartolano PF, Seltzer R, Patel AR (1991) US Pat 5.004, 770
146. Carette L, Gay M, Lavault S, Mur G (1991) Eur Pat Appl 421, 890
147. Bauer DR (1990) In: Patsis AV (ed) 12th Intern Confer on stabilization and degradation of polymers, 25–27 May 1990. Luzern, Proc p 11
148. Böhnke H, Avar L, Hess E (1991) J Coat Technol 63(799): 53
149. Böhnke H, Hess E (1991) Advan Org Coat, Sci Technol Ser 13: 49
150. Neri C, Grattani F, Landoni G, Constanzi S (1991) Eur Pat Appl 448, 163
151. Mielewski DF, Bauer DR, Gerlock JL (1990) Polym Mater Sci Eng 63: 642
152. Mielewski DF, Bauer DR, Gerlock JL (1993) Polym Degrad Stab 4: 323
153. Bechtold K, Hess E, Ligner G (1993) Farbe + Lack 99: 25
154. Schirmann PJ, Dexter M (1987) In: Galbo JL (ed) Handbook of coatings additives. Dekker, New York, Chap 8
155. Valet A (1990) Farbe + Lack 96: 689
156. Valet A (1992) Eur Polym Paint Colour J 182(4311): 406
157. Ligner G, Hess E (1991) Eur Coat J 10: 622
158. Patel GA, Trapp MA (1993) US Pat 5. 214, 085
159. Bauer DR, Gerlock JL, Mielewski DF (1990) Polym Degrad Stab 28: 115
160. Bauer DR, Gerlock JL, Mielewski DF (1992) Polym Degrad Stab 36: 9
161. Rozantsev EG, Kagan ES, Sholle ES, Ivanov VB (1985) In: Klemchuk PP (ed) Polymer stabilization and degradation. ACS Symp Ser 280: 11
162. Raspanti G (1993) Eur Pat Appl 521, 372
163. Raymor RJ (1988) US Pat 4.751, 073
164. Chirinos-Padron AJ (1990) J Macromol Sci, Rev Macromol Chem Phys C10: 107
165. Sedlář J (1990) In: Pospíšil J, Klemchuk PP (eds) Oxidation inhibition in organic materials, vol 2. CRC Press, Boca Raton, p 1
166. Gugumus F (1990) Angew Makromol Chem 182: 111
167. Bauer DR, Dean MJ, Gerlock JL (1988) Ind Eng Chem, Research 27: 65
168. Gijsman P, Hennekens J, Tummers D (1993) Polym Degrad Stab 39: 225
169. Gijsman P, Hennekens J, Tummers D (1994) Angew Makromol Chem 216: 37
170. Faucitano A, Buttafawa A, Martinotti F, Bortolus P (1984) J Phys Chem 88: 1187
171. Toda T, Mori E, Horiuchi H, Murayama K (1972) Bull Chem Soc Japan 42: 1802
172. Malatesta F, Ingold KU (1971) J Am Chem Soc 95: 6400
173. Kurumada T, Ohsawa H, Fujita T, Toda T, Yoshioka T (1985) J Polym Sci, Polym Chem Ed 23: 2747
174. Kurumada T, Ohsawa H, Fujita T, Toda T, Yoshioka T (1984) J Polym Sci, Polym Chem Ed 22: 277
175. Belluš D (1978) In: Ranby B, Rabek JF (eds) Singlet oxygen. Wiley, Chichester, p 80
176. Dagonneau M, Kagan ES, Mikhailov VI, Rozantsev EG, Sholle VD (1984) Synthesis –: 895
177. Ingold KU (1984) In: Fischer H (ed) Landolt-Börnstein, New series, vol 13, part C. Springer, Berlin, p 181
178. Beckwith ALJ, Bowry VW, Ingold KU (1992) J Am Chem Soc 114: 4983
179. Neri C, Malatesta V, Constanzi S, Riva R (1994) Angew Makromol Chem 216: 135
180. Keana JFW (1979) In: Berliner LJ (ed) Spin labeling, vol 2. Academic Press, New York, p 115, 347

181. Hodgeman DKC (1981) J Polym Sci, Polym Chem Ed 19: 807
182. Allen NS, Parkinson A, Gardette J-L, Lemaire J (1984) Polymer 25: 235
183. Dudler V (1993) Polym Degrad Stab 42: 205
184. Beckwith ALJ, Bowry VW, Moad G (1988) J Org Chem 53: 1633
185. Chateauneuf J, Lusztyk J, Ingold KU (1988) J Org Chem 53: 1629
186. Hinsken H, Moss S, Pauquet JR, Zweifel H (1991) Polym Degrad Stab 34: 279
187. Bowry VW, Ingold KU (1992) J Am Chem Soc 114: 4992
188. Marchal J (1991), Radiat Phys Chem 37: 53
189. Gugumus F (1993) In: 3rd Intern. Confer. on polymer photochemistry, 5–10 Sept, 1993. Sestri-Levante
190. Carlsson DJ, Chan KH, Garton A, Wiles DM (1980) Pure Appl Chem 52: 389
191. Gande ME, Cordola E, Klemchuk PP (1988) Polym Mater Sci Eng 58: 445
192. Denisov ET (1989) Polym Degrad Stab 25: 209
193. Grattan A, Carlsson DJ, Wiles DM (1979) Can J Chem 57: 2834
194. Neri C, Constanzi S, Farris R, Malatesta V (1991) In: Patsis AV (ed) 13th Intern. Confer. on stabilization and degradation of polymers, 22–24 May 1991. Luzern, Proc p 111
195. Klemchuk PP, Gande ME (1989), Makromol Chem, Macromol Symp 28: 117
196. Klemchuk PP, Gande ME, Cordola E (1990) Polym Degrad Stab 27: 65
197. Step NE, Turro NJ, Klemchuk PP, Gande ME (1994) In: Patsis AV (ed) 16th Intern. Confer. on stabilization and degradation of polymers, 20–22 June 1994. Luzern, Proc p 73
198. Kysel O, Mach P (1993) Polym Degrad Stab 40: 31, 42: 117
199. Shilov YA, Battalova RM, Denisov ET (1972) Dokl Akad Nauk USSR 207: 388
200. Carlsson DJ, Chan KH, Durmis J, Wiles DM (1982) J Polym Sci, Polym Chem Ed 20: 575
201. Feng X (1992) Makromol Chem, Macromol Symp 63: 1
202. Sedlář J, Petrůj J, Marchal J (1982) Polym Photochem 2: 175
203. Geuskens J, Debie F, Kabamba MS, Nedelkos G (1984) Polym Photochem 5: 313
204. Toda T, Mori E, Murayama K (1972) Bull Chem Soc Japan 42: 1640
205. Falicki S, Gosciniak DJ, Cooke JM, Carlsson DJ (1993) Polym Degrad Stab 41: 205
206. Gijsman P, Hennekens J, Vincent JAJM (1989) In: Patsis AV (ed) 11th Intern Confer on stabilization and degradation of polymers, 24–26 May 1989. Luzern, Proc p 33
207. Crouzet C, Marchal J (1992) Radiat Phys Chem 39: 359, 40: 233
208. Darmanyan AP, Tatikolov AS (1986) J Photochem 32: 157
209. Monroe BM (1977) J Phys Chem 81: 1861
210. Ballardini R, Beggiato G, Bortolus P, Faucitano A, Buttafawa A, Grattani F (1984) Polym Degrad Stab 7: 41
211. Furukawa F, Ogryzlo EA (1972/1973) J Photochem 1: 163
212. Yang YY, Lucki J, Rabek JF, Ranby B (1983) Polym Photochem 3: 47, 97
213. Bortolus P, Camoioni N, Flamigni L, Minto F, Monti A, Faucitano A (1992) J Photochem Photobiol 68A(2): 239
214. Allen NS, Fatinikun KU, Gardette J-L, Lemaire J (1981) Polym Degrad Stab 3: 243
215. Lucki J, Rabek JF, Ranby B, Dai GS (1984) Polym Photochem 5: 385
216. Fairgrieve SP, McCallum JR (1986) Polym Degrad Stab 15: 81
217. McCallum JR (1989) In: Patsis AV (ed) Advances in the stabilization and degradation of polymers, vol 1. Technomic, Lancaster, p 52
218. Yang XZ, Chen Y, Dickinson JCh, Chien JCW (1988) Polym Degrad Stab 20: 1
219. Gugumus F (1990) Angew Makromol Chem 176/177: 241
220. Chmela Š, Carlsson DJ, Wiles DM (1989) Polym Degrad Stab 26: 185
221. Pan J, Cui S (1993) Polym Degrad Stab 40: 375
222. Shtsukin GI, Grigorjev IA, Volodarskij AV (1980) Izv Akad Nauk, ser khim –: 1421
223. Lucki J, Rabek JF, Ranby B, Watanabe Y (1988) J Appl Polym Sci 36: 1067
224. Hodgeman DKC (1982) In: Grassie N (ed) Developments in polymer degradation, vol 4. Applied Science, London, p 189
225. Moad G, Rizzardo E, Solomon DH (1981) Tetrahedron Lett 22: 1165
226. Falicki S, Carlsson DJ, Cooke JM, Gosciniak DJ (1992) Polym Degrad Stab 38: 265
227. Falicki S, Gosciniak DJ, Cooke JM, Carlsson DJ (1994) Polym Degrad Stab 43: 1, 117
228. Carlsson DJ, Yang C, Wiles DM (1987) J Appl Polym Sci 33: 875
229. Dulog L, Bleher R (1986) Makromol Chem 187: 2357
230. Gray RL (1991) Plast Eng 47(6): 21
231. Valet A (1991) Polym Paint Colour J 181(4276): 38

232. Evans E, Chasan D, Seltzer R (1991) US Pat 5.019, 285
233. Winter RAE, Galbo JP, Mar A, Behrens RA, Malherbe RF (1989) Eur Pat Appl 309, 401
234. Bramer D, Holt MS (1991) In: Patsis AV (ed) 13th Intern. Confer. on stabilization and degradation of polymers, 22–24 May, 1991. Luzern, Proc p 23
235. Bramer D, Holt MS, Mar A (1990) Polym Mater Sci Eng 63: 647
236. Gatechair LR, Evers H, Hyun JL (1992) Eur Pat Appl 467, 851
237. Henninger F, Pedrazetti E (1988) Plasticulture (80): 5
238. Chmela Š, Hrdlovič P (1990) Polym Degrad Stab 27: 159
239. Kikkawa K, Nakahara Y, Ohkatsu Y (1987) Polym Degrad Stab 18: 237
240. Lucki J, Jian SZ, Rabek JF, Ranby B (1986) Polym Photochem 7: 27
241. Lind H, Loeliger H (1979) In: Mechanisms of degradation and stabilization of hydrocarbon polymers, 8–12 July 1979. Prague, Paper M44
242. Ray JW, Reynolds AB (1990) Nuclear Technol 91: 394
243. Nishimoto S, Chaisupakitsin M, Inuni T (1992) Radiat Phys Chem 39: 413
244. David C, Zabeau F (1985) Eur Polym J 21: 343
245. Wheeler EL (1991) PCT Int Appl WO 91 05, 773
246. Miller DE, Dessent RW, Kuczkowski JA (1985) Rubber World 193(1): 31
247. Lattimer RP, Layer RW, Hooser ER, Rhee CK (1991) Rubber Chem Technol 64: 780
248. Lattimer RP, Hooser ER, Layer RW, Rhee CK (1983) Rubber Chem Technol 56: 431
249. Lattimer RP, Layer RW, Rhee CK (1984) Rubber Chem Technol 57: 1023
250. Pospíšil J (1990) In: Pospíšil J, Klemchuk PP (eds) Oxidation inhibition in organic materials, vol 1. CRC Press, Boca Raton, p 173
251. Pospíšil J (1993) Polym Degrad Stab 39: 103, 40: 217
252. Kasaikina OT, Kartasheva ZS, Mazaletskij AB (1992) Izv Akad Nauk, ser khim –: 417
253. Tsvetkovskij IB, Korennaya AB, Andreeva AB (1991) Zh Prikl Khim 64: 387, 393
254. Raymond S, Patel AR, Stewart PW (1985) Eur Pat Appl 138, 767
255. Paisner MJ (1990) US Pat 4.822, 839
256. Ono H, Yoshitsugu T (1988) Jpn Kokai Tokkyo Koho 01 225, 697
257. Evans S, Schumacher R (1989) Eur Pat Appl 346, 283
258. Roling PV (1990) US Pat 4.929, 778
259. Hawkins WL, Hansen RH, Matreyek W, Winslow FA (1959) Rubber Chem Technol 32: 1164
260. Pospíšil J, Rotschová J (1993) In: 2nd Intern. Confer. on carbon black, 27–30 Sept 1993. Mulhouse, Proc p 161
261. Gugumus F (1985) Angew Makromol Chem 137: 189
262. Vyprachtický D, Pospíšil J, Sedlář J (1990) Polym Degrad Stab 27: 227
263. Scheim K, Pospíšil J , Habicher W (1992) In: 34th IUPAC Intern. Symposium on macromolecules, 13–18 July 1992. Prague, Book of abstracts 8-P-16
264. Scheim K, Habicher W, Pospíšil J (1992) In: 34th IUPAC Intern. Symposium on macromolecules, 13–18 July 1992. Prague, Book of abstracts 8-P-47
265. Allen NS, Gardette J-L, Lemaire J (1982) J Appl Polym Sci 27: 2761
266. Ohkatsu Y, Yamaguchi K, Yamaki A, Mantuyama Y, Kusano T (1991) Sekiyu Gakkaishi 34: 280
267. Carloni P, Greci L, Stipa P, Rizzoli C, Sgarabotto P, Ugozzoli F (1993) Polym Degrad Stab 39: 73
268. Shibata T, Matsuda M, Ishida M (1992) Jpn Kokai Tokkyo Koho 04 159, 353
269. Kletecka G, Lai JTY, Son PN (1991) Eur Pat Appl 443, 328
270. Mori H, Yamazaki S, Okada K (1993) Jpn Kokai Tokkyo Koho 05 86, 230
271. Wirth HO, Sander HJ (1992) Ger Offen 4.141, 661
272. Bergner KD, Pfahler G (1991) Eur Pat Appl 457, 228
273. Drake WO, Pauquet JR, Todesco RV (1989) In: Intern. Confer. PRI, 3–5 Nov 1989. Madrid
274. Hepburn C, Alun A (1984) Rubber World 190(2): 49
275. Kuczkowski JA (1978) US Pat 4.124, 565
276. Hähner U, Habicher WD, Chmela Š (1993) Polym Degrad Stab 41: 197
277. Chmela Š, Habicher WD, Hähner U, Hrdlovič P (1993) Polym Degrad Stab 39: 367
278. Nesvatba P (1993) Eur Pat Appl 530, 139
279. Avar L, Böhnke H, Hess E (1988) In: 19th FATIPEC congress, vol 1. Proc p 317
280. Allen NS, Edge M, He J, Chen W, Kikawa K, Minagawa M (1994) Polym Degrad Stab 44: 99
281. Slongo M, Birnbaum JL, Rody J, Valet A (1991) Eur Pat Appl 453, 405
282. Ravichandran R, Galbo JP (1990) Eur Pat Appl 389, 427

283. Kazmierczak RT, MacLay RE (1990) Eur Pat Appl 366, 057
284. Malík J, Hrivík A, Tomová E (1992) Polym Degrad Stab 35: 61, 125
285. Kato K, Kadoi S, Higaki Y (1992) Jpn Kokai Tokkyo Koho 04 103, 607
286. Rytz G, Slongo M (1989) Eur Pat Appl 344, 114
287. Seltzer R, Winter RA, Schirmann PJ (1990) Eur Pat Appl 389, 423
288. Ravichandran R (1990) Eur Pat Appl 389, 429
289. Vilén CE, Auer M, Näsman JH (1992) J Polym Sci, Part A, Polym Chem 30: 1163
290. Nithianandam VS, Erhan S (1991) Polymer 32: 1146
291. Foster GN, Enelly PH, King IR (1990) Eur Pat Appl 358, 190
292. Friedrich H, Jansen I, Rühlmann K (1993) Polym Degrad Stab 42: 127
293. Galbo JP, Ravichandran R, Schirmann PJ, Mar A (1990) Eur Pat Appl 389, 419
294. Al-Mehdave MS, Stuckey JE (1989) Rubber Chem Technol 62: 13
295. Braun D, Rettig R, Rogler W (1993) Angew Makromol Chem 211: 165
296. Haubold W, Taplick P, Seifferth K, Roehr Ch, Steinert V, Rätzsch M (1988) Ger (East) Pat 264, 699
297. Hahnfeld JL, Devore DD (1993) Polym Degrad Stab 39: 241
298. Al-Malaika S, Scott G, Wirjosentona B (1993) Polym Degrad Stab 40: 233
299. Schulz GO, Parker DK (1992) Eur Pat Appl 470, 042
300. Hrdlovič P, Chmela Š (1990) Int J Polym Mater 13: 249
301. Chmela Š, Hrdlovič P (1993) Polym Degrad Stab 42: 55
302. Pan J, Cheng W, Song Y, Hu X (1993) Polym Degrad Stab 39: 85
303. Lee CS, Lau WWY, Lee SY, Goh SH (1992) J Polym Sci, Part A, Polym Chem 30: 983
304. Chmela Š, Hrdlovič P, Maňásek Z (1985) Polym Degrad Stab 11: 233, 339
305. Malatesta V, Neri C, Raughino G, Montanari R, Fantucci P (1993) Macromolecules 26: 4287
306. Leimgruber R (1990) In: Gächter R, Müller H (eds) Plastics additives handbook, 3rd edn. Hanser, Munich, p 877
307. Code of Federal regulations, Foods and drugs, vol 21, parts 177–199, April 1, 1992. Washington
308. Clemens MR (1991) Klin Wochenschau 69: 1123
309. Turesky RJ, Lang NP, Butler MA, Teitel CH, Kadlubar FF (1991) Carcinogenesis 12: 1839
310. Nutt AR (1984) Toxic hazards of rubber chemicals. Elsevier, London
311. Munday R (1992) Chem Biol Interactions 82: 165
312. Borman S (1992) Chem Eng News 70(48): 27
313. Hard GC, Neal GE (1992) Fundam Appl Toxicol 18: 278
314. Voest EE, van Faassen E, van Asbeck BS, Neijt JP, Marx JJM (1992) Biochem Biophys Acta 1136: 113
315. Sun Y (1990) Free Radic Biol & Med 8: 583

Editor : Prof. K. Dušek
Received: August 1994

Application of Nitrogen-15 NMR to Polymers

M. Andreis[1] and J.L. Koenig
Department of Macromolecular Science, Case Western Reserve University,
Cleveland, OH 44106-7202/USA

A large number of polymers for commercial use as well as biological macromolecules are based on nitrogen containing monomers. Thus, the understanding of their structure and molecular dynamics is of great importance. Practical observation of low natural abundance and sensitivity of ^{15}N isotope was made possible with the development of Fourier Transform techniques, although precise structural analysis and relaxation measurements require isotopic enrichment. The review summarizes all aspects of solution and solid state ^{15}N NMR techniques for the study of structure, conformation and molecular dynamics in linear and crosslinked polymers such as polyamides (nylons), synthetic polypeptides, conducting polymers, urea and melamine based resins. The application of recently developed NMR techniques significantly improves our understanding of structural and dynamical properties of nitrogen containing polymers at the molecular level.

1   Introduction 192

2   ^{15}N NMR in Solutions.................................. 193
    2.1   Broad Band and Inverse Gated Decoupling 194
        2.1.1   Polyamides (Nylons)........................... 195
        2.1.2   Synthetic Polypeptides 197
        2.1.3   Other Polymers 198
    2.2   Spin-Lattice and Spin-Spin Relaxation 205
    2.3   Cross polarization................................... 206

3   ^{15}N NMR in the Solid State 208
    3.1   Gated High Power Decoupling......................... 211
    3.2   Cross Polarization.................................. 211
        3.2.1   Polyamides (Nylons).......................... 213
        3.2.2   Conducting Polymers.......................... 217
        3.2.3   Other Polymers 219
    3.3   Double Cross Polarization and Dipolar Rotational Spin-Echo   226
    3.4   Spin-Lattice Relaxation............................. 230
    3.5   Interrupted Decoupling.............................. 233

4   References 235

---

[1]Permanent address: Ruder Boskovic Institute, Zagreb, Croatia

Advances in Polymer Science, Vol. 124
© Springer-Verlag Berlin Heidelberg 1995

1 Introduction

The nuclear magnetic resonance of nitrogen isotopes ^{14}N and ^{15}N was first reported in 1950 [1]. However, the nitrogen NMR in the 1950s and early 1960s was limited mainly to the ^{14}N isotope because of the very low natural abundance and sensitivity of ^{15}N. The major disadvantage of ^{14}N NMR is that the nucleus has a spin quantum number I = 1 associated with a quadrupole moment which provides a very efficient relaxation mechanism. In most cases the spin-spin relaxation times are very short, and consequently the linewidths are very broad. Although the nitrogen chemical shift range covers several hundred parts per million, linewidths of several hundred Hz limits the precise determination of small differences in chemical shifts.

The low natural abundance of ^{15}N nuclei can be overcome by isotopic enrichment. However, the enrichment is not a straight-forward procedure and it is relatively expensive. The spectra of ^{15}N enriched samples were reported beginning from 1964, but the major advance in nitrogen NMR was made in the early 1970s with the development of Fourier Transform (FT) techniques in NMR spectroscopy. The application of FT NMR to nitrogen nuclei in solution made possible the observation of natural abundance nitrogen spectra within practical experimental time limits.

The application of NMR to low natural abundance nuclei in the solid state was made in 1976 by combining high power decoupling, magic angle spinning and cross polarization techniques [2]. The first practical applications of solid state ^{15}N NMR to both natural abundance and ^{15}N enriched macromolecules were reported in 1982 [3, 4]. Since then ^{15}N NMR has become an established technique for studying polymers in the solid state which contain nitrogen.

One of the major advantages of ^{15}N NMR is the relatively large range of chemical shifts compared to the chemical shift ranges for other commonly observed nuclei in polymers. Although the full range of chemical shifts covers approximately 1000 ppm, about half of the range is covered with various nitroso groups and azo-bridges which are usually not present in polymer systems. The typical chemical shift scale for nitrogen shieldings in polymers covers about 400 ppm which leads to a smaller likelihood of overlapping resonances compared to ^{13}C NMR spectra [5, 6].

In addition, fewer chemically distinct types of nitrogen moieties, compared to the number of carbon moieties, in polymers should result in simpler spectra, and easier identification of resonances and their quantitative analysis. These properties of ^{15}N NMR spectra should also enable the detection of byproducts and minor products which would probably be undetected in the relatively complicated ^{13}C spectra. However, the small gyromagnetic ratio (about 40% of that of ^{13}C) and low natural abundance of the ^{15}N isotope (Table 1) result in about 30 times less sensitivity compared to ^{13}C NMR. Therefore, isotopic enrichment is usually required, particularly for solid state experiments.

Table 1. Properties of hydrogen, carbon and nitrogen nuclei

	^1H	^{13}C	^{14}N	^{15}N
Natural abundance/%	99.99	1.1	99.64	0.36
Spin quantum number	1/2	1/2	1	1/2
NMR frequency at 7.047 T/MHz	300.0	75.47	21.67	30.40
Gyromagnetic ratio, $\gamma/2\pi/T^{-1}s^{-1}$	$267.4\cdot10^6$	$67.2\cdot10^6$	$19.3\cdot10^6$	$-27.1\cdot10^6$
Relative sensitivity	1.0	0.01	0.001	0.002

2 ^{15}N NMR in Solutions

Changes in signal intensities which occur in double resonance experiments arise from the Nuclear Overhauser Effect (η), i.e. from the ratio of the proton-decoupled signal intensity to the proton-coupled intensity. The broad band decoupling yields up to a 200% increase in ^{13}C signal intensities. However, since the gyromagnetic ratio of ^{15}N nuclei is negative the Nuclear Overhauser Enhancement Factor (NOEF $= 1 + \eta$) for complete dipolar $^{15}N-^1H$ relaxation is -3.9 (for ^{15}N nuclei $\eta = \gamma H/2\gamma N = -4.93$). This results in about four times the enhanced inverted signals compared to the signals observed with ^1H irradiation. Furthermore, any η between 0 and -2 lowers the intensity of the observed signal, and in very unfavorable cases ($\eta = -1$) no ^{15}N signal can be observed. The dependence of the Nuclear Overhauser Effect on the molecular motion (rotational correlation time), resonance frequency, solvent, concentration, and temperature for polyglycine (nylon-2) is shown in Fig. 1. Although the commonly used broad band decoupling could enhance signal intensities up to 300%, the signal intensities in some cases can approach zero. The loss of nitrogen resonances can be suppressed by using the gated decoupling technique.

Nitrogen chemical shifts are more sensitive to the type and acidity of the solvent compared to the chemical shifts of protons and carbons. For example, the chemical shifts of 1 mol/l, 2 mol/l, and 10 mol/l aqueous nitric acid solutions are 375.8, 367.6 and 362.0 ppm downfield from anhydrous liquid ammonia at 25°, respectively. Although more than ten common standards are used for chemical shift references, the standards most frequently used for referencing nitrogen chemical shifts in solutions are nitromethane and liquid ammonia which absorb at the lowest and highest fields of the commonly used 380 ppm ^{15}N chemical shift scale, respectively (only benzofuran, some triazene-, terminal diazo-, azo- and nitroso-groups which are not present in polymers absorb at fields lower than that of nitromethane) [5]. The reference standards are usually used externally in order to avoid interactions between reference standard, solvent, and sample. The interactions can result in changes of chemical shifts of both reference standard and nitrogen signals from the sample.

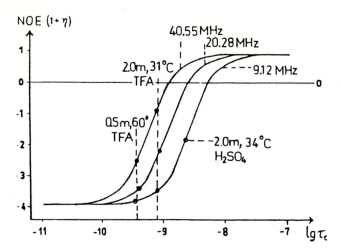

Fig. 1. Plot of calculated NOE values vs correlation time (τ) for three resonance frequencies and NOE measurements of various polyglycine ($P_n = 50$) solutions (reprinted from [4])

2.1 Broad Band and Inverse Gated Decoupling

The broad band proton decoupling technique is widely used for nuclei with positive gyromagnetic ratios. However, the negative NOEF in proton decoupled ^{15}N NMR experiments can yield zero intensities for some resonances. Since the relaxation parameters (spin-lattice relaxation time and NOEF) for all the ^{15}N nuclei present in the sample are generally not known, the use of the gated decoupling technique is preferable, particularly for the quantitative determination of peak areas. In the gated decoupling technique the decoupler is gated on during a short free induction decay accumulation and then is gated off during the relatively long pulse intervals. Suppression of the Nuclear Overhauser Effect depends on the relative magnitudes of the ^{15}N T_1 relaxation times and the decoupler on/off duty cycle. However, if the immediate vicinity of the observed group is not affected by various substituents, such as amide groups in polyaminamides, reliable quantitative results can be obtained even when measured with continuous broad band proton decoupling [7]. On the contrary, the three types of amine triad broad band decoupled signals in poly(vinylamine), with nearly equal spin-lattice relaxation times, have NOEFs ranging from -3.7 to -2.8, and require the normalization of peak areas for quantitative analysis [8].

Although ^{15}N chemical shifts are substantially more sensitive to the sequence lengths compared to ^{13}C chemical shifts (for example, the 2.5 ppm ^{13}C chemical shift range between nylon-2 and nylon-5 in trifluoroacetic acid is approximately four times smaller than the corresponding ^{15}N chemical shift range) there is no general rule as to whether nitrogen or carbon resonance will yield more information on the sequence structure. Various types of polymers containing carbonyl and amide/amine groups exhibit different behavior. Carbonyl signals of aromatic polyamides are sensitive to the sequence effects and have smaller linewidths compared to the corresponding amide nitrogen (100–200 Hz), while in the case of aliphatic polyamides the amide linewidths are in the range of 6–15 Hz, making ^{15}N NMR spectra more suitable for sequence analysis than ^{13}C spectra. Carbonyl signals of polyamides are sensitive to the neighboring residue effects (to a lesser extent than amide nitrogen) while the carbonyl signals of polyureas are not. Furthermore, sequences consisting of ω-amino acids of different chain lengths are best characterized by ^{13}C NMR, while ^{15}N is more useful when sequences consist of monomers which differ in their side chains. In general, the ^{13}C NMR spectra are more suitable for quantitative analysis, while the sensitivity of ^{15}N nuclei to the sequence distribution and stereoisomerism makes the nitrogen spectra more useful in sequence and tacticity analysis (particularly in the case of synthetic peptides and copolypeptides, linear polyamides, polyureas and polyurethanes).

From the point of view of structural analysis of nitrogen containing polymers, simultaneous application of both ^{13}C and ^{15}N NMR techniques will result in a more complete qualitative and quantitative definition of the polymer system, because the two nuclei/spectroscopies behave complementarily.

2.1.1 Polyamides (Nylons)

The application of ^{15}N NMR is limited mainly to the aliphatic polyamides. Less mobile aromatic polyamides have the unfavorable Nuclear Overhauser Enhancement Factor (signal nulling effect), and the broad signals obtained by inverse gated proton decoupling are not suitable for sequence analysis. The ^{15}N chemical shifts for ω-aminoacyl homopolyamides (nylon-2 to nylon-7) exhibit a downfield shift with increasing length of the aliphatic chains and solvent acidity (Fig. 2). Since amide groups are highly sensitive to both hydrogen bonds and protonation at the carbonyl oxygen, the shifts of the chemical shift as a function of chain length strongly depend on the solvent acidity and the basicity of polymer carbonyl and amino groups. Although increasing the acidity of a solvent causes a downfield chemical shift for all the polyamides, the effect of chain length decreases exponentially with the increased number of methylene groups. The short distance between positively charged amide groups in protonated nylon-2 (polyglycine) makes a quantitative protonation of nylon-2 much more difficult compared to nylon-8. Consequently, a significant upfield shift of nylon-2 and nylon-3 followed by a small shift difference between nylon-4 and nylon-8 is observed in the extreme cases of nearly quantitative protonation (in fluorosulfonic acid) and entirely absent protonation (in trifluoroethanol and formic acid). The best differentiation of the amide groups with different basicity is obtained in trifluoroacetic acid, where polyamides are partially protonated. The low viscosity of trifluoroacetic acid compared to all other useful solvents is also of great importance for having an acceptable signal to noise ratio of

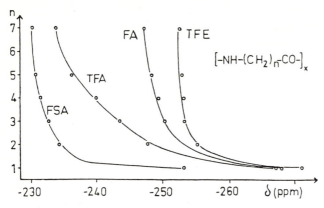

Fig. 2. ^{15}N NMR chemical shifts (δ, upfield from external NO_3^-) of various polyamides measured in fluorosulfonic acid (FSA), trifluoroacetic acid (TFA), formic acid (FA) and trifluoro-ethanol (TFE) (reprinted from [4])

naturally abundant ^{15}N nuclei in a reasonable time interval [9, 10]. The shift effects of various solvents on nitrogen groups are very helpful in structural analysis of soluble polyamides and polypeptides. Moreover, a variation of solvents is found to be more appropriate for structural investigation than the effects caused by paramagnetic shift reagents. Additional disadvantages of using paramagnetic shift reagents are choosing a solvent for the polymer which does not destroy metal complexes, and ensuring that the solvents themselves are not complexed with shift reagents [11].

A similar, but less pronounced chain length vs chemical shift dependence is observed for a series of nylon-6,6 type polyamides based on aliphatic or aromatic α,ω-diamines and aliphatic or aromatic diacids in trifluoroacetic acid. However, the downfield chemical shift caused by increasing the chain length of aliphatic diacid residues is less pronounced relative to that of aliphatic diamines [12].

Strong neighboring residue effects are found in alternating copolyamides based on two different ω-amino acids [4], alternating copolyamides containing ω-amino sulfonic acid [13], and in copolyamides containing β-alanine and {β-aminobutyric/β-amino-isobutyric acid} (recorded with the INEPT pulse sequence) [14]. On the basis of this effect it is possible to distinguish alternating sequences from mixtures of the corresponding homopolymers or random copolyamides. However, the ^{15}N NMR sequence analysis is generally not suitable for all types of copolyamides. For example only two or three broad signals (linewidths in the range from 30–60 Hz) instead of the expected four sharp signals were observed in nylon-m/-n type copolyamides obtained by anionic copolymerization of lactams. The broad signals arise from unresolved long range neighboring residual effects (up to 2–3 ppm), i.e. from monomer units not directly linked to the observed nitrogen [4, 14].

2.1.2 Synthetic Polypeptides

The sequence analysis of synthetic polypeptides (which are used as models for proteins) by ^{13}C NMR techniques is not successful except for some alanine based copolypeptides and some homopolypeptides of β-amino acids. Although the ^{13}C NMR is more suitable for quantitative analysis, the main advantage of ^{15}N NMR is the high sensitivity of ^{15}N nuclei to structural and conformational changes of the macromolecules as well as to solvent effects.

One approach to the detection of neighboring residue effects on polypeptide acid$_1$–acid$_1$ type bonds is based on isotopic enrichment of one of the acid units. Different chemical shifts of the two glycine units, dependent on the nature of X (X = alanine, valine and other aminoacids), indicate a strong direct neighboring residue effect. Following this model a number of polymers containing X-glycine with ^{15}N enriched glycine units were analyzed (15, 16).

A second method for assigning the A–B and B–A bonds in copolypeptides is by the so-called guest–host polypeptides which are prepared from a 1:50 ratio of ^{15}N enriched monomer A*: natural abundance monomer B. Chemical shifts of the ^{15}N enriched 'guest-monomers' (A*) flanked by blocks of B monomers represent the B–A bonds in copolypeptides. Numerous copolypeptides having the general structure (B_n)-A*-(B_m) were synthesized and analyzed as models for syndiotactic sequences in the spectra of poly (D,L-amino acids) (4, 17).

The sequence analysis, tacticity, primary and secondary structure, influences of (a) cis-trans isomerization, (b) neighboring residue effects, and (c) type and concentration of solvent on chemical shifts for a number of synthetic polypeptides have been reported by Kricheldorf and coworkers. The investigated polypeptides (excluding homopolyglycine which is usually considered as a polyamide) include various poly-, oligo-, and copolypeptides (10, 13, 18–19, 20–23), and particularly polymers based on glycyl-glycine units, (15, 16, 21, 24) D- and L-amino acids (17, 25–28), and L-lysine and iso-L-lysine (29).

^{15}N NMR was also used to study the reaction site of the α- and ε-nitrogen of lysine in a lysine-formaldehyde-urea polymer. The unchanged resonance position of α-nitrogen, accompanied by the upfield shift of ε-nitrogen of about 7 ppm for lysine hydrochloride and for the lysine-formaldehyde-urea polymer, respectively, indicate that the ε-nitrogen is bound to a carbon derived from formaldehyde and linked to urea. However, the intensity of the α-nitrogen signal in a lysine-formaldehyde-urea polymer is only half that of the ε-nitrogen (the intensity ratio α:ε in a lysine hydrochloride spectrum is approximately 1:1) because the ε-nitrogen bound in the matrix has intramolecular dipole-dipole interactions, whereas the α-nitrogen has only intermolecular interactions which are limited to the polymer. Although the NOE effect arising from different relaxation mechanisms makes quantitative analysis of a broad band proton decoupled spectra difficult, the absence of a resonance for unincorporated ε-nitrogen in the polymer suggests that the incorporation of lysine in the polymer is virtually complete (30).

2.1.3 Other Polymers

A downfield shift of ^{15}N signals with the increasing chain length n found in polyamides is also observed in linear homopolyureas, $[-NH-(CH_2)_n-NH-CO-]$ (Table 2). The maximum shift for a polymer with 12 methylene groups is 6 ppm. The corresponding carbonyl ^{13}C signals exhibit an upfield shift with increasing chain length with approximately four times smaller shifts effects ($\Delta\delta_{max} = 1.57$ ppm). The nitrogen groups in copolyureas, $[-NH-(CH_2)_m-NH-CO-NH-(CH_2)_n-NH-CO-]$, are sensitive to neighboring residue effects, i.e. the monomer unit, which influences chemical shifts of two directly attached nitrogens, also influences the shifts of neighboring residues. For a series of copolyureas based on diaminohexane (m = 6) the chemical shifts of nitrogen attached to the $-(CH_2)_6-$ unit are shifted downfield from -293.5 to -286.5 ppm (external $NH_4NO_3 = 0.0$ ppm) with increasing n (the signals are slightly shifted upfield compared to the corresponding homopolyureas). The influence of the $-(CH_2)_6-$ units on the nitrogen attached to $-(CH_2)_n-$ units results in their slight upfield shift from -285.5 to -287.4, with n increasing from 2 to 12, respectively (31). Nitrogen in linear polyurethanes shows similar downfield shifts with increasing chain length of the diamine unit (Table 2; polyurethane A). The downfield shift/chain length relationship is a consequence of both direct substituent effects of the methylene groups in the aliphatic chains and the protonation of urethane groups which become more basic with increasing chain length of the diamine unit. The effect of increased diol chain length on the diamine nitrogen is similar to the neighboring residue effect in linear polyureas ($\Delta\delta_{max}$ max = 1.8 ppm), but it is opposite in direction (Table 2; polyurethane B). However, chemical shifts of urethane groups in polymers based on aromatic diisocyanates are independent of the chain length of the diol units (32).

Contrary to the high sensitivity of ^{15}N chemical shifts to tacticity in polypeptides, the nitrile residues in polymethacrylonitrile prepared by radical polymerization of unlabelled monomer show only a single resonance signal. The use of ^{15}N enriched azoisobutyronitrile (AIBN) in copolymers of styrene with methyl methacrylate resulted in a broad unresolved multiplet at 244.9 ppm (in $CDCl_3$ solution), intermediate between the signal positions for the two homopolymers. The fusion of the signal from initiator fragments attached to the two types of monomer unit indicates that the use of ^{15}N enriched initiator does not seem to be a promising method for detailed examination of initiator fragments in polymers and copolymers (33).

The natural abundance ^{15}N NMR spectra of urea-formaldehyde and melamine-formaldehyde adducts and resins (recorded with approximately 20 000 pulses) can be used for direct determination of different types of amino groups. The spectra of a urea-formaldehyde resin (1:1 formaldehyde/urea ratio) after 15 and 60 min of reaction are shown in Fig. 3. On the basis of chemical shifts of urea based model compounds (monomethylol-, N,N'-dimethylol- and N,N,N'-trimethylol-ureas) the following assignment of nine ^{15}N resonances is proposed

Table 2. ^{13}C and ^{15}N chemical shifts, δ_2 and $\Delta\delta_{n,2}$ (ppm), of polyureas and polyurethanes in trifluoroacetic acid[a]

		δ_2 $n=2$	$\Delta\delta_{3,2}$ $n=3$	$\Delta\delta_{4,2}$ $n=4$	$\Delta\delta_{6,2}$ $n=6$	$\Delta\delta_{8,2}$ $n=8$	$\Delta\delta_{12,2}$ $n=12$
Linear polyurea							
$-NH-(CH_2)_n-NH-CO-$	^{15}N	$-292.6 = 0.0$	-3.2	-5.0	-5.6	-5.8	-6.0
$-NH-(CH_2)_n-NH-CO-$	^{13}C	$160.89 = 0.00$	0.62	1.03	1.39		1.57
Alternating linear copolyurea							
$-NH-(CH_2)_n-NH-CO-NH-(CH_2)_6-NH-CO-$	^{15}N	$-293.5 = 0.0$	-3.5	-5.2		-6.9	-7.0
$-NH-(CH_2)_n-NH-CO-NH-(CH_2)_6-NH-CO-$	^{15}N	$-285.5 = 0.0$	1.0	1.2		1.7	1.9
$-NH-(CH_2)_n-NH-CO-NH-(CH_2)_6-NH-CO-$	^{13}C	$160.24 = 0.00$	0.43	0.59		0.80	0.86
Linear polyurethane A							
$-NH-(CH_2)_n-NH-CO-O-(CH_2)_2-O-CO-$	^{15}N	$-293.5 = 0.0$	-3.8	-5.6	-6.8	-7.6	-7.7
$-NH-(CH_2)_n-NH-CO-O-(CH_2)_2-O-CO-$	^{13}C	$159.70 = 0.00$	0.03	0.00	0.21	0.37	0.30
Linear polyurethane B							
$-NH-(CH_2)_6-NH-CO-O-(CH_2)_n-O-CO-$	^{15}N	$-286.7 = 0.0$	-0.7	-1.1	-1.6	-1.7	-1.8
$-NH-(CH_2)_6-NH-CO-O-(CH_2)_n-O-CO-$	^{13}C	$159.91 = 0.00$	0.51	0.82	1.09		1.30

[a] For polymers with alkyl chain length $n = 2$ ^{13}C chemical shifts (δ_2) are relative to external TMS and ^{15}N chemical shifts (δ_2) are relative to external NO_3^-. For polymers with $n > 2$ chemical shifts are relative to $\delta_2 = 0.0$ (adapted from [31, 32])

(chemical shift scale is calibrated to NH_4^+ in ammonium nitrate):

$HOCH_2NHCONH_2$	80.3 ppm
$HOCH_2NHCONHCH_2OH$	80.3 ppm
$HOCH_2NHCONH-$	79.8; 79.2 ppm
$NH_2CONHCH_2NHCONH_2$	73.5 ppm
$-NHCH_2NH-$	73.0 ppm
$-NHCH_2OCH_2NH-$	72.4 ppm
$NH_2CONHCH_2OH$	55.3 ppm
NH_2CON-	54.8; 54.2 ppm

As the reaction proceeds the intensity of the group of signals around 55 ppm, associated with the primary amino nitrogen in monomers and dimers and endgroups in oligomers, decreases. The signal at 55.3 ppm arising from urea or monomethylol-urea disappears most rapidly. The three signals around 80 ppm attributed to methylol-substituted secondary amines also decrease in intensity with the reaction time. Analogously to the group of primary nitrogen resonances, the highest reduction in signal intensity is associated with the mono- and di-methylol-urea (80.3 ppm), and indicates a loss of low-molecular weight components with reaction time. The relative increase in intensity for three signals at around 72 ppm, arising from secondary amino nitrogen adjacent to the methylene and methylene ether linkages, is a direct measure of chain extension during the polycondensation. Furthermore, the sensitivity of the nitrogen resonances to the neighboring group effects makes it possible to distinguish $-NH-CH_2-NH-$ signals near chain ends (73.5 ppm) from those within the chain (73.0 ppm). This is not possible by ^{13}C NMR spectroscopy. According to the proposed assignments it can be concluded that the methylene linkages are formed to a greater extent compared to the methylene ether linkages (highest field component). The absence of the expected resonances in the 93–100 ppm range arising from tertiary nitrogen atoms suggests that chain branching does not take place in the low molecular weight resins. A similar analysis of a melamine-formaldehyde system (formaldehyde/melamine ratio 2.3:1) also confirmed the absence of tertiary nitrogen atoms in resins cured for up to 45 min. In addition, the melamine ring nitrogens are found to be sensitive to the amine substituents through three bonds (i.e. the hydroxymethylation of melamine amines causes a 1.3 ppm upfield shift of melamine triazine nitrogen resonances (34)).

Phenolic polymers based on p-benzoquinone and amino compounds are proposed as models for humic acids. The ^{15}N NMR spectra of a polymer based on p-benzoquinone and ^{15}N enriched ammonium chloride (molar ratio 1:1.47) exhibit seven signals arising from a variety of nitrogen types incorporated in the polymer structure. However, besides the most intense high field signal, which is assigned to the NH_4^+ (a pentate in the undecoupled spectrum), and a signal in the aniline absorption range, the origin of the five resonances in the low field range is not clear. These signals could arise from polymer nitrogen forms such as amines, lactames, pyrrole and indole-like groups incorporated by formation of covalent bonds between amino nitrogen and aromatic carbon (35).

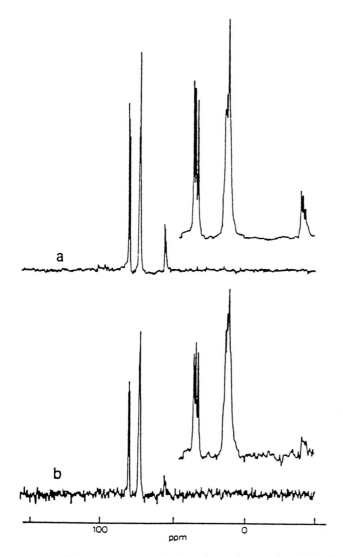

Fig. 3a,b. ^{15}N NMR spectra (with expansions) of a urea-formaldehyde resin made with a 1:1 formaldehyde/urea ratio; **a** after 15 min reaction; **b** after 1 h reaction. Spectra were referenced with respect to an external standard consisting of the NH_4^+ ion in a solution of ammonium nitrate (in d_6-DMSO) (reprinted from [34])

The formation of cyclic structures and polymerization of aryl cyanates was followed from the changes in signal intensities of the reaction products. All intermediate spectra consist of two well separated signals arising from cyanate and triazine groups, respectively. Since a chromium acetylacetonate is added in order to eliminate the Nuclear Overhauser Effect and reduce the spin-lattice relaxation time, the signal intensities are directly proportional to the concentration of groups present in the mixture. It was found that the degrees of conversion of monofunctional and difunctional cyanates to the corresponding triazines, determined from differential scanning calorimetry, ^{15}N NMR, ^{13}C NMR, and Fourier transform infrared spectroscopy, are in good agreement (36). However, the addition of relaxation agents for shortening the T_1 relaxation time of tertiary nitrogen causes line broadening of the faster relaxing protonated nitrogen, if they are present in the same molecule (depending on the paramagnetic ion concentration) (11).

The mechanism of the curing reaction of cyanate polymer resins based on bisphenol A dicyanate (BPACN) has been investigated by both ^{13}C and ^{15}N NMR techniques. The ^{15}N NMR spectrum of soluble species after heating of a ^{15}N enriched BPACN exhibits only two new nitrogen signals associated with the triazine ring and amide nitrogen, respectively. This clearly indicates that (a) there are no substantial amounts of any long-lived dimeric or other intermediate species prior to triazine ring formation and (b) the side products (of the general form $R-O-CONH_2$) are formed by the reaction of the cyanate functionalities with trace water present in the solvent (37).

Natural abundance ^{15}N NMR spectra of the commercial poly(ethylene imine) in alkaline H_2O solution (collected with 60 000 45° pulses) reveal one, two, and three signals in the primary, secondary, and tertiary amine resonance regions, respectively. From the signal intensities it can be concluded that approximately every third nitrogen in the polymer backbone belongs to a tertiary amine group. This result indicates that the polymer is highly branched. However, an exact calculation of the degree of branching is not possible because of different Nuclear Overhauser Effects on the intensities of primary, secondary and tertiary amine signals obtained without inverse gated proton decoupling (38).

The effect of pH on the nitrogen charge density of poly(ethylene imine) based polymers is shown in Fig. 4. Although a detailed ^{15}N chemical shift analysis was not made it is obvious that the increasing acidity of the solution removes almost all the resonances arising from nonprotonated nitrogen-containing species. Only the signals of protonated secondary and tertiary nitrogen can be observed (Fig. 4b) (39).

Detection of tacticity in D_2O solutions of poly (vinylamine) depends on the configurational sensitivity to the solution acidity. Based on the results obtained from ^{13}C NMR data the six well resolved peaks observed at pH 10.5 are assigned at the triad level to the mm (18%-1 signal), mr (56%-2 signals) and rr (26%-3 signals) triads. It should be noted that the three high field signals associated with the rr triads arise from the pentad sensitivity of ^{15}N NMR compared to the ^{13}C spectra. However, the expected three and four signals for mm and rm pentads, respectively, are not observed. The low pH range (pH 2-8) is not suitable for configurational assignments since a single signal shifted to a higher field with increasing acidity is observed, while the rr and mr triads are clearly resolved only at pH > 9. The ^{15}N NMR titration (protonation) curves over a wide pH range (2.2 to 12.6) are fitted assuming a two-stage titration process with pK_1 about 5.2

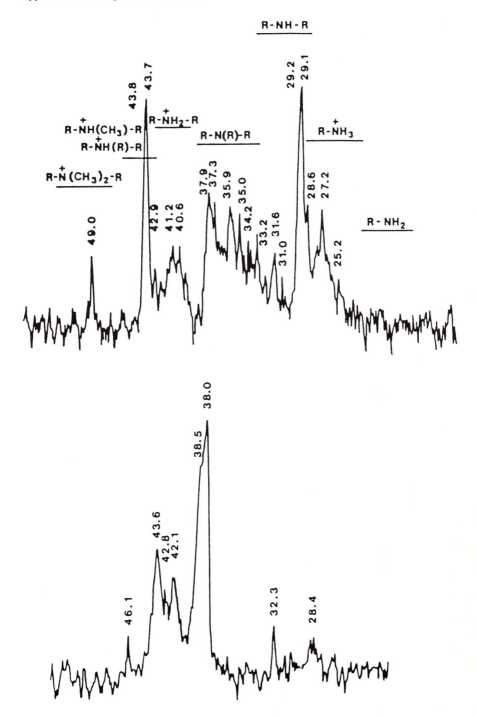

Fig. 4a, b. ^{15}N NMR spectrum of poly(ethylene imine); **a** at pH = 7.3; **b** pH = 1 (adapted from [39])

and pK_2 about 10. A model proposed for this titration process involves a conformation change of the polymer mediated by hydrogen bond formation near half neutralization (8, 40).

The titration of natural abundance ^{15}N polyelectrolytes based on poly (amine adipamides) with HCl in the pH range from 13 to 1 was also studied. For a polymer based on diethylene triamine the protonation results in a downfield amine shift of 9.7 ppm and an upfield amide shift of 4.4 ppm (the point of inflection is at pH 7.9). The unexpected observed upfield shift of amide signal with increasing protonation could arise from an electric field effect combined with a solvent effect. These shifts allow one to distinguish substitutions involving exchange of electric charges from other substitutions without the isolation of the polymers. Other reactions such as alkylation and acylation of polyaminamides were carried out, and the structures of the resulting polyelectrolytes were determined from the NMR spectra (7).

The sensitivity of ^{15}N chemical shifts to the polymer chain stereoregularity is observed for the succinimide unit in ^{15}N enriched alkene/N-methylmaleimide copolymers. The ^{15}N NMR spectra of both ethene and propene copolymers exhibit two signals separated by 2.5 ppm. By comparing chemical shifts with those of model compounds the most intense signal at lower field is assigned to the *trans*-succinimide units, while the smaller signal at higher fields is assigned to the *cis*-succinimide units. In addition, a slight splitting of the trans signal of the propene copolymer is explained by the sensitivity of the ^{15}N chemical shift to the different chirality (tacticity) of the tertiary propyl carbon three bonds removed from the nitrogen. The same assignment could be applied to the spectra of styrene copolymers which show two signals of nearly the same intensity. However, the two equally intense signals observed in the spectra of a styrene copolymer prepared by the polymer-analogous reaction cannot be assigned to the trans and cis methylmaleimide units. Since these polymers have a strong tendency to epimerize by forming trans units, both signals are assigned to the trans configuration of the succinimide unit; they are split by a random tacticity of the methine carbon of the styryl unit (41). Four nuclei (^1H, ^{13}C, ^{15}N, and ^{29}Si) are investigated by NMR for the characterization of a new polymer synthesized by the ring opening polymerization of cyclosilazanes. All spectra confirmed a linear polymer structure, with the absence of significant branching on nitrogen and silicon atoms or through the vinyl groups. The single resonance in ^{15}N NMR spectra of the polymer with chain structure $(-Si(CH_3)_2-N(CH_3)-)_n$ is observed 2 ppm downfield from the signal of the polymer with chain structure $(-Si(CH_3)(vinyl)-N(CH_3)-Si(CH_3)_2-N(CH_3)-)_n$ (42).

The ^{15}N NMR spectra of bisphenol-A-diglycidylether (DGEBA)-aniline addition polymers and their acetyl derivatives in THF (using ^{15}N enriched aniline) exhibit only a single signal, attributed to the tertiary phenylamino group (NR_3). The observed splitting in some samples (0.3 ppm) indicates the sensitivity of nitrogen nuclei to diastereomers. From the signal intensities it can be concluded that both erythro- and threo- aminodiol units are formed in nearly equal concentrations. Telechelic polymers synthesized with an excess of aniline show a high concentration of secondary phenylamino endgroups (NHR_2), which absorb approximately 1 ppm downfield from the NH_3 groups (43).

2.2 Spin-Lattice and Spin-Spin Relaxation

The main mechanism for the relaxation of ^{15}N nuclei with directly bonded protons is the $^{15}N-^1H$ dipolar interaction. Under the extreme spectral narrowing condition, $(\omega_N + \omega_H) \tau_C \ll 1$, where ω_N and ω_H are corresponding resonance frequencies, the nuclear magnetic relaxation is given by:

$$\frac{1}{T_1} = \frac{\gamma_H^2 \gamma_N^2 h^2}{r_{NH}^6} \tau_c \qquad (1)$$

where γ_H and γ_N are the ^1H and ^{15}N gyromagnetic ratios, n is the number of protons attached to the nitrogen atom, r_{NH} is the nitrogen–proton internuclear distance, and τ_c is the effective correlation time.

If dipolar relaxation is the main relaxation mechanism, the ratio of ^{15}N and ^{13}C relaxation times under the conditions of extreme spectral narrowing provides information on the relative motions of these substituents.

With the assumption that similar (isotropic) motions cause relaxation of these nuclei, the ratio $^{15}N/^{13}C$ nT_1 should be 4.4 (for bond distances r_{NH} and r_{CH} of 1.03 Å and 1.09 Å, respectively). The ratios of dipolar spin-lattice relaxation in aqueous solution of poly (ethyleneimine) are in the range from 2.6 to 3.6 indicating that the motions of backbone NH (or NH_2^+) groups are restricted. This is a result of intramolecular interactions and hydrogen bonding to the solvent. The motion of amine side-groups in poly (vinylamine) strongly depends on the acidity of solution. It can be estimated that the average number of protons attached to the nitrogen decreases from NH_3 at low pH values (2.8–3.8) to $NH_{2.5}$ at pH 6.7–8.6, and NH_2 at pH = 10.1. Using these values of n the ratio nT_1 at low pH is about 16, which is significantly higher than the expected value. With decreasing fraction of protonated amine the ratio decreases sharply to about 6 at neutral pH and reaches a value close to the value predicted at pH 10.1 ('rigid' NH_2 group). A large amount of amine group rotation in strong acidic media (20% methane-sulfonic acid) is also confirmed with a low energy barrier (less than 11.7 kJ/mol compared to 16.3 kJ/mol at pH 2.8) as calculated by assuming that the rotation is between three equivalent positions (44).

^{15}N relaxation is particularly useful for studies of molecular motion in polypeptides when the relaxation of the carbonyl carbon cannot provide direct information on molecular dynamics. Since carbonyl carbons do not have directly attached protons, their relaxation can be governed by other relaxation mechanisms, and the interpretation of relaxation data assuming dipolar relaxation may not be appropriate. The effect of γ-irradiation on a hydrated polypentapeptide (PPP) is studied in the 10–70 °C temperature range. A single resonance observed in both uncrosslinked and crosslinked PPP arises from ^{15}N enriched glycine (glycine[5]) in a pentapeptide repeating unit: valine-proline-glycine[3]-

valine-glycine[5]. Although the spin-spin (T_2) relaxation time for a typical polymer increases with increasing temperature, or reaches a constant 'plateau' value in the case of crosslinked polymers (45), a broadening of the lines (decrease of T_2) with increasing temperature is observed for both samples in the 10–70 °C range ('inverse temperature transition'). However, the broader line widths for irradiated PPP compared to the nonirradiated sample in this temperature range indicate that the motional restrictions in the former are introduced by crosslinking. The rotational correlation times, τ_c, calculated from the T_1 data (using Eq. 1) and from the T_2 data are presented in Fig. 5. In all cases the mobility increases (τ_c decreases) on raising the temperature from 10 to 20 °C; further increase of temperature to 40 or 50 °C results in decreasing mobility, while increased mobility resumes on raising the temperature above 50 °C. This defines an inverse temperature transition in the 25–40 °C temperature range. The similar slopes of τ_c vs $1/T$ (calculated from the T_1 data) for both samples before and after the transition yield similar activation energies of about 5.8 kJ/mol, expected for motional processes giving rise to entropic elasticity (46).

2.3 Cross Polarization

The relatively low sensitivity of natural abundance ^{15}N nuclei can be enhanced by using the cross polarization technique (47). The main advantage of this technique is that signal enhancement depends only on the ratio of the gyromagnetic ratio magnitudes (γ_H/γ_M) and not on their signs. Furthermore, the pulse repetition delays are governed by proton T_1 relaxation times, which are shorter than the nitrogen T_1 relaxation times. The combined effect of cross polarization and shorter recycle delay results in a 100-times shorter experiment for a 10-times smaller sample compared to the standard FT experiment without NOE. Since the magnetization is coherently exchanged between ^{15}N nuclei and coupled protons with a frequency comparable to the coupling constant, J_{NH}, (contrary to the solids) the cross polarization technique in liquids is called J cross polarization (JCP). The application of this technique to nylons (polyamides and copolyamides) confirmed the deshielding of ^{15}N nuclei with both increasing solvent acidity and decreasing polymer concentration. It was found that the degree of polymerization has little effect on nitrogen chemical shifts at lower nylon-66 concentrations. However, at higher nylon concentrations the degree of polymerization affects the extent of protonation and hydrogen bonding which results in increased shielding at the amide ^{15}N nuclei. In the case of copolyamides a larger shift in chemical shift vs temperature dependence is observed only for a nylon-6,12 component of the copolyimide with nylon-6 between 49 and 56 °C, ($\Delta\delta$ = 0.6 ppm). Since the data observed after a high temperature measurement were in good agreement with the data collected before elevation of the temperature to 56 °C a relatively large shift may indicate a conformational change rather than a degradation (48, 49).

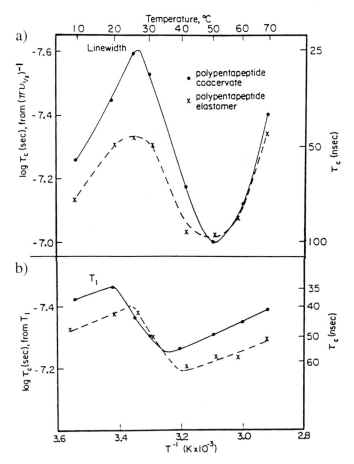

Fig. 5a, b. Plots of (a) linewidth and (b) spin-lattice relaxation time, T_1 vs the inverse temperature for polypentapeptides (adapted from [46])

3 ^{15}N NMR in Solid State

Although the same basic principles of solid state NMR work as well for ^{15}N as for ^{13}C nuclei, the lower sensitivity of nitrogen nuclei is the major problem in natural abundance experiments of complex polymer systems, particularly for polypeptides and biologically interesting polymers where nitrogen content is generally between 5% and 10%. The natural abundance spectra involve accumulations of several thousand to several hundred thousand scans for satisfactory signal to noise ratios. Thus, the use of ^{15}N enriched samples is required in order to achieve a reasonable experimental time required for a good quality spectrum. As an example, the spectra of commercially available and ^{15}N enriched nylon-12 are shown in Fig. 6. Approximately 27 000 transients (23 h) were needed for acceptable signal to noise at natural abundance levels (spectra A, B and C), while only 256 transients (approximately 13 min) were required for a very good signal to noise ratio for ^{15}N enriched samples (spectra A', B' and C') (50).

A number of polymers have been studied by various ^{15}N NMR techniques in the solid state. Although the low natural abundance of ^{15}N is a major disadvantage of nitrogen compared to the carbon nuclear magnetic resonance, in many cases the ^{15}N NMR of ^{15}N enriched samples in the solid state provides information on molecular structure and dynamics which cannot be obtained from ^{13}C NMR spectra. An example is the hexamethylenetetramine curing of phenolic resins. The relative simplicity of nitrogen compared to the carbon NMR spectra of isocyanate based resins also allows a quantitative structural analysis of these polymer systems. However, in some cases (such as urea-formaldehyde and melamine-formaldehyde resins) ^{15}N NMR does not provide a straightforward approach to the structural analysis, and ^{13}C NMR spectra are more suitable for the analysis of these polymers. The sensitivity of particular nitrogen and carbon atoms to structural changes differs from polymer to polymer. For example, the carbonyl carbons in ^{13}C NMR spectra of nylons are nearly insensitive to the crystal and amorphous forms (173.0, 173.4 and 174.5 ppm for α, γ and amorphous phase, respectively), while the two signals in ^{15}N NMR spectra of nylon-6 and nylon-12 (see Fig. 6A[1]) arising from γ and α crystal forms are clearly separated by approximately 5 ppm.

There is still no unique standard for the calibration of ^{15}N NMR spectra. Six solid standards as well as liquid ammonia are frequently used as external references for the calibration of nitrogen shieldings in the solid state. Although chemical shifts in solids do not depend on the temperature and concentration (as the chemical shifts of liquid standards) there are discrepancies in the literature data on the exact chemical shifts for particular standards. In order to avoid errors created by conversion of literature data from different chemical shift scales, precise resonance positions of chemical shift standards are required. Since the difference in the resonance frequency between proton decoupled spectra

Table 3. [15]N chemical shifts for common standards

Standard	δ ($CH_3NO_2 = 0$) ppm	δ ($NH_3 = 0$) ppm
NH_3 (liquid)	− 380.0	0.0
NH_4NO_3 (solid)	− 358.4	21.6
$(NH_4)_2SO_4$ (solid)	− 355.8	24.2
Glycine (solid)	− 347.5	32.5
NH_4Cl (solid)	− 341.2	38.8
$HCONH_2$ (solid)	− 266.7	113.3
CH_3NO_2 (liquid)	0.0	380.0

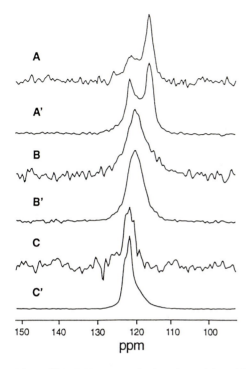

Fig. 6. ^{15}N NMR spectra of nylon-12: precipitated (α plus γ forms; A and A′); quenched (γ' form; B and B′) and annealed (γ form; C and C′). Commercial samples are shown in spectra A,B and C; ^{15}N enriched samples are shown in spectra A′, B′ and C′ (adapted from [50])

fluctuates up to 0.20 ppm and the difference between the coupled spectra of two standards are always constant, the centers of coupled spectra should be taken as the shift standard. Table 3 displays chemical shifts for common standards precisely determined from the resonance position of the magic angle spinning spectra of liquid nitromethane (51). Chemical shifts of nitrogen absorptions in solids are relatively constant compared to those in solutions. For example, amide group in aliphatic polyamides in solutions is shifted by approximately 30 ppm downfield of the solid resonances because of the strong interactions of nitrogen groups with solvent. Chemical shifts of typical nitrogen containing groups in various solid polymers are presented in Table 4. In order to compare the data presented in the literature using various chemical shift scales, all the data are converted to the liquid NH_3 chemical shifts scale ($^{15}NH_3$, 0 ppm) according to Table 3. The advantage of NH_3 over the nitromethane scale is that all the nitrogen resonances found in polymers are (de)shielded (chemical shifts are positive relative to ammonium signal), analogously to the ^{1}H and ^{13}C chemical shift scale relative to TMS.

Table 4. ^{15}N chemical shifts for polymers in solid state

$\delta(NH_3 = 0)$ ppm	Group	Polymer
46	$-C_6H_4-N=C=O$	MDI-polyisocyanurate resins: endgroups[77]
51.6	$-C_6H_4-NH_2$	polyaniline hydrochloride: endgroups[68]
53	$-C_6H_4-NH_2$	MDI-polyisocyanurate resins: endgroups[77]
58.1–60.1	$-C_6H_4-NH_2$	polyaniline: endgroups[68]
~73	$-C_6H_4-OCONH_2$	cyanate resins: amide endgroups[37]
78.4	$-C_6H_4-NH-C_6H_4-$	polyaniline: homopolymeric diads[66]
85.1	$-C_6H_4-NH-C_6H_4-$	polyaniline hydrochloride[68]
86.5	$-C_6H_4-NH-C_6H_4-$	polyaniline: copolymeric diads[66]
98.7	$> C_3N_3-NH-CH_2OR$	melamine-formaldehyde resins[53]
89.5–90.6	$-C_6H_4-NH-C_6H_4-$	polyaniline (low and high molecular weights)[68]
104	$-C_6H_4-NH-CO-NH-C_6H_4-$	MDI-polyisocyanurate resins: urea linkages[77]
106.3	$-NH-CH_2-CO-$	polyglycine (nylon-2, (β)helix-form)[79]
106.3	$> C_3N_3-N(CH_2OCH_3)CH_2OR$	melamine-formaldehyde resins[53]
110.8	$-NH-CH_2-CO-$	polyglycine (nylon-2, (α)helix-form)[79]
113.6; 114.2	$> C_3N_3-N(CH_2OR)_2$	melamine-formaldehyde resins[53]
116.5	$-NH-(CH_2)_n-CO-$	nylon-6 (n = 5), nylon-11 (n = 10) and nylon-12 (n = 11) (α form)[50, 58, 64]
117.5	$-NH-(CH_2)_3-CO-$	nylon-4 (α form)[56]
119.1	$-NH(CH_2)_n-CO-$	nylon-6 (n = 5) and nylon-12 (n = 11) (amorphous phase)[50,58]
119.8	$-NH-(CH_2)_5-CO-$	3-arm star polycaproamide (nylon-6)[54]
119.8	$-NH-(CH_2)_{12}-CO-$	nylon-13 (crystal form 'B')[61]
120.4	$-NH-(CH_2)_2-CO-$	poly(β-alanine) (nylon-3, α-phase)[79]
121.0; 121.2	$-NH-(CH_2)_{12}-CO-$	nylon-13 (crystal form 'A')[61]
121.3	$-NH-(CH_2)_{11}-CO-$	nylon-12 (γ form)[50]
121.4	$-NH-CH(CH_3)-CO-$	poly(α-alanine) ((α)helix form)[79]
121.8	$-NH-(CH_2)_n-CO-$	nylon-6 (n = 5) and nylon-12 (n = 11) (γ form)[50, 58]
123.2	$-NH(CH_2)_5-CO-$	nylon-6/I_2/KI complex[60]
126.6	$-C_6H_4=N\cdot HCl-C_6H_4-$	polyaniline hydrochloride[68]
128.2	$-NH-C_6H_4-CO-$	poly(p-benzamide)[54]
130.2	$-NH-CH(CH_3)-CO-$	poly(α-alanine) ((β)helix form)[79]
141	$-C_6H_4-NH-CO-N(C_6H_5)-$ $CO-NH-C_6H_4-$	MDI-polyisocyanurate resins: biuret linkages[77]
145	$-C_6H_4-N < ^{CO}_{CO} > N-C_6H_4-$	MDI-polyisocyanurate resins: uretidione linkages[77]
145.5	$-C_4H_2N-C_4H_2N-$	polypyrrole: imino aromatic form (deconvoluted)[71]
150	$-C_6H_4-N < ^{CO-N <}_{CO-N <} ^{C_6H_4-}_{^{CO}_{C_6H_4-}}$	MDI-polyisocyanurate resins: isocyanurate linkages[77]
158.8	$-C_4H_2NC_4H_2N-$	polypyrrole and related polymers containing pyrrole ring (imino form) (broad unresolved signal)[69, 70]
161.5	$-C_4H_2N=C_4H_2N-$	polypyrrole: imino quinoid form (deconvoluted)[71]
171.6	$-HN-C_3N_3 <$	melamine formaldehyde resins: triazine ring nitrogen in the vicinity of primary amines[74]
177.4	$> N-C_3N_3 <$	melamine formaldehyde resins: triazine ring nitrogen in the vicinity of secondary amines[53]
~200.3	$-O-C_3N_3 < ^{O-}_{O-}$	cyanate resins: triazine ring nitrogen[37]
258.8	$-C_4H_2N-CH=C_4H_2N-$	polypyrrilenemethine and polyfurylenepyrrylenemethine[70]
326–357	$-C_6H_4-N=C_6H_4-$	polyaniline: homo- and co-polymeric diads[66]
350–351.1	$-C_6H_4-N=C_6H_4-$	polyaniline (low and high molecular weights)[68]
358.8	$-N=CH-CH=N-$	polyazine[65]
363.6	$-N=C(CH_3)-C(CH_3)=N$	polyazine[65]

3.1 Gated High Power Decoupling

The gated high power decoupling (GHPD) sequence involves a 90° nitrogen pulse and the observation of the FID under the conditions of high power proton decoupling. Direct polarization of observed nuclei is useful when the spin-lattice relaxation time does not exceed a few seconds and the nuclei under observation are highly mobile. For example, direct application of a 90° nitrogen pulse to the hexamethylenetetramine cured phenolic resins results in a spectrum revealing only a sharp line at 38 ppm, which confirmed its origin in a highly mobile ammonium ion present in the cured resin (52). The application of both direct and cross-polarization techniques to hexamethylmethoxylated melamine-formaldehyde resins enables the detection of two motionally different portions in the sample (Fig. 7). Highly cured samples exhibit only two broad signals arising from nonprotonated and protonated amines, respectively, indicating that motionally mobile and rigid parts of crosslinked resins have similar chemical structure. Only the spectra of uncured resin at elevated temperatures exhibit sufficiently narrow ^{15}N signals to allow the detection of amine nitrogens in slightly different chemical moieties. However, besides the similar chemical structure, relaxation measurements based on direct and cross-polarization pulse sequences clearly show the difference in molecular dynamics for the two regions (53).

Since the ^{15}N T_1 relaxation times in solids are in the 10^0–10^3 s range, this technique is generally not suitable for analysis of solid polymers. However, the GHPD pulse sequence can be used for selective saturation of signals with longer T_1 times if the difference in T_1 times between two overlapping signals is more than an order of magnitude and if the shorter T_1 has a value less than a few seconds (54–56). By using the GHPD technique with the recycle delay of 5 s the magnetization of ^{15}N nuclei in the rigid crystalline region of the polymer ($T_{1l} > 100$ s) is quickly saturated and the signals from the faster relaxing amorphous region ($T_{1s} > 1$ s) are observed directly (56).

3.2 Cross-Polarization

Cross-polarization is a double-resonance experiment in which the energy levels of the ^{1}H and ^{15}N spins are matched to the Hartman-Hahn condition in the rotating frame [47]. Under this condition the energy between the two coupled spin systems may be exchanged and the ^{15}N magnetization (S) develops as a function of time at the expense of the proton magnetization according to the relation:

$$\frac{dS}{dt} = \frac{S_0 \exp(-t/T_{1\rho N}) - S}{T_{NH}} - \frac{S}{T_{1\rho N}} \tag{2}$$

where S_0 is the maximum nitrogen polarization which would be obtained in the absence of dissipative processes, $T_{1\rho N}$ and $T_{1\rho H}$ are ^{15}N and ^{1}H relaxation times

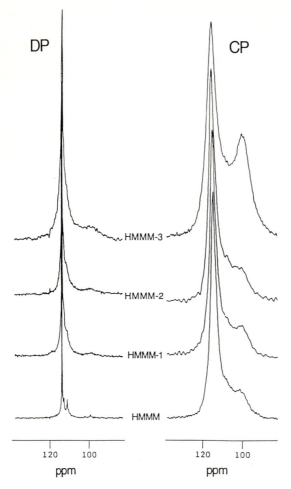

Fig. 7. Gated high power decoupling (DP) and cross-polarization (CP) spectra of ^{15}N enriched hexamethylmethoxylated melamine-formaldehyde resin (HMMM). HMMM-1, HMMM-2 and HMMM-3 are resins cured for 1, 3 and 24 h respectively. DP spectra were recorded at 343 K; CP spectra were recorded at 298 K with a contact time of 2 ms (adapted from [53])

in the rotating frame, respectively, and T_{NH} is the $^1H–^{15}N$ cross-polarization time constant. Due to proton relaxation in the rotating frame only a fraction, S, of the theoretical nitrogen intensity S_0 is observed. The signal intensity is determined by the time constants of two opposing relaxation mechanisms, $T_{1\rho H}$ and T_{NH}. The cross-polarization rate constant is inversely proportional to the sixth power of the $^{15}N–^1H$ internuclear distance. Thus, nitrogens with directly attached protons are expected to cross-polarize more rapidly than nitrogens without direct interactions, if the nitrogens are subjected to the same motions (rapid molecular motion attenuates the cross-polarization mechanism). In general, shorter cross-polarization contact times (less than 1 ms) will optimize the dectection of nitrogens tightly coupled to hydrogens (amine and amide) while the signals of nitrogens bearing no directly attached hydrogens (imide, isocyanate, isocyanurate, tertiary nitrogens etc.) will have the highest intensity at the longer contact times (more than 1 ms). However, strong hydrogen bonding in uncured melamine-formaldehyde resins significantly shortens the transfer of polarization, and observed T_{NH} constants for nonprotonated nitrogens are in the range 0.5–0.8 ms [53].

3.2.1 Polyamides (Nylons)

Aliphatic polyamides are extensively studied by natural abundance ^{15}N NMR spectroscopy in solution. However, characterization of polyamides in solution is limited by the insolubility of many (particularly aromatic) polyamides. On the other hand, chemical shifts of amide nitrogens are strongly dependent on the nature of a solvent, and for a particular polyamide, could cover approximately 20 ppm, as in the case of fluorosulfonic acid and trifluoroethanol (see Fig. 2). Since the important properties of solid polyamides such as crystalline structure and hydrogen bonding cannot be studied by solution spectra, the various solid state ^{15}N NMR techniques have been used for structural and dynamical characterization of these polymeric materials.

Despite a low natural abundance of ^{15}N nuclei, the solid state ^{15}N NMR spectra of ^{15}N nonenriched polyamides recorded with 20 000 to 50 000 scans could yield valuable information on composition and conformation in the solid state. However, if the concentration of particular nitrogens is very low, no signals are observed. In the case of 3-arm star nylon-6 there were no signals arising from less than 1% of trifunctional initiator species. Even a signal from 20% of p-benzamide in block copolymers with nylon-6 is not observed, although at the same time the original signal of a nylon-6 component is shifted 5–6 ppm downfield due to the influence of the p-benzamide unit. On the contrary, an alternating copolymer of nylon-6 and poly(p-benzamide) exhibits two absorptions with equivalent areas, separated by 20.7 ppm (it should be noted that the separation is solution is only about 5 ppm because of solvent effects). The absorption of the aromatic amide component is shifted about 5 ppm downfield from the signal of the corresponding homopolymer, while nearly the same chemical shift of the aliphatic amide signal is in the opposite direction [54].

The resolutions and signal to noise ratios of natural abundance ^{15}N NMR spectra of nylons are sufficient to detect individual absorptions arising from morphologically different phases in nylons. In addition, the NMR data, which give immediate qualitative identification of the type of crystallinity, do not depend as much on size and perfection of the crystalline regions as X-ray analysis [56, 57]. However, the broad unresolved peaks and shoulders on sharp peaks of various annealed nylons can arise from rigid interphases or transition zones between triclinic α and γ regions as well as from the amorphous regions. In order to obtain high-quality spectra suitable for spectral deconvolution the use of ^{15}N enriched samples is required [56, 58, 59]. By using ^{15}N enriched nylon-6 the sharp absorptions arising from α (116.5 ppm) and γ (121.8 ppm) crystalline forms as well as from the amorphous phase (119.1 ppm), can be detected [58]. Deconvolution of the overlapping signals yield the line widths at half heights, 2.4 and 6.3 ppm for α-crystalline and amorphous phases respectively [59]. Detection of the crystalline phase only with the cross-polarization technique is possible when the mobility of amorphous regions is high so cross-polarization is inefficient. The CP spectra of nylon-11 obtained at 388 K exhibit only a signal arising from the pseudohexagonal δ crystalline form (temperature is well above α and δ transition temperature). On the contrary, the slow relaxing nuclei from the crystalline phase will not be detected if a 90° pulse is applied directly to the nitrogen nuclei and the recycle delay is relatively short. Thus, by subtracting the GHPD spectrum from the CP spectrum, which also contains absorptions from more mobile amorphous phase, it is possible to obtain a 'spectrum' from the rigid crystalline α phase only. The chemical shift differences arise from conformational as well as substituent effects. The semi-empirical MO calculations confirm qualitatively the conformational relationship and predict that the nitrogens in the predominant crystal forms of polyamides α and γ are most and least shielded, respectively [56]. The γ form of nylon-6 can also be obtained by desorption of I_2/KI from iodinated nylon-6/I_2/KI complexes. Iodinated nylon-6 film exhibits ^{15}N signal at 123.2 ppm, which is downfield shifted compared to those for α, γ and amorphous nylon-6. This shift, accompanied with the analogous downfield shift of corresponding carbonyl groups in ^{13}C NMR spectra, indicates that hydrogen bonding networks are created upon complexation and that the carbonyl is no longer hydrogen bonded to an –NH– group on a neighboring chain. Parallel X-ray, ^{13}C and, ^{15}N NMR data suggest that the γ form evolves from the complex without going through an identifiable, intermediate, native amorphous phase [60].

The improved sensitivity of ^{15}N enriched nylon-12 samples allows the clear identification of a shoulder appearing at 122.8 ppm (see Fig. 6C'). Although the exact identity is unclear, this peak is assigned on the basis of spin-lattice relaxation data to material epitaxially crystallized on the surface of γ crystallites or to an ordered interphase region. The spectra of quenched samples (Fig. 6B, 6B') are very similar in shape. However, the broader base of natural abundance sample indicates that the residual monomer (lauryllactam) may be acting as a plasticizer and increasing the size of conformational mobility of the amorphous region. The residual monomer is probably responsible for the difference in the relative intensities between natural abundance and ^{15}N enriched samples (Fig. 6A, 6A') [50].

However, two different crystal forms in nylon-13 do not correspond to those reported for any other nylon and they are believed to be unique to this nylon. Moreover, the data obtained from ^{15}N NMR spectra are not consistent with the

X-ray data. The 'A' form, obtained by extended annealing just below the melting temperature and by film casting from 1,1,1,3,3,3-hexafluoroisopropanol, has the X-ray peaks similar to α crystals, while the ^{15}N resonances at 121.0 and 121.2 are very close to that of the γ form in nylon-11 and nylon-12. The 'B' form, developed on precipitating from m-cresol into methanol or film casting from m-cresol, has the X-ray peak positions and overall shape virtually identical to the γ form of nylon-12, while ^{15}N NMR data are contradictory or inconclusive. A very sharp signal at 119.8 ppm is approximately half-way between the values for α and γ crystals, corresponding to both the high temperature δ form and the mesomorphic pseudohexagonal phase observed for nylon-11 above 95 °C (61).

^{15}N NMR spectra of polyamide/polyether block copolymers based on poly(ethylene oxide) and either nylon-6 or nylon-12 are dominated by peaks at approximately 119 ppm, arising from the amorphous nylon phase. In addition, both polymers also exhibit a shoulder. Downfield (about 112 ppm) and upfield (about 117 ppm) shoulders indicate that polyamide phases exist as α (nylon-6) and γ (nylon-12) forms, respectively (62).

Only one chemically distinct nitrogen in the polyamide repeating unit makes ^{15}N NMR superior to ^{13}C NMR in the study of chemical shift anisotropies. Static cross-polarization spectra of nylon-6, and the best fit to the nonaxial powder pattern obtained at 27 °C, are shown in Fig. 8. Calculated values for anisotropic tensor $\sigma_{11} = 37.5$ ppm, $\sigma_{22} = 92.5$ ppm and $\sigma_{33} = 212.5$ ppm yield the isotropic chemical shift σ_{iso} of 114.1 ppm, which is in good agreement with the value of 116.7 obtained through the magic angle spinning spectra. A motionally narrowed amorphous resonance at about 116.5 ppm, which is clearly seen above 100 °C, becomes more intense with increasing temperature. However, the σ_{33} component of the powder pattern becomes smaller with the increase of temperature, and above 115 °C cannot be accurately defined. On the contrary, the σ_{22} and σ_{11} components remain essentially unchanged from the spectrum at room temperature. The retention of nonaxial symmetry and the loss of the σ_{33} component indicate the onset of anisotropic motion or libration associated with the σ_{33} tensor component, which lies nearly parallel to the NH bond in amides. The proposed explanations of the 'loss' of the σ_{33} component include (a) a change or transition in intermolecular hydrogen bonding at elevated temperatures that apparently does not destroy the crystal structure, and (b) a change in cross-polarization efficiency which affects only the intensity of the σ_{33} component (55, 59, 63). The temperature dependence of chemical shift anisotropy spectra for nylon-11 generally exhibit similar behavior as the spectra of nylon-6. However, all three nonaxial shielding tensor components for the crystal region remain unchanged upon approaching the polymer melting point, and the crystalline component shows no evidence of narrowing at the 95 °C α–δ transition. Although previous studies suggested that the α–δ transition involves the onset of rapid hydrogen bond disruption and re-formation within the crystal lattice, the ^{15}N static NMR spectra show that the hydrogen bonded amide units remain conformationally rigid in the crystal lattice despite increasing librational motion (64).

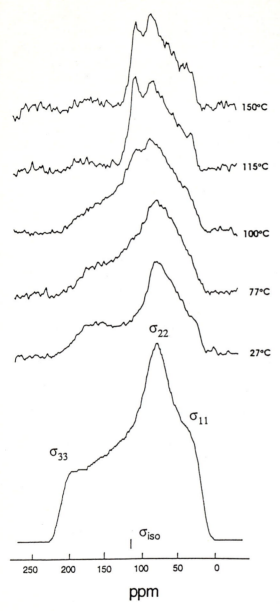

Fig. 8. Static ^{15}N NMR spectra of annealed nylon-6 sample obtained with cross-polarization and high-power decoupling at temperatures indicated. The bottom spectrum shows the best fit of the spectrum recorded at 27 °C (adapted from [59, 63])

3.2.2 Conducting Polymers

Polymers with conjugated π-electron backbones have characteristic electronic properties such as low energy optical transitions, low ionization potentials, and high electron affinity. This class of polymers, which can be oxidized and reduced more easily than conventional polymers, is called electrically conductive polymers. The ^{13}C NMR spectra of conducting aromatic polymers and polymers of heterocycles usually exhibit unresolvable broad aromatic ^{13}C signals arising from slightly magnetically nonequivalent carbons which are difficult to assign. Moreover, from the ^{13}C NMR spectra, it cannot be concluded whether nitrogen atoms are protonated or doped (oxidized). Thus the direct observation of ^{15}N nuclei in conducting polymers can provide valuable information on molecular structure and electrochemical reactivity of these polymers.

Polyazines ($-N = CR_1-CR_2 = N-$) may contain various numbers of (usually) alkyl groups along the polymer backbone. A series of nine polyazines in the range from permethyl ($R_1 = R_2 = CH_3$) to unsubstituted polyazines ($R_1 = R_2 = H$), prepared from corresponding natural abundance ^{15}N α,β-dihydrazones and α,β-dicarbonyls, were investigated by both ^{13}C and ^{15}N NMR in order to determine the defect structures. Some partially methylated as well as unsubstituted polyazines show, besides the imine adsorptions in the 300 ppm range, a broad ^{15}N peak in the region at about 139 ppm. This signal, arising from tertiary nitrogens (including cyclic structures), is associated with the defect structures in the polymer. An additional narrow signal observed in some samples at approximately 113 ppm is attributed to a hydrazone endgroup (65).

The general formula of polyaniline polymers is $-(-A_y -B_{1-y})_x-$ where A is the reduced form (–Ph–NH–Ph–NH–) and B is the oxidized form (–Ph–N = Ph = N–). The relative ratio of A and B units can be controlled through electrochemical modification. A conducting polymer (called emeraldine) consists of an equal number of alternating reduced and oxidized units (y = 0.5). If y $\neq$ 0.5 the chemical shifts of both –NH– and –N = signals are the same as the homopolymeric (imine-imine or amine-amine sequences) and copolymeric (amine-amine) diads. The structure of polymers prepared by various methods and having various relative ratios are studied, but the end groups are not detected (66, 67). However, a signal at about 58 ppm arising from primary amine groups, presumably on the ends of the chains, is detected in some high and low molecular weight polyanilines. The ratios of peak areas for primary and secondary amines at different contact times are similar indicating that the intensity of cross-polarization at the two amine groups is equivalent. Thus, it is possible to calculate the number of repeating units and the molecular weight. With the assumption that polymer chains have only one –NH$_2$ end group (strictly p-linked phenyleneamine-imine units) the calculated molecular weights from NMR data are found to be five times smaller compared to the molecular weights from gel permeation chromatography data (68).

Pyrrole based polymers have potential use as conducting materials because of their higher chemical and thermal stability compared to other commercial conducting polymers. However, since the polymers are amorphous and in-soluble their exact molecular structure is not known and some basic physical properties are not explained on the molecular level. The possible structures of polypyrrole include nonprotonated and protonated nitrogen atoms as well as aromatic and quinoid forms. ^{13}C NMR spectra of polypyrroles exhibit an unresolved broad ^{13}C aromatic absorption arising from signals with slightly different chemical shifts and are not suitable for structural analysis. Various ^{15}N enriched polypyrrole samples prepared electrochemically in different ways (charged and discharged (69) and polypyrrylenemethine (70) show only one broad nitrogen resonance centered at about 158 ppm, which is characteristic of protonated imino nitrogens in pyrrole rings. The polypyrrole broad absorption is decomposed by computer fitting into at least four signals centered at 177.5, 161.5, 145.5 and 122.5 ppm, respectively (Fig. 9). On the basis of shielding constant calculations and using model compounds for aromatic and quinoid polypyrroles, the two major peaks are assigned to nitrogen atoms in quinoid form (γ, 161.5 ppm, doped state) and aromatic form (β, 145.5 ppm, undoped state). The minor peak δ probably arises from nitrogens bonded to exchanging hydrogens, while the origin of the other minor peak α cannot be explained with quantum mechanical calculations on model compounds (71).

Contrary to polypyrrole and polypyrrylenemethine, the spectra of poly-pyrrylfurylmethine exhibit absorptions in two spectral regions. The signal at about 288 ppm (typical for azomethine $-N =$ form) and the two peaks in the

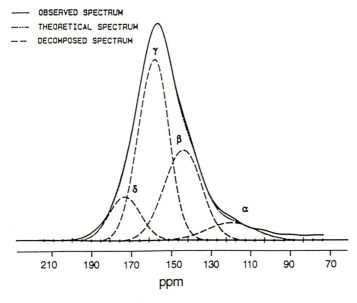

Fig. 9. ^{15}N NMR spectrum (contact time $= 800\,\mu s$) and a simulated spectrum of polypyrrole. The four components were decomposed by computer-fitting (adapted from [71])

158 ppm region (165 and 143 ppm, respectively) indicate the presence of non-protonated and protonated (which probably include doubly protonated pyrrole units) nitrogen atoms. The appearance of a new signal at about 298 ppm after treatment of polypyrrylenemethine with ammonia indicates that protonated nitrogen atoms exist in both –NH– and –NH$^+$= form, the latter being converted by NH_3 to the –N= form. Although the line intensities of CP MAS spectra generally do not correspond to the concentrations of particular chemical structures, an estimate that only 20% (out of theoretically 50%)|of the –NH$^+$= units are converted to –N= units was made on the basis that the increase of cross-polarization time did not really affect the line intensities (70).

A substantial change in the spectra of polymeric ^{15}N enriched μ-oxo-phthalocyaninatosilicon ([PcSiO]$_n$) is observed after doping the polymer with iodine ([(PcSiO)I$_{1.1}$]$_n$). The two signals in the spectra of ([PcSiO]$_n$) arise from outer and inner core nitrogens, respectively. On doping ([PcSiO$_n$]) with iodine the nitrogen signals are shifted between −160 and −180 ppm upfield. Since the polymer is not completely oxidized by doping with iodine, the presence of both unshifted and shifted signals indicates the coexistence of conducting and insulating regions (72).

3.2.3 Other Polymers

The first application of natural-abundance solid state ^{15}N NMR was on the complex solids from the reaction of hydrogen cyanide and ammonia. Signals in the solid state spectrum of the cold water soluble fraction are attributed to the NH_4^+ salt, amine-, urea-, peptide-, nitrile-, and pyrrole-like nitrogens. The spectrum of the cold water insoluble fraction shows two broad resonances indicating a diversity of nitrogen functionality, possibly including crosslinked aromatics and purine-like compounds (3).

The ^{15}N NMR spectra of cured urea-formaldehyde resins based on ^{15}N enriched urea show only two main signals centered at 114 and 94 ppm, although at least five different signals were expected due to the complexity and variety of the urea-formaldehyde resin structure. The signal at higher field is attributed to the secondary amides, while the broader lower shielding peak, which is more pronounced in highly cured resins, has contributions from at least two absorptions arising from tertiary and secondary amides. Thus, nitrogen spectra seem to be less useful than carbon spectra in providing structural analysis of these resins, although ^{15}N data can provide a useful supplement (73).

^{15}N NMR spectra of natural abundance (74) and ^{15}N enriched (53) melamine-formaldehye resins were reported. The natural abundance spectra of cured resins exhibit three broad resonances arising from the triazine nitrogens, mono- and disubstituted amines, respectively. The ^{15}N spectra of samples isotopically enriched at amine sites show only the two broad amine resonances (see Fig. 7; CP spectra) although ten different $>$ C–NR$_1$R$_2$ structures (four protonated and six nonprotonated nitrogen signals) were theoretically expected with the assumption that R_1 and R_2 are either –H, –CH$_2$OH, –CH$_2$OCH$_3$, or –CH$_2$OCH$_2$– units (53). Analogously to urea-formaldehyde resins, the ^{15}N cross-polarization spectra do not provide a straight-forward structural analysis of melamine based resins (directly polarized nitrogen nuclei in less cured resins at higher temperatures have narrower signals which make possible the assignment of directly observed resonances to the corresponding structural units).

A parallel ^{13}C and ^{15}N NMR study of the hexamethylenetetramine (HMTA) curing of phenolic resins using isotope enriched ^{13}C and ^{15}N HMTA was found to be a powerful method for identification of intermediates and establishing the course of the curing process. Figure 10 shows the ^{13}C (10a) and ^{15}N NMR spectra (10b and 10c) of the networks originating from ^{13}C and ^{15}N enriched HMTA respectively (52). The most dominant ^{13}C signal at lower temperatures

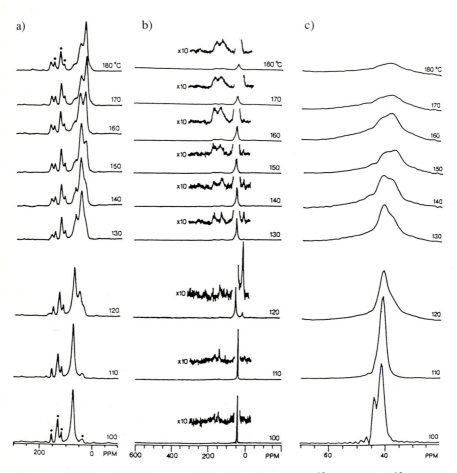

Fig. 10a–c. ^{13}C (**a**) and ^{15}N (**b**) spectra of phenolic resin cured with ^{13}C-labelled and ^{15}N-labelled HMTA, respectively, for 1 h at the temperature indicated. *Asterisks* in the top and bottom spectra (**a**) denote natural-abundance peaks. Spectra (**c**) present expansions of spectra (**b**) (adapted from [52])

(100–120 °C) arising from ^{13}C enriched HMTA (75 ppm) gradually decreases with increasing curing temperature, while the spectral intensity around 33 ppm significantly increases. This indicates that the methylene crosslinks (para-para, para-ortho, and ortho-ortho) formed during the crosslinking originate from the HMTA. The new resonances appearing with increased temperature at 83 and 56 ppm show a decrease in intensity at the later stages of reaction, indicating their origin in a reaction intermediate. The parallel ^{15}N NMR spectra show less dramatic changes compared to the carbon spectra. A broadening and decreasing intensity of the main group of signals around 40 ppm is observed as well as a few additional low intensity nitrogen signals which occur during the reaction. The appearance of a relatively intense signal at 3 ppm in the spectrum of the sample heated at 120 °C is assigned to NH_3 which initially becomes trapped in the resin upon curing and is released at higher temperatures. The main group of signals around 40 ppm at the onset of the cure split into resonances at 44 and 41 ppm, and are attributed to the crystalline and amorphous HMTA involved in hydrogen bonding with phenolic OH groups, respectively. Following the initial curing period at 100 °C the signal at 44 ppm disappears, the signal at 41 ppm decreases, while an additional shoulder at 38 ppm assignable to the NH_4^+ ion, whose intensity is increasing with the increasing curing time, appears. On the basis of both carbon and nitrogen spectra (including the interrupted decoupling spectra) the intermediates are identified as benzoxazine and tribenzylamine type species. However, it should be mentioned that not all of the signals appearing in either nitrogen or carbon spectra have a counterpart in the other spectra. For example, the loss of crystallinity of HMTA results in a 3 ppm downfield shift in ^{15}N NMR spectra while such a change is not observed in the ^{13}C spectra. On the contrary, the signal arising from tribenzylamine structures which is observed in ^{13}C spectra at 57 ppm, does not have a corresponding large intensity around 52 ppm in the ^{15}N NMR spectrum. The latter anomaly is explained in terms of the extreme sensitivity of ^{15}N nuclei to the geometrical environment, i.e. the structural heterogeneity and the resulting chemical shift dispersion (52).

Various ^{13}C and ^{15}N NMR techniques were simultaneously applied for the characterization (75), ageing (76) and thermal degradation (77) of 4,4'-methylenebis(phenylisocyanate) (MDI) based resins. The cross-polarization spectra of resins prepared at different cure temperatures and obtained with various contact times are shown in Fig. 11 (spectra 'A'). At short contact time (0.4 ms) the dominant signals in the spectral region between 100 and 120 ppm arise from nitrogens with directly attached protons. The assignment of two major resonances to –NH– groups in urea (104 ppm) and biuret (114 ppm) linkages is based on model compounds as well as on the parallel ^{13}C spectra and interrupted decoupling data. These signals are strongly attenuated when the contact time of 6 s is used for building the nitrogen magnetization. On the contrary, the two most intense signals at long contact times arise from nitrogen nuclei with no directly attached protons such as isocyanurate (150 ppm) and isocyanate (46 ppm) groups; a small resonance at 141 ppm corresponds to the imide linkage in the biuret-type network, while the resonance at 145 ppm in the

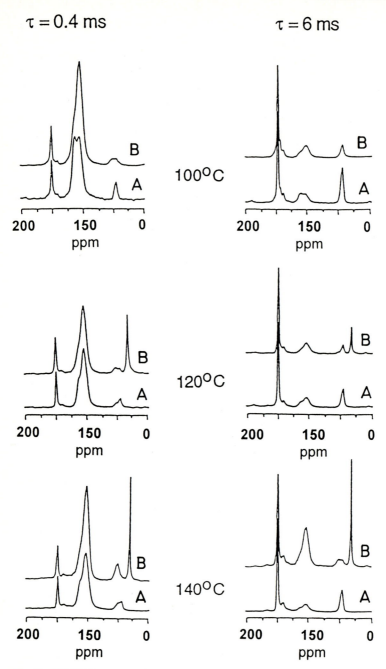

Fig. 11. ^{15}N NMR spectra of MDI-polyisocyanurate resins initially cured at the temperatures indicated; *left* and *right* spectra are recorded with contact times (τ) of 0.4 and 6 ms, respectively; spectra A present the initially cured samples (recorded at 20.3 MHz); spectra B present the resins after a 7-month exposure to air (recorded at 26.4 MHz). The resonance at 32 ppm corresponds to glycine (adapted from [76])

spectrum of a resin cured at $100\,°C$ is ascribed to the uretidione linkage, whose formation is known to be favored under certain conditions.

Although there is minor overlapping of signals in the ^{15}N NMR spectra of MDI-polyisocyanurate resins, the nitrogen absorptions arising from the seven structural units are more suitable for quantitative study than the corresponding ^{13}C data. Since single cross-polarization spectra cannot yield quantitative data because of different cross-polarization rates for each nitrogen nuclei, the signal intensities observed at particular contact times, $S(\tau)$, should be corrected to the theoretical intensity S_0 according to relation:

$$S(\tau) = S_0 \frac{T_{1\rho H}}{T_{1\rho H} - T_{NH}} \exp(-t/T_{1\rho H}) - \exp(-t/T_{NH})) \tag{3}$$

which is derived from Eq. (2) under the assumption that both $T_{1\rho N}$ and $T_{1\rho H}$ are much longer than T_{NH}. The constants T_{NH} and $T_{1\rho H}$ can be determined from the variable contact time cross-polarization experiment. The $T_{1\rho H}$ value of a homogeneous solid polymer is typically an average over all the protons in the sample because of proton spin diffusion. However, the $T_{1\rho H}$ values corresponding to urea-type signals (about 4.6 ms; 104 ppm) are shorter by a factor of approximately 3 than $T_{1\rho H}$ values obtained for isocyanate and isocyanurate nitrogens. This finding suggests the existence of a phase separation between isocyanurate/isocyanate-rich environments and polyurea-rich environments which have increased molecular motion. The isocyanurate:isocyanate ratio calculated from the corrected peak intensities indicates that the formation of isocyanurate crosslinks is most favoured at $120\,°C$ (mol ratio 2.8:1), followed by the resins prepared at $160\,°C$ (mol ratio 2.3:1) and $100\,°C$ (mol ratio 1.6:1).

Spectra of samples initially cured at 100, 120, and $160\,°C$ after a 7-month exposure to air are also shown in Fig. 11 (spectra 'B'). The major changes in the spectra of samples after a 7-month exposure compared to the initial samples are: (a) decreased intensity of the isocyanate nitrogen resonance (46 ppm), (b) increased intensity of the amine nitrogen resonance (53 ppm), and (c) increased intensity of the urea nitrogen resonance (104 ppm). The changes in intensity indicate the consumption of isocyanate to form the amine and the formation of urea condensation products from the reaction of amine and isocyanate groups during the 7-month period. In addition, no significant reaction of urea linkages with isocyanate groups to form biuret linkages is observed from the nitrogen spectra. The changes in particular resonances, which differ for short and long contact times, occur to different extents for the three samples. A quantitative analysis is performed on the basis of signal intensities obtained with 0.4 ms contact time and corrected according to Eq. (3). Insignificant changes in the relative concentration of the isocyanurate moieties (150 ppm) indicate that isocyanurate linkages are hydrolytically stable. However, consumptions and initial relative intensities (i.r.i.) of isocyanate groups of 86% (i.r.i. = 16%), 50% (i.r.i. = 22%), and 43% (i.r.i. = 15%) for samples cured at 160, 100, and $120\,°C$, respectively, indicate a different postcure reaction chemistry and overall isocyanate reactivities for the three resins. Higher reactivity of the sample initially cured at 160 °C is explained in terms of a more inhomogeneous curing process resulting in more irregularly crosslinked resins, with larger pockets or regions of unreacted MDI monomer which might be trapped or occluded in the solid matrix. This irregular structure may also result in a larger number of voids or imperfections in the cured product which makes this sample undergo more accelerated degradation (easier diffusion of water vapour into the sample) compared to polymers undergoing a more homogeneous curing process (76).

Thermal decomposition of an MDI-polyisocyanurate resin initially cured at 100 °C, exposed to air for 18 months, and heated for 30 min under a nitrogen atmosphere was also followed by both ^{15}N and ^{13}C NMR spectra. From the ^{15}N NMR spectra (Fig. 12) the following changes in the chemical structure of the resin upon increasing heating temperature were observed: (a) the degradation of biuret and uretidione linkages by 230 °C, (b) a decrease in residual isocyanate groups at heating temperatures up to 240 °C, (c) increase (up to 240 °C) and decrease (between 250 and 260 °C) in concentration of urea linkages, (d) a steady increase in amine group concentration, and (e) decrease in isocyanurate concentration for heating temperatures of 280 °C and above (77). In order to obtain quantitative information on the relative concentrations of major components known to exist in the resin during the thermal degradation process the relative cross-polarization signal intensities were corrected according to Eq. (3) assuming that the values for relaxation parameters T_{NH} and $T_{1\rho H}$ for all the resins are not dramatically different. The semiquantitative results obtained with this approximate method are presented in Table 5. The results support some of the conclusions of previous investigations in terms of ranking of thermal stabilities of the various linkages which fall into the following order: biuret, uretidione < urea < isocyuanurate < urea', where the prime indicates that certain urea linkages are more thermolitically stable than isocyanurate linkages (77).

The curing of cyanate polymer resins based on bisphenol A dicyanate (BPACN) was studied by both ^{13}C and ^{15}N NMR spectroscopies. Spectra of the ^{15}N enriched BPACN monomer cured in the solid state and in solution indicate that the solid, cured resin is composed mainly of triazine ring linkages, whether the reaction is carried out in the solid state or the product is obtained by precipitation from a solution polymerization. The curing reaction in the solid state is shown to be nearly quantitative (contribution of triazine signals is estimated to be more than 98%), while the isolated resins from solution curing contain a certain amount of side products. The cross-polarization experiment with variable contact times confirmed that all the signals associated with side products arise from proton bearing nitrogens (including amide endgroups which are also found in the solution ^{15}N NMR spectra) (37).

The hydrophilic surface structures of plasma polymerized acrylonitrile are also investigated by both ^{13}C and ^{15}N NMR spectroscopies. In the samples of plasma polymerized acrylonitrile and a polymer produced near the electrode only amine signals were detected. However, the spectra of a film produced from acrylonitrile under nitrogen gas show a number of amide/amine signals in a wide chemical shift range (80 ppm). In addition, the polymer formed near the

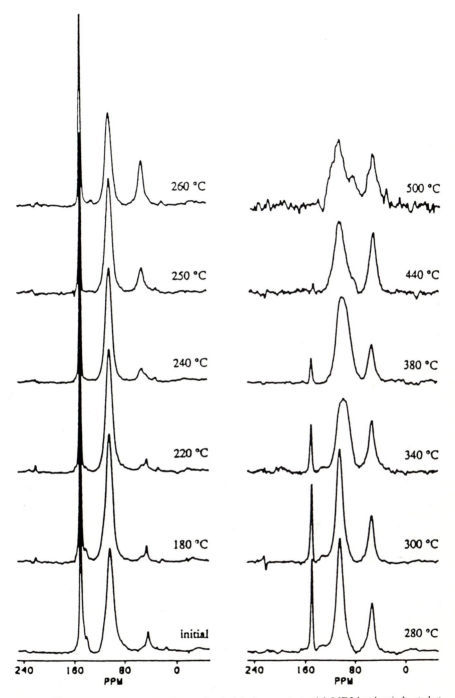

Fig. 12. ^{15}N NMR spectra (contact time = 6 ms) of the isocyanurate-rich MDI-basd resin heated at the temperatures indicated (reprinted from [77])

Table 5. Corrected relative intensities of some nitrogen absorptions for various thermal decomposition temperatures of MDI-based polymers[a]

Thermal decomposition temperature	corrected relative intensities (%) of nitrogen structures			
	isocyanate 46 ppm	amine 53 ppm	urea 104 ppm	isocyanurate 150 ppm
initial	22	–	22	34
initial (after 7 month)	11	2.2	32	35
initial (after 18 month)	3.3		77	
180 °C	2.1		83	14
220 °C	1.1	1.3	84	14
240 °C	0.7	4.7	80	15
250 °C	–	12	75	13
260 °C	–	21	62	17
280 °C	–	20	73	7
300 °C	–	20	74	5.9
340 °C	–	24	74	3.4
360 °C	–	21	76	3.0
380 °C	–	16	82	1.6
440 °C	–	29	71	0.4
500 °C	–	28	72	

[a] Data from [76, 77]

electrode also contains heterocyclic nitrogen structures. This indicates that the plasma polymerization of acrylonitrile with nitrogen gas promotes the formation of nitrogen radical species and, consequently, a preparation of various types of amines and amides (78).

The solid state ^{15}N NMR spectra of natural abundance and ^{15}N enriched synthetic polypeptides are dominated by their secondary structure rather than by their chemical shifts. The nitrogen signals of α-helices generally absorb about 1.2–10.0 ppm upfield from the signal of the β-sheet structure (the polyglycine helix absorbs downfield). However, if the differences in chemical shifts are too small they might be obscured by the relatively large line widths (150–300 Hz). Contrary to the solution spectra, where differences between dissolved coil and helices are only in the range of 1.0 to 1.2 ppm and neighboring residue effects may reach 10 ppm, the neighboring residue effects in the solid state are negligible. Although ^{13}C chemical shift differences for the carbonyl, C_α and C_β signals between the α-helix and β-sheet form are nearly identical (around 4–5 ppm), the variations of the ^{13}C chemical shift difference are very small (1.0–1.5 ppm). The variations of the ^{15}N chemical shifts for various types of homopolypeptides are around 2.5 ppm for the α helix form and around 7.5 ppm for the β-sheet form indicating that nitrogen chemical shifts are also dependent on the sidechain structure of individual amino acid residues (4, 79, 80).

3.3 Double Cross Polarization and Dipolar Rotational Spin-Echoes

Two pulse sequences were used to characterize hydrogen cyanide (HCN) polymers. Characterization of polymerization products of HCN is of great importance for prebiotic and extraterrestrial chemistry since the hypothesis was made that the original heteropolypeptides on primitive earth may have been formed spontaneously from HCN and water without the prior formation of α-amino acids. Double cross-polarization (DCP) can be performed in several ways. For direct observation of ^{15}N spectra with ^{1}H to ^{13}C to ^{15}N polarization transfer a knowledge of $T_{1\rho C}$ and $T_{1\rho N}$ relaxation parameters is required. However, if the polarization is transferred from ^{1}H to ^{15}N to ^{13}C (Fig. 13), and if during the second contact the ^{13}C radiofrequency field is phase modulated to prevent buildup of ^{13}C polarization, the double cross-polarization difference is maximised by preventing transfer of ^{13}C polarization back to the ^{15}N spin system. In this case no knowledge of the carbon and nitrogen spin-lattice relaxation times in the rotating frame is required. By using the additional cross-polarization step for transferring the magnetization from ^{15}N to ^{13}C nuclei, which requires strong dipolar coupling between the spins, only the ^{15}N–^{13}C chemical bonds are detected compared with the standard CPMAS ^{15}N NMR spectra. The draining away of ^{15}N polarization occurs at a rate which is dependent on the specific ^{15}N–^{13}C bond (it may be determined from model compounds). The DCP experiment was performed on HCN labeled with both ^{13}C and ^{15}N which are introduced in the polymer by a reaction either between H^{13}C^{15}N and HC^{15}N or between H^{13}C^{15}N and HCN. Both CP and DCP ^{15}N NMR spectra before and after hydrolysis for a sample prepared from a 1:1 mixture of H^{13}C^{15}N and HCN are shown in Fig. 14. The nitrogen functionalities before hydrolysis are assigned to amines (64 ppm), olefinic and aromatic nitrogens (184–244 ppm), and nitrile nitrogen (264 ppm). After exposure to water all sharp nitrogen resonances have disappeared from the spectrum of the insoluble fraction, and a new broad resonance at 119 ppm (the region associated with amide and amidine sp^3 nitrogens) appears. The independence of nitrogen–carbon polarization transfer rates on isotopic composition (1:1,

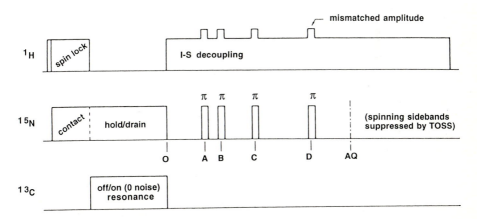

Fig. 13. Pulse sequence for double-cross-polarization

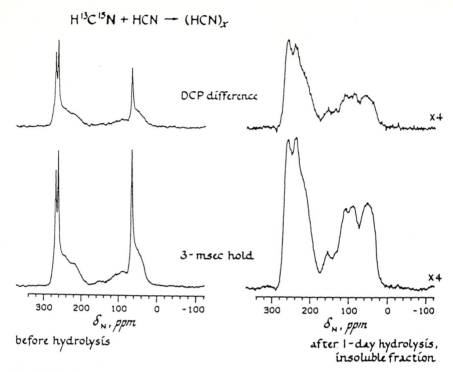

$$H^{13}C^{15}N + HCN \longrightarrow (HCN)_x$$

before hydrolysis

after 1-day hydrolysis, insoluble fraction

Fig. 14. ^{15}N NMR spectra of the dried reaction products of a 1:1 mixture of H^{13}C^{15}N and HCN. The spectrum of the total solids before hydrolysis appears at the *lower left of the figure*. A ^{15}N spin-lock of 3 ms was used in obtaining this spectrum. The corresponding double-cross-polarization difference spectrum is also shown (*upper left*). Analogous ^{15}N NMR spectra of the water-insoluble fraction after a 1-day hydrolysis are shown at the *right of the figure* (adapted from [81])

0.2:0.8 and 0.05:0.95 H^{13}C^{15}N:HCN) observed for amines and nitriles indicates that there is no evidence for bond breaking, isotopic exchange with an unreacted label pool, and bond reformation. On the contrary, the strong dependence of the polarization transfer rate on isotopic composition observed for the signal at 224 ppm indicates the formation of chemically new C–N bonds. In addition, the strong DCP difference signal (which arises from all ^{15}N–^{12}C nitrogen and a fraction of ^{15}N–^{13}C nitrogens) indicates the long term stability of labeled pairs to mild water conditions. The spectra of an insoluble fraction after the hydrolysis of HCN polymers prepared from a separate double labeled experiment (1:1 H^{13}CN:HC^{15}N mixture) clearly show only signals arising from newly formed C–N bonds, i.e. absorptions at 224 and 119 ppm, respectively. With the assumption that the signal at 119 ppm is potentially a sum from two types of nitrogen (peptides and primary amides) an estimate that about 80% of the intensity at 119 ppm represents nitrogens in the peptide linkage was made. However, the complexity of the spectrum indicates that other structures are present in the insoluble fraction after hydrolysis, particularly purines, cross-linked ladder structures, and the polymeric extensions of diaminomaleonitrile. These results, together with the finding that the water soluble fraction gives no DCP difference spectrum, support the hypothesis that heteropolypeptides can be synthesized directly from hydrogen cyanide and water according to the following equation (81–83):

$$H^{13}C^{14}N + H^{12}C^{15}N \xrightarrow[H_2O]{} -^{12}CH(R)-^{15}NH-^{13}C(O)- \qquad (4)$$

The direct measurement of the average number of protons attached to the amide nitrogens (a possible peptide bond signal at 119 ppm) can be made by using the dipolar rotational spin-echo technique. This is a two dimensional experiment (Fig. 15) in which, during the additional time dimension, the nitrogen magnetization is allowed to evolve under the influence of N–H coupling while H–H coupling is suppressed by proton multiple pulse irradiation. The Fourier transform of intensity vs evolution time for resolved resonances yields a dipolar spectrum consisting of $^{15}N-^{1}H$ Pake doublets scaled by the multiple pulse decoupling and broken up into sidebands by magic angle spinning. Since this experiment measures nitrogen-proton dipolar coupling it can be used to count the average number of protons attached to a nitrogen. Two types of HCN polymers are studied by this pulse sequence: (a) solids formed by base catalyzed polymerization of HCN and (b) solids formed by the electrical discharge of methane and ammonia (Fig. 16). The experimental dipolar rotational sideband pattern (Fig. 16b) for the resonance at 119 ppm (Fig. 16a, indicated by the arrow) matches that of a simulated pattern produced by adding calculated rigid lattice NH and NH_2 patterns in a 66/34 ratio (Fig. 16c) (84). Although the majority of the amidine or amide nitrogens are secondary, i.e. peptide-like, none of the amidine or amide nitrogens observed are bonded to aliphatic carbons. Thus, the amides in base-catalyzed HCN polymers do not

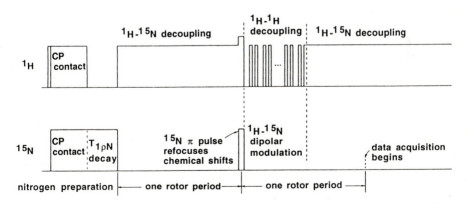

Fig. 15. Pulse sequence for dipolar rotational spin-echo ^{15}N NMR experiment

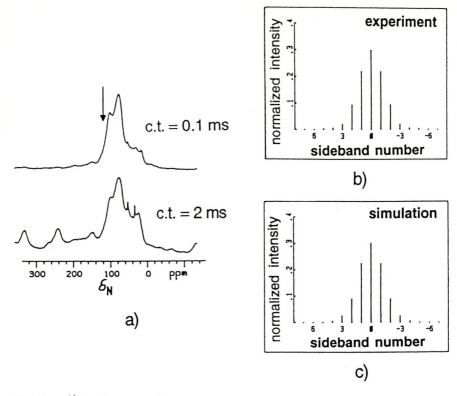

Fig. 16a–c. ^{15}N NMR spectra of the anhydrous solids obtained from electrical discharge through an atmosphere of methane and ammonia (25% enriched in ^{15}N). Figs. 16b and 16c refer to the contact time (c.t.) of 0.1 ms when only the protonated nitrogens have significant intensity. (adapted from [84])

indicate the direct formation of heteropolypeptides from HCN. The anhydrous solids formed by electrical discharge have aliphatic carbons but little or no amidine functionality. Consequently they too are unlikely precursors for the direct formation of heteropolypeptides (83, 84).

3.4 Spin-Lattice Relaxation

Spin-lattice relaxation times, nT_1, for different phases of various nylon samples are listed in Table 6. The α crystal form of both nylon-6 and nylon-11 exhibit a single exponential decay function in a plot of signal intensity vs delay τ, with corresponding nT_1 values in the 10^3 s range. The sample prepared in situ has the shortest T_1 relaxation time which indicates more rapid motions due either to plasticization by residual caprolactam or differences in the crystalline regions.

Table 6. ^{15}N nT_1 relaxation times (s) and chemical shifts^a for nylon-6, nylon-11 and nylon-12^b

	122.8 ppm^c (90.3 ppm)	γ form 121.8 ppm (89.3 ppm)	γ' form 121.3 ppm (88.8 ppm)	amorphous phase 119.7 ppm (87.2 ppm)	amorph.phase and δ(δ') form 119.1 ppm (86.6 ppm)	α form 116.7 ppm (84.2 ppm)
Nylon-6						
in situ (23% cryst.)				–; 26		111
melt quenched (29% cryst.)				2.7; 19		230
annealed (160 °C) (45% cryst.)				1.9; 29.1		416
Nylon-11						
from trifluoroacetic acid			84.1^d		4.2; 19.5	—
quenched in dry ice/i-PrOH			—		2.3; 13.4 (δ')	—
annealed			—		4.9; 68.1	178
annealed (115 °C)			—		7.2; 112.2 (δ)	—
Nylon-12						
precipitated	—	9; 160			2; 107^e	2; 323
quenched	—	—	3; 34		4; 79^e	—
annealed	10; 319	20; 314	—		4; 41^e	—

^{a} Chemical shifts in parenthesis are relative to solid glycine at 0 ppm
^{b} Data from [50, 64, 85]
^{c} The exact identity of this signal is unclear (see text)
^{d} This signal is assigned to γ (not to γ' form)
^{e} The resonance at 119.1 ppm in nylon-12 is assigned to amorphous phase only

These samples also exhibit a single exponential relaxation curve for the amorphous phase. However, there is a single exponential decay function for the
amorphous signal of both samples quenched from the melt, but the slowly
cooled and annealed sample exhibits a two component decay with a short
component in the $10°$ s range (about 2.3 s) and a long component in a 10^2 s
range (from 19 to 29 s). The existence of two nT_1s indicates the presence of two
motionally different noncrystalline regions. These are ascribed to a bulk
amorphous fraction with liquidlike mobility and a noncrystalline interphase
region with restricted motion. Biexponential relaxation functions with the short
nT_1 component times similar to that obtained for the bulk amorphous phase of
nylon-6 are also found in the relaxation of the 119.1 ppm signal in the samples of
nylon-11 and nylon-12. However, the long nT_1 component varies greatly from
sample to sample and may depend on crystallite size. The nT_1s of 112 s and 13 s
for the δ and δ' component (119.1 ppm), respectively, indicate that while the
amides in these crystals may be similar chemically, the relative mobility of the δ'
form is much greater than that of the δ form (55, 64, 85). On the contrary, two-
component decay with nT_1s of 4 and 41 s, observed for a signal at 119.1 ppm in
nylon-12, is assigned to the amorphous phase only (both components have short
T_1 values compared to the crystalline γ form (50). On the basis of long ^{15}N nT_1
times for a biexponential decay of the shoulder observed at 122.8 ppm (see Fig.
6C'), which are similar to those for the γ form signal, it was concluded that the
residues associated with this peak are probably in rigid environments as in the γ
crystalline form. This peak may belong to material epitaxially crystallized on the
surface of γ crystallites or to an ordered interphase region (50).

Spin-lattice relaxation times for the individual chemical structures of three
MDI-polyisocyanurate resins after seven month exposure to air depend on the
number of directly bonded protons. In principle, the largest nT_1 values are
observed for isocyanurate and isocyanate nitrogens (102–140 s), followed by the
–NH– groups in urea and biuret units (61–83 s) and the –NH$_2$ endgroups
(20–32 s). The smaller nT_1 values indicate that the effective relaxations of amino
nitrogens are dominated by modulation of the ^{15}N–^{1}H dipolar interaction due
to the motion of the –NH$_2$ group (analogously to the methyl group motions in
^{13}C NMR). Larger values of nT_1 for isocyanurate nitrogens (116–140 s) compared to the isocyanate nitrogens (102–121 s) in the same samples suggest that
the isocyanurate crosslink sites are part of the most rigid environments in each
sample. In addition, the single exponential nT_1 behavior for isocyanate signals
confirmed that substantial MDI monomer is not present in the system (76).

The observation of rigid and mobile phases in ^{15}N enriched melamine-
formaldehyde resins makes it possible to measure the spin-lattice relaxation
times in both regions. By using standard inversion recovery and cross-polarization (86) T_1 pulse sequences the values in the range 6–22 s and 25–69 s,
respectively, were obtained for the secondary amine resonance. Shorter relaxation times in more mobile portions of the sample, as well as longer times in more
rigid parts, show a similar temperature dependence with a possible minimum
(around 343 K) associated with relaxation processes in the resin (53).

The spin-lattice relaxations in the rotating frame ($T_{1\rho}$) for the two signals of nylon-6 increase with the increasing degree of crystallinity, although large differences are not evident. The α form has longer relaxation times (13–58 ms) and the amorphous phase shorter (0–40 ms). However, the relaxation in the rotating frame has contributions from both spin-lattice and spin-spin processes due to rapid proton diffusion between the different phases. Although the $T_{1\rho}$ in highly crystalline systems is completely dominated by spin-spin relaxation, it was found that the spin-spin diffusion in nylon-6 does not dominate $T_{1\rho}$. Nevertheless, the two T_1 times observed in the amorphous phase are not evident from the $T_{1\rho}$ data for nylon-6 (55, 59). On the contrary, two component decay of $T_{1\rho}$ follows that of T_1 decay in nylon-12. Within the respective samples, both short and long values for all components are similar, suggesting that similar motions in the kHz range occur throughout each sample (50).

Two different $T_{1\rho}$ times for nonprotonated nitrogen in uncured and cured ^{15}N enriched melamine-formaldehyde resins were observed by direct- (DP) and cross-polarization (CP) pulse sequences, respectively. DP $T_{1\rho}$ times increase with increasing temperature indicating a relaxation process at or below 298 K. At 298 K they are very similar to the CP $T_{1\rho}$ values which are temperature independent. Shorter CP $T_{1\rho}$ times are probably dominated by spin-spin relaxation, and do not represent a motional spin-lattice relaxation (53).

3.5 Interrupted Decoupling

By turning off the proton rf field (decoupler) for a time T_D the [15]N–[1]H static dipolar interactions cause the dephasing of nitrogen magnetization (87). Since the dipolar relaxation rate is proportional to the number of directly attached protons, internuclear distances r_{NH} (as r_{NH}^6), and rotational correlation times, only rigid nitrogens with directly attached protons ($r_{NH} = 1.03$ Å) will dephase rapidly. However, tertiary nitrogens with no directly attached protons, as well as highly mobile protonated nitrogen atoms whose motions are effective in averaging dipolar couplings to reduced values, will not dephase during the appropriate interrupted decoupling period T_D. The two amine resonances at 114 and 98.7 ppm observed in cured melamine-formaldehyde resins (see Fig. 7) are assigned to secondary and primary amines, respectively. While the interrupted-decoupling period of 100 μs almost completely suppresses the protonated nitrogen signal, an additional signal at 106.3 ppm arising from secondary amines can be directly detected (Fig. 17). By using this technique the assignment of the completely overlapped signal, which cannot be observed directly, was possible (53).

A T_D of 20 μs is sufficiently short for nitrogen magnetizations in nylon-6 to be intact, while a T_D of 100 μs at 300 K reduces the rigid crystalline resonance into the base line noise. At 370 K only a small signal remains with a T_D of 120 μs, indicating that enough motion is present to weaken the dipolar interaction such that the signal is not completely lost. Since the remaining resonance at approx-

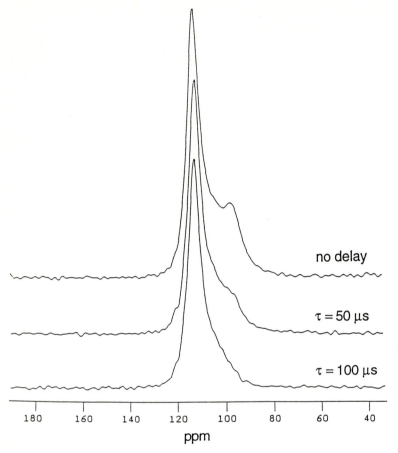

no delay

$\tau = 50\ \mu s$

$\tau = 100\ \mu s$

Fig. 17. ^{15}N interrupted decoupling spectra of a cured ^{15}N enriched melamine-formaldehyde resin with the delays (τ) indicated (adapted from [53])

imately 116.5 ppm is correlated with the resonance of the α crystal form and not with the amorphous portion, this result supports the conclusions from chemical shift anisotropy experiments that significant motion at higher temperatures is occurring in the crystalline region of the sample (55, 59).

The observation of nitrogen signals in hexamethylenetetramine cured phenolic resins (50–38 ppm range) after an interrupted decoupling period as long as 125 μs confirmed that the resonances cannot arise from nitrogens either with one or two directly attached protons. This finding eliminates many of the intermediates proposed for the curing process and confirmed the presence of a highly mobile NH_4^+ ion in the resin (52). The interrupted decoupling period of 70 μs almost completely suppresses the resonances arising from urea and amide linkages in the MDI-polyisocyanurate based resins leaving only isocyanurate, isocyanate, imide, and uretidione resonances, which contain nonproton bearing nitrogens (spectra are nearly identical to the spectra 'A' obtained with a contact time of 6 ms and shown in Fig. 11, except the intensity of signals in the spectral region between 100 and 120 ppm is almost reduced into the baseline) (75). Since the magnetization of nitrogens with directly attached protons dephases more rapidly than the magnetization of nitrogen spins with no directly attached proton, a decrease in intensity at long T_Ds is expected for overlapping signals which arise from both protonated and nonprotonated nitrogens. Consequently, the intensity of a signal arising only from secondary amides in urea-formalde-hyde resins decreases with increasing T_D up to 50 µs and vanishes at $T_D = 120$ µs, while the overlapping signal arising from both tertiary and secondary amides is still present at T_D of 120 µs with the intensity proportional to the ratio of tertiary vs secondary amides (75).

The interrupted dephasing pulse sequence is also used in order to determine the number of hydrogen atoms bonding to nitrogen atoms in doped and undoped polypyrroles. The same relative intensities of signals in doped and undoped polymers obtained with T_D of 20, 100 and 200 µs indicate that the number of bonded hydrogens and the bond lengths in both samples are approximately equal (71).

4 References

1. Proctor WG, Yu FC (1950) Phys Rev 77: 716
2. Schaefer J, Stejskal EO (1976) J Am Chem Soc 98: 1031
3. Schaefer J, Stejskal EO, Jacob Gary S, McKay RA (1982) Appl Spectr 36: 179
4. Kricheldorf HR (1982) Pure Appl Chem 54: 467
5. Levy GC, Lichter RL (1979) Nitrogen-15 nuclear magnetic resonance spectroscopy, J Wiley, New York
6. Witanowski W, Stefaniak L, Webb GA (1986) Ann Rep NMR Spectr 18: 1
7. Kricheldorf HR, (1981) J Polym Sci Polym Chem Ed 19: 2195
8. Chang C, Muccio DD, St. Pierre T, Chen CC, Overberger CG (1986) Macromolecules 19: 913
9. Kricheldorf HR, Schilling G (1978) Makromol Chem 179: 2667
10. Kricheldorf HR (1978) Makromol Chem 179: 2675
11. Kricheldorf HR (1980) Hull WE, Makromol Chem 181: 507
12. Kricheldorf HR (1978) Makromol Chem 179: 2687
13. Kricheldorf HR, Hull WE (1978) J Macromol Sci Chem A12, 51
14. Kricheldorf HR, Joshi SV, Hull WE (1982) J Polym Sci Polym Chem Ed 20: 2791
15. Kricheldorf HR, Hull WE (1978) J Polym Sci Polym Chem Ed
16. Kricheldorf HR, Hull WE (1979) Makromol Chem 180: 161
17. Kricheldorf HR, Hull WE (1982) Biopolymers 21: 359
18. Kricheldorf HR, Hull WE, Formacek V (1977) Biopolymers 16: 1609
19. Kricheldorf HR (1979) Makromol Chem 180: 147
20. Kricheldorf HR, Hull WE (1980) Macromolecules 13: 87
21. Kricheldorf HR (1981) Org Magn Reson 15: 162
22. Kricheldorf HR, Haupt E (1983) Int J Biol Macromol 5: 237
23. Kricheldorf HR, Mueller D, Hull WE (1985) Biopolymers 24: 2113
24. Kricheldorf HR, Hull WE (1980) Org Magn Reson 13: 335
25. Kricheldorf HR, Hull WE (1977) Makromol Chem 178: 583
26. Kricheldorf HR, Hull WE (1978) Liebigs Ann Chem 1817

27. Kricheldorf HR, Hull WE (1979) Makromol Chem 180: 1707
28. Kricheldorf HR, Hull WE (1979) Makromol Chem 180: 1715
29. Hull WE, Kricheldorf HR, Fehrle M (1978) Biopolymers 17: 2427
30. Barton FE, Himmelsbach DS, Amos HE (1981) J Agric Food Chem 29: 669
31. Kricheldorf HR (1980) J Macromol Sci Chem A14: 959
32. Kricheldorf HR, Hull WE (1981) Makromol Chem 182: 1177
33. Bevington JC, Huckerby TN, Hutton NWE (1982) Eur Polym J 18: 963
34. Ebdon JR, Heaton PE, Huckerby TN, O'Rourke WTS, Parkin J (1984) Polymer 25: 821
35. Preston CM, Rauthan BS, Rodger C, Ripmeester JA (1982) Soil Sci 134: 277
36. Barton JM, Greenfield DCL, Hameron I, Jones JR (1991) Polym Bull 25: 475
37. Fyfe CA, Niu J, Rettig SJ, Burlinson NE, Reidsema CM, Wang DW, Poliks M (1992) Macromolecules 25: 6289
38. Kricheldrof HR (1980) Polym Bull 3: 53
39. Axelson DE, Blake SL (1985) J Polym Sci Polym Chem Ed 23: 2507
40. Chang C, Fish F, Muccio DD, St. Pierre T (1987) Macromolecules 20: 621
41. Komber H, Jakisch L, Zschoche S, Mobus H, Ratzsch M (1991) Makromol Chem Rapid Commun 12: 547
42. Duguet E, Schappacher M, Soum A (1992) Macromolecules 25: 4835
43. Horhold HH, Klee J, Shutz H, Radeglia R (1986) Angcw Makrom Chem 144: 1
44. Rinaldi PL, Yu C, Levy GC (1981) Macromolecules 14: 551
45. Andreis M, Koenig JL (1989) Adv Polym Sci 89: 69
46. Urry DW, Trapane TL, McMichens RB, Iqbal N, Harris RD, Prasad KU (1986) Biopolymers 25: S209
47. Pines A, Gibby MG, Waugh JS (1973) J Chem Phys 59: 569
48. Holmes BS, Chingas GC, Moniz WB, Ferguson RC (1981) Macromolecules 14: 1785
49. Holmes BS, Moniz WB, Ferguson RC (1982) Macromolecules 15: 129
50. Mathias LJ, Johnson CG (1991) Macromolecules 24: 6114
51. Hayashi S, Hayamizu K (1991) Bull Chem Soc Jpn 64: 688
52. Hatfield GR, Maciel GE (1987) Macromolecules 20: 608
53. Andreis M, Koenig JL, Gupta M, Ramesh S submitted to J Polym Sci Polym Phys Ed
54. Powell DG, Sikes AM, Mathias LJ (1988) Macromolecules 21, 1533
55. Powell DG, Mathias LJ (1990) J Am Chem Soc 112: 669
56. Powell DG, Sikes AM, Mathias LJ (1991) Polymer 32: 2523
57. Mathias LJ, Powell DG, Sikes AM (1988) Polymer Comm 29: 192
58. Hatfield GR, Glans JH, Hammond WB (1990) Macromolecules 23: 1654
59. Mathias LJ, Powell DG, Autran JP, Porter RS (1990) Mater Sci Eng A126: 253
60. Murthy NS, Hatfield GR, Glans JH (1990) Macromolecules 23: 1342
61. Johnson CG, Mathias LJ (1993) Polymer 34: 4978
62. Hatfield GR, Guo Y, Killinger WE, Andrejak RA, Roubicek PM (1993) Macromolecules 26: 6350
63. Powell DG, Mathias LJ (1990) Polym comm 31: 58
64. Mathias LJ, Powell DG, Autran JP, Porter RS (1990) Macromolecules 23: 963
65. Chaloner-Gill B, Euler WB, Roberts JE (1991) Macromolecules 24: 3074
66. Richter AF, Ray A, Ramanathan KV, Manohar SK, Furst GT, Opella SJ, Macdiarmid AG (1989) Synth Metals 29: E243
67. Wehrle B, Limbach HH, Zipplies T, Hanack M (1989) Angew Chem Int Ed Engl Adv Mater 28: 1741
68. Adams PN, Apperley DC, Monkman AP (1993) Polymer 34: 328
69. Wehrle B, Limbach HH, Mortensen J, Heinze J (1990) Synth Metals 38: 293
70. Wehrle B, Limbach HH, Braunling H (1991) Synth Metals 39: 319
71. Kikuchi M, Kurosu H, Ando I (1992) J Mol Struc 269: 183
72. Wehrle B, Limbach HH, Zipplies T, Hanack M (1989) Angew Chem Int Ed Engl Adv Mater 28: 1743
73. Chung I-S, Hawkins BL, Maciel GE, Myers GE (1985) Macromolecules 18: 1482
74. Ebdon JR, Hunt BJ, O'Rourke WTS, Parkin J (1988) Br Polym J 20: 327
75. Duff DW, Maciel GE (1990) Macromolecules 23: 3069
76. Duff DW, Maciel GE (1991) Macromolecules 24: 387
77. Duff DW, Maciel GE (1991) Macromolecules 24: 651
78. Tajima I, Suda T, Yamamoto M, Satta K, Morimoto H (1988) Polym J 20: 919
79. Forster HG, Muller D, Kricheldorf HR (1983) Int J Biol Macromol 5: 101

80. Shoji A, Ozaki T, Fujito T, Deguchi K, Ando I (1987) Macromolecules 20: 2441
81. McKay RA, Schaefer J, Stejskal EO, Ludicky R, Matthews CN (1984) Macromolecules 17: 1124
82. Matthews CN, Ludicky R, Schaefer J, Stejskal EO, McKay RA (1984) Origin Life 14: 243
83. Matthews CN, Ludicky R (1991) in Solid State NMR of Polymers (L. Mathias Ed.), Plenum Press, New York
84. Garbow JR, Schaefer J, Ludicky R, Matthews CN (1987) Macromolecules 20: 305
85. Powell DG, Mathias LJ (1989) Macromolecules 22: 3812
86. Torchia DA (1978) J Magn Reson 30: 613
87. Opella SJ, Frey MH (1979) J Am Chem Soc 101: 5854

Edited: Prof. Koenig
Received: July 1994

Author Index Volumes 101-124

Author Index Vols. 1-100 see Vol. 100

Adolf, D. B. see Ediger, M. D..: Vol. 116, pp. 73-110.

Aharoni, S. M. and *Edwards, S. F.*: Rigid Polymer Networks. Vol. 118, pp. 1-231.

Améduri, B. and *Boutevin, B.*: Synthesis and Properties of Fluorinated Telechelic Monodispersed Compounds. Vol. 102, pp. 133-170.

Amselem, S. see Domb, A. J.: Vol. 107, pp. 93-142.

Andreis, M. and *Koenig, J. L.*: Application of Nitrogen-15 NMR to Polymers. Vol. 124, pp. 191-238.

Angiolini, L. see Carlini, C.: Vol. 123, pp. 127-214.

Anseth, K. S., Newman, S. M. and *Bowman, C. N.*: Polymeric Dental Composites: Properties and Reaction Behavior of Multimethacrylate Dental Restorations. Vol. 122, pp. 177-218.

Arnold Jr., F. E. and *Arnold, F. E.*: Rigid-Rod Polymers and Molecular Composites. Vol. 117, pp. 257-296.

Arshady, R.: Polymer Synthesis via Activated Esters: A New Dimension of Creativity in Macromolecular Chemistry. Vol. 111, pp. 1-42.

Bahar, I., Erman, B. and *Monnerie, L.*: Effect of Molecular Structure on Local Chain Dynamics: Analytical Approaches and Computational Methods. Vol. 116, pp. 145-206.

Baltá-Calleja, F. J., González Arche, A., Ezquerra, T. A., Santa Cruz, C., Batallón, F., Frick, B. and *López Cabarcos, E.*: Structure and Properties of Ferroelectric Copolymers of Poly(vinylidene) Fluoride. Vol. 108, pp. 1-48.

Barshtein, G. R. and *Sabsai, O. Y.*: Compositions with Mineralorganic Fillers. Vol. 101, pp.1-28.

Batallán, F. see Baltá-Calleja, F. J.: Vol. 108, pp. 1-48.

Barton, J. see Hunkeler, D.: Vol. 112, pp. 115-134.

Bell, C. L. and *Peppas, N. A.*: Biomedical Membranes from Hydrogels and Interpolymer Complexes. Vol. 122, pp. 125-176.

Berry, G.C.: Static and Dynamic Light Scattering on Moderately Concentraded Solutions: Isotropic Solutions of Flexible and Rodlike Chains and Nematic Solutions of Rodlike Chains. Vol. 114, pp. 233-290.

Bershtein, V. A. and *Ryzhov, V. A.*: Far Infrared Spectroscopy of Polymers. Vol. 114, pp. 43-122.

Bigg, D. M.: Thermal Conductivity of Heterophase Polymer Compositions. Vol. 119, pp. 1-30.

Binder, K.: Phase Transitions in Polymer Blends and Block Copolymer Melts: Some Recent Developments. Vol. 112, pp. 115-134.

Biswas, M. and *Mukherjee, A.*: Synthesis and Evaluation of Metal-Containing Polymers. Vol. 115, pp. 89-124.

Boutevin, B. and *Robin, J. J.*: Synthesis and Properties of Fluorinated Diols. Vol. 102. pp. 105-132.

Boutevin, B. see Amédouri, B.: Vol. 102, pp. 133-170.
Bowman, C. N. see Anseth, K. S.: Vol. 122, pp. 177-218.
Boyd, R. H.: Prediction of Polymer Crystal Structures and Properties. Vol. 116, pp. 1-26.
Bruza, K. J. see Kirchhoff, R. A.: Vol. 117, pp. 1-66.
Burban, J. H. see Cussler, E. L.: Vol. 110, pp. 67-80.

Candau, F. see Hunkeler, D.: Vol. 112, pp. 115-134.
Capek, I.: Kinetics of the Free-Radical Emulsion Polymerization of Vinyl Chloride. Vol. 120, pp. 135-206.
Carlini, C. and *Angiolini, L.*: Polymers as Free Radical Photoinitiators. Vol. 123, pp. 127-214.
Casas-Vazquez, J. see Jou, D.: Vol. 120, pp. 207-266.
Chen, P. see Jaffe, M.: Vol. 117, pp. 297-328.
Choe, E.-W. see Jaffe, M.: Vol. 117, pp. 297-328.
Chow, T. S.: Glassy State Relaxation and Deformation in Polymers. Vol. 103, pp. 149-190.
Chung, T.-S. see Jaffe, M.: Vol. 117, pp. 297-328.
Connell, J. W. see Hergenrother, P. M.: Vol. 117, pp. 67-110.
Criado-Sancho, M. see Jou, D.: Vol. 120, pp. 207-266.
Curro, J.G. see Schweizer, K.S.: Vol. 116, pp. 319-378.
Cussler, E. L., Wang, K. L. and *Burban, J. H.*: Hydrogels as Separation Agents. Vol. 110, pp. 67-80.

Dimonie, M. V. see Hunkeler, D.: Vol. 112, pp. 115-134.
Dodd, L. R. and *Theodorou, D. N.*: Atomistic Monte Carlo Simulation and Continuum Mean Field Theory of the Structure and Equation of State Properties of Alkane and Polymer Melts. Vol. 116, pp. 249-282.
Doelker, E.: Cellulose Derivatives. Vol. 107, pp. 199-266.
Domb, A. J., Amselem, S., Shah, J. and *Maniar, M.*: Polyanhydrides: Synthesis and Characterization. Vol.107, pp. 93-142.
Dubrovskii, S. A. see Kazanskii, K. S.: Vol. 104, pp. 97-134.
Dunkin, I. R. see Steinke, J.: Vol. 123, pp. 81-126.

Economy, J. and *Goranov, K.*: Thermotropic Liquid Crystalline Polymers for High Performance Applications. Vol. 117, pp. 221-256.
Ediger M. D. and *Adolf, D. B.*: Brownian Dynamics Simulations of Local Polymer Dynamics. Vol. 116, pp. 73-110.
Edwards, S. F. see Aharoni, S. M.: Vol. 118, pp. 1-231.
Erman, B. see Bahar, I.: Vol. 116, pp. 145-206.
Ezquerra, T. A. see Baltá-Calleja, F. J.: Vol. 108, pp. 1-48.

Fendler, J.H.: Membrane-Mimetic Approach to Advanced Materials. Vol. 113, pp. 1-209.
Fetters, L. J. see Xu, Z.: Vol. 120, pp. 1-50.
Förster, S. and *Schmidt, M.*: Polyelectrolytes in Solution. Vol. 120, pp. 51-134.
Frick, B. see Baltá-Calleja, F. J.: Vol. 108, pp. 1-48.
Fridman, M. L.: see Terent'eva, J. P.: Vol. 101, pp. 29-64.

Ganesh, K. see Kishore, K.: Vol. 121, pp. 81-122.
Geckeler, K. E. see Rivas, B.: Vol. 102, pp. 171-188.
Geckeler, K. E.: Soluble Polymer Supports for Liquid-Phase Synthesis. Vol. 121, pp. 31-80.

Gehrke, S. H.: Synthesis, Equilibrium Swelling, Kinetics Permeability and Applications of Environmentally Responsive Gels. Vol. 110, pp. 81-144.

Godovsky, D. Y.: Electron Behavior and Magnetic Properties Polymer-Nanocomposites. Vol. 119, pp. 79-122.

González Arche, A. see Baltá-Calleja, F. J.: Vol. 108, pp. 1-48.

Goranov, K. see Economy, J.: Vol. 117, pp. 221-256.

Grosberg, A. and *Nechaev, S.*: Polymer Topology. Vol. 106, pp. 1-30.

Grubbs, R., Risse, W. and *Novac, B.*: The Development of Well-defined Catalysts for Ring-Opening Olefin Metathesis. Vol. 102, pp. 47-72.

van Gunsteren, W. F. see Gusev, A. A.: Vol. 116, pp. 207-248.

Gusev, A. A., Müller-Plathe, F., van Gunsteren, W. F. and *Suter, U. W.*: Dynamics of Small Molecules in Bulk Polymers. Vol. 116, pp. 207-248.

Guillot, J. see Hunkeler, D.: Vol. 112, pp. 115-134.

Guyot, A. and *Tauer, K.*: Reactive Surfactants in Emulsion Polymerization. Vol. 111, pp. 43-66.

Hadjichristidis, N. see Xu, Z.: Vol. 120, pp. 1-50.

Hall, H. K. see Penelle, J.: Vol. 102, pp. 73-104.

Hammouda, B.: SANS from Homogeneous Polymer Mixtures: A Unified Overview. Vol. 106, pp. 87-134.

Hedrick, J. L. see Hergenrother, P. M.: Vol. 117, pp. 67-110.

Heller, J.: Poly (Ortho Esters). Vol. 107, pp. 41-92.

Hemielec, A. A. see Hunkeler, D.: Vol. 112, pp. 115-134.

Hergenrother, P. M., Connell, J. W., Labadie, J. W. and *Hedrick, J. L.*: Poly(arylene ether)s Containing Heterocyclic Units. Vol. 117, pp. 67-110.

Hirasa, O. see Suzuki, M.: Vol. 110, pp. 241-262.

Hirotsu, S.: Coexistence of Phases and the Nature of First-Order Transition in Poly-N-isopropylacrylamide Gels. Vol. 110, pp. 1-26.

Hunkeler, D., Candau, F., Pichot, C., Hemielec, A. E., Xie, T. Y., Barton, J., Vaskova, V., Guillot, J., Dimonie, M. V., Reichert, K. H.: Heterophase Polymerization: A Physical and Kinetic Comparision and Categorization. Vol. 112, pp. 115-134.

Ichikawa, T. see Yoshida, H.: Vol. 105, pp. 3-36.

Ilavsky, M.: Effect on Phase Transition on Swelling and Mechanical Behavior of Synthetic Hydrogels. Vol. 109, pp. 173-206.

Inomata, H. see Saito, S.: Vol. 106, pp. 207-232.

Irie, M.: Stimuli-Responsive Poly(N-isopropylacrylamide), Photo- and Chemical-Induced Phase Transitions. Vol. 110, pp. 49-66.

Ise, N. see Matsuoka, H.: Vol. 114, pp. 187-232.

Ivanov, A. E. see Zubov, V. P.: Vol. 104, pp. 135-176.

Jaffe, M., Chen, P., Choe, E.-W., Chung, T.-S. and *Makhija, S.*: High Performance Polymer Blends. Vol. 117, pp. 297-328.

Jou, D., Casas-Vazquez, J. and *Criado-Sancho, M.*: Thermodynamics of Polymer Solutions under Flow: Phase Separation and Polymer Degradation. Vol. 120, pp. 207-266.

Kaetsu, I.: Radiation Synthesis of Polymeric Materials for Biomedical and Biochemical Applications. Vol. 105, pp. 81-98.

Kammer, H. W., Kressler, H. and *Kummerloewe, C.*: Phase Behavior of Polymer Blends - Effects of Thermodynamics and Rheology. Vol. 106, pp. 31-86.

Kandyrin, L. B. and *Kuleznev, V. N.*: The Dependence of Viscosity on the Composition of Concentrated Dispersions and the Free Volume Concept of Disperse Systems. Vol. 103, pp. 103-148.

Kaneko, M. see Ramaraj, R.: Vol. 123, pp. 215-242.

Kang, E. T., Neoh, K. G. and *Tan, K. L.*: X-Ray Photoelectron Spectroscopic Studies of Electroactive Polymers. Vol. 106, pp. 135-190.

Kazanskii, K. S. and *Dubrovskii, S. A.*: Chemistry and Physics of „Agricultural" Hydrogels. Vol. 104, pp. 97-134.

Kennedy, J. P. see Majoros, I.: Vol. 112, pp. 1-113.

Khokhlov, A., Starodybtzev, S. and *Vasilevskaya, V.*: Conformational Transitions of Polymer Gels: Theory and Experiment. Vol. 109, pp. 121-172.

Kilian, H. G. and *Pieper, T.*: Packing of Chain Segments. A Method for Describing X-Ray Patterns of Crystalline, Liquid Crystalline and Non-Crystalline Polymers. Vol. 108, pp. 49-90.

Kishore, K. and *Ganesh, K.*: Polymers Containing Disulfide, Tetrasulfide, Diselenide and Ditelluride Linkages in the Main Chain. Vol. 121, pp. 81-122.

Klier, J. see Scranton, A. B.: Vol. 122, pp. 1-54.

Kobayashi, S., Shoda, S. and *Uyama, H.*: Enzymatic Polymerization and Oligomerization. Vol. 121, pp. 1-30.

Koenig, J. L. see Andreis, M.: Vol. 124, pp. 191-238.

Kokufuta, E.: Novel Applications for Stimulus-Sensitive Polymer Gels in the Preparation of Functional Immobilized Biocatalysts. Vol. 110, pp. 157-178.

Konno, M. see Saito, S.: Vol. 109, pp. 207-232.

Kopecek, J. see Putnam, D.: Vol. 122, pp. 55-124.

Kressler, J. see Kammer, H. W.: Vol. 106, pp. 31-86.

Kirchhoff, R. A. and *Bruza, K. J.*: Polymers from Benzocyclobutenes. Vol. 117, pp. 1-66.

Kuleznev, V. N. see Kandyrin, L. B.: Vol. 103, pp. 103-148.

Kulichkhin, S. G. see Malkin, A. Y.: Vol. 101, pp. 217-258.

Kuchanov, S. I.: Modern Aspects of Quantitative Theory of Free-Radical Copolymerization. Vol. 103, pp. 1-102.

Kummerloewe, C. see Kammer, H. W.: Vol. 106, pp. 31-86.

Kuznetsova, N. P. see Samsonov, G. V.: Vol. 104, pp. 1-50.

Labadie, J. W. see Hergenrother, P. M.: Vol. 117, pp. 67-110.

Laschewsky, A.: Molecular Concepts, Self-Organisation and Properties of Polysoaps. Vol. 124, pp. 1-86.

Laso, M. see Leontidis, E.: Vol. 116, pp. 283-318.

Lazár, M. and *Rychlý, R.*: Oxidation of Hydrocarbon Polymers. Vol. 102, pp. 189-222.

Lenz, R. W.: Biodegradable Polymers. Vol. 107, pp. 1-40.

Leontidis, E., de Pablo, J. J., Laso, M. and *Suter, U. W.*: A Critical Evaluation of Novel Algorithms for the Off-Lattice Monte Carlo Simulation of Condensed Polymer Phases. Vol. 116, pp. 283-318.

Lesec, J. see Viovy, J.-L.: Vol. 114, pp. 1-42.

Liang, G. L. see Sumpter, B. G.: Vol. 116, pp. 27-72.

Lin, J. and *Sherrington, D. C.*: Recent Developments in the Synthesis, Thermostability and Liquid Crystal Properties of Aromatic Polyamides. Vol. 111, pp. 177-220.

López Cabarcos, E. see Baltá-Calleja, F. J.: Vol. 108, pp. 1-48.

Majoros, I., Nagy, A. and *Kennedy, J. P.*: Conventional and Living Carbocationic Polymerizations United. I. A Comprehensive Model and New Diagnostic Method to Probe the Mechanism of Homopolymerizations. Vol. 112, pp. 1-113.

Makhija, S. see Jaffe, M.: Vol. 117, pp. 297-328.

Malkin, A. Y. and *Kulichkhin, S. G.*: Rheokinetics of Curing. Vol. 101, pp. 217-258.

Maniar, M. see Domb, A. J.: Vol. 107, pp. 93-142.

Matsumoto, A.: Free-Radical Crosslinking Polymerization and Copolymerization of Multivinyl Compounds. Vol. 123, pp. 41-80.

Matsuoka, H. and *Ise, N.*: Small-Angle and Ultra-Small Angle Scattering Study of the Ordered Structure in Polyelectrolyte Solutions and Colloidal Dispersions. Vol. 114, pp. 187-232.

Mays, W. see Xu, Z.: Vol. 120, pp. 1-50.

Mikos, A. G. see Thomson, R. C.: Vol. 122, pp. 245-274.

Miyasaka, K.: PVA-Iodine Complexes: Formation, Structure and Properties. Vol. 108. pp. 91-130.

Monnerie, L. see Bahar, I.: Vol. 116, pp. 145-206.

Morishima, Y.: Photoinduced Electron Transfer in Amphiphilic Polyelectrolyte Systems. Vol. 104, pp. 51-96.

Müllen, K. see Scherf, U.: Vol. 123, pp. 1-40.

Müller-Plathe, F. see Gusev, A. A.: Vol. 116, pp. 207-248.

Mukerherjee, A. see Biswas, M.: Vol. 115, pp. 89-124.

Mylnikov, V.: Photoconducting Polymers. Vol. 115, pp. 1-88.

Nagy, A. see Majoros, I.: Vol. 112, pp. 1-113.

Nechaev, S. see Grosberg, A.: Vol. 106, pp. 1-30.

Neoh, K. G. see Kang, E. T.: Vol. 106, pp. 135-190.

Newman, S. M. see Anseth, K. S.: Vol. 122, pp. 177-218.

Noid, D. W. see Sumpter, B. G.: Vol. 116, pp. 27-72.

Novac, B. see Grubbs, R.: Vol. 102, pp. 47-72.

Novikov, V. V. see Privalko, V. P.: Vol. 119, pp. 31-78.

Ogasawara, M.: Application of Pulse Radiolysis to the Study of Polymers and Polymerizations. Vol.105, pp.37-80.

Okada, M.: Ring-Opening Polymerization of Bicyclic and Spiro Compounds. Reactivities and Polymerization Mechanisms. Vol. 102, pp. 1-46.

Okano, T.: Molecular Design of Temperature-Responsive Polymers as Intelligent Materials. Vol. 110, pp. 179-198.

Onuki, A.: Theory of Phase Transition in Polymer Gels. Vol. 109, pp. 63-120.

Osad'ko, I.S.: Selective Spectroscopy of Chromophore Doped Polymers and Glasses. Vol. 114, pp. 123-186.

de Pablo, J. J. see Leontidis, E.: Vol. 116, pp. 283-318.

Padias, A. B. see Penelle, J.: Vol. 102, pp. 73-104.

Penelle, J., Hall, H. K., Padias, A. B. and *Tanaka, H.*: Captodative Olefins in Polymer Chemistry. Vol. 102, pp. 73-104.

Peppas, N. A. see Bell, C. L.: Vol. 122, pp. 125-176.

Pichot, C. see Hunkeler, D.: Vol. 112, pp. 115-134.

Pieper, T. see Kilian, H. G.: Vol. 108, pp. 49-90.

Pospíšil, J.: Functionalized Oligomers and Polymers as Stabilizers for Conventional Polymers. Vol. 101, pp. 65-168.

Pospíšil, J.: Aromatic and Heterocyclic Amines in Polymer Stabilization. Vol. 124, pp. 87-190.

Priddy, D. B.: Recent Advances in Styrene Polymerization. Vol. 111, pp. 67-114.

Priddy, D. B.: Thermal Discoloration Chemistry of Styrene-co-Acrylonitrile. Vol. 121, pp. 123-154.

Privalko, V. P. and *Novikov, V. V.:* Model Treatments of the Heat Conductivity of Heterogeneous Polymers. Vol. 119, pp 31-78.

Putnam, D. and *Kopecek, J.:* Polymer Conjugates with Anticancer Acitivity. Vol. 122, pp. 55-124.

Ramaraj, R. and *Kaneko, M.:* Metal Complex in Polymer Membranc as a Model for Photosynthetic Oxygen Evolving Center. Vol. 123, pp. 215-242.

Rangarajan, B. see Scranton, A. B.: Vol. 122, pp. 1-54.

Reichert, K. H. see Hunkeler, D.: Vol. 112, pp. 115-134.

Risse, W. see Grubbs, R.: Vol. 102, pp. 47-72.

Rivas, B. L. and *Geckeler, K. E.:* Synthesis and Metal Complexation of Poly(ethyleneimine) and Derivatives. Vol. 102, pp. 171-188.

Robin, J. J. see Boutevin, B.: Vol. 102, pp. 105-132.

Roe, R.-J.: MD Simulation Study of Glass Transition and Short Time Dynamics in Polymer Liquids. Vol. 116, pp. 111-114.

Rusanov, A. L.: Novel Bis (Naphtalic Anhydrides) and Their Polyheteroarylenes with Improved Processability. Vol. 111, pp. 115-176.

Rychlý, J. see Lazár, M.: Vol. 102, pp. 189-222.

Ryzhov, V. A. see Bershtein, V. A.: Vol. 114, pp. 43-122.

Sabsai, O. Y. see Barshtein, G. R.: Vol. 101, pp. 1-28.

Saburov, V. V. see Zubov, V. P.: Vol. 104, pp. 135-176.

Saito, S., Konno, M. and *Inomata, H.:* Volume Phase Transition of N-Alkylacrylamide Gels. Vol. 109, pp. 207-232.

Samsonov, G. V. and *Kuznetsova, N. P.:* Crosslinked Polyelectrolytes in Biology. Vol. 104, pp. 1-50.

Santa Cruz, C. see Baltá-Calleja, F. J.: Vol. 108, pp. 1-48.

Scherf, U. and *Müllen, K.:* The Synthesis of Ladder Polymers. Vol. 123, pp. 1-40.

Schmidt, M. see Förster, S.: Vol. 120, pp. 51-134.

Schweizer, K. S.: Prism Theory of the Structure, Thermodynamics, and Phase Transitions of Polymer Liquids and Alloys. Vol. 116, pp. 319-378.

Scranton, A. B., Rangarajan, B. and *Klier, J.:* Biomedical Applications of Polyelectrolytes. Vol. 122, pp. 1-54.

Sefton, M. V. and *Stevenson, W. T. K.:* Microencapsulation of Live Animal Cells Using Polycrylates. Vol.107, pp. 143-198.

Shamanin, V. V.: Bases of the Axiomatic Theory of Addition Polymerization. Vol. 112, pp. 135-180.

Sherrington, D. C. see Lin, J.: Vol. 111, pp. 177-220.

Sherrington, D. C. see Steinke, J.: Vol. 123, pp. 81-126.

Shibayama, M. see Tanaka, T.: Vol. 109, pp. 1-62.

Shoda, S. see Kobayashi, S.: Vol. 121, pp. 1-30.

Siegel, R. A.: Hydrophobic Weak Polyelectrolyte Gels: Studies of Swelling Equilibria and Kinetics. Vol. 109, pp. 233-268.

Singh, R. P. see Sivaram, S.: Vol. 101, pp. 169-216.

Sivaram, S. and *Singh, R. P.*: Degradation and Stabilization of Ethylene-Propylene Copolymers and Their Blends: A Critical Review. Vol. 101, pp. 169-216.

Starodybtzev, S. see Khokhlov, A.: Vol. 109, pp. 121-172.

Steinke, J., Sherrington, D. C. and *Dunkin, I. R.*: Imprinting of Synthetic Polymers Using Molecular Templates. Vol. 123, pp. 81-126.

Stenzenberger, H. D.: Addition Polyimides. Vol. 117, pp. 165-220.

Stevenson, W. T. K. see Sefton, M. V.: Vol. 107, pp. 143-198.

Sumpter, B. G., Noid, D. W., Liang, G. L. and *Wunderlich, B.*: Atomistic Dynamics of Macromolecular Crystals. Vol. 116, pp. 27-72.

Suter, U. W. see Gusev, A. A.: Vol. 116, pp. 207-248.

Suter, U. W. see Leontidis, E.: Vol. 116, pp. 283-318.

Suzuki, A.: Phase Transition in Gels of Sub-Millimeter Size Induced by Interaction with Stimuli. Vol. 110, pp. 199-240.

Suzuki, A. and *Hirasa, O.*: An Approach to Artifical Muscle by Polymer Gels due to Micro-Phase Separation. Vol. 110, pp. 241-262.

Tagawa, S.: Radiation Effects on Ion Beams on Polymers. Vol. 105, pp. 99-116.

Tan, K. L. see Kang, E. T.: Vol. 106, pp. 135-190.

Tanaka, T. see Penelle, J.: Vol. 102, pp. 73-104.

Tanaka, H. and *Shibayama, M.*: Phase Transition and Related Phenomena of Polymer Gels. Vol. 109, pp. 1-62.

Tauer, K. see Guyot, A.: Vol. 111, pp. 43-66.

Terent'eva, J. P. and *Fridman, M. L.*: Compositions Based on Aminoresins. Vol. 101, pp. 29-64.

Theodorou, D. N. see Dodd, L. R.: Vol. 116, pp. 249-282.

Thomson, R. C., Wake, M. C., Yaszemski, M. J. and *Mikos, A. G.*: Biodegradable Polymer Scaffolds to Regenerate Organs. Vol. 122, pp. 245-274.

Tokita, M.: Friction Between Polymer Networks of Gels and Solvent. Vol. 110, pp. 27-48.

Uyama, H. see Kobayashi, S. : Vol. 121, pp. 1-30.

Vasilevskaya, V. see Khokhlov, A., Vol. 109, pp. 121-172.

Vaskova, V. see Hunkeler, D.: Vol. 112, pp. 115-134.

Verdugo, P.: Polymer Gel Phase Transition in Condensation-Decondensation of Secretory Products. Vol. 110, pp. 145-156.

Viovy, J.-L. and *Lesec, J.*: Separation of Macromolecules in Gels: Permeation Chromatography and Electrophoresis. Vol. 114, pp. 1-42.

Volksen, W.: Condensation Polyimides: Synthesis, Solution Behavior, and Imidization Characteristics. Vol. 117, pp. 111-164.

Wake, M. C. see Thomson, R. C.: Vol. 122, pp. 245-274.

Wang, K. L. see Cussler, E. L.: Vol. 110, pp. 67-80.

Wunderlich, B. see Sumpter, B. G.: Vol. 116, pp. 27-72.

Xie, T. Y. see Hunkeler, D.: Vol. 112, pp. 115-134.

Xu, Z., Hadjichristidis, N., Fetters, L. J. and *Mays, J. W.*: Structure/Chain-Flexibility Relationships of Polymers. Vol. 120, pp. 1-50.

Yannas, I. V.: Tissue Regeneration Templates Based on Collagen-Glycosaminoglycan Co-polymers. Vol. 122, pp. 219-244.

Yamaoka, H.: Polymer Materials for Fusion Reactors. Vol. 105, pp. 117-144.

Yaszemski, M. J. see Thomson, R. C.: Vol. 122, pp. 245-274.

Yoshida, H. and *Ichikawa, T.*: Electron Spin Studies of Free Radicals in Irradiated Polymers. Vol. 105, pp. 3-36.

Zubov, V. P., Ivanov, A. E. and *Saburov, V. V.*: Polymer-Coated Adsorbents for the Separation of Biopolymers and Particles. Vol. 104, pp. 135-176.

Subject Index

N-Acylhydroxylamines 143
Acylperoxy radicals 133
Ageing 24
Aggregation number 34, 43-48, 55, 58
Alkyl radicals, scavenging of 114, 116, 123, 131
- -, selfreactivity 131
O-Alkylhydroxylamine 107, 133, 135, 147
Alkylperoxy radicals 100, 132, 135
Amine oxide 137, 153
Amines, reductive regeneration 113, 115
-, toxic risk assessment 174
3-Anilino-1,5-diphenylpyrazole 123
Anti-polyelectrolyte 31
Antifatigue activity 155
Arborol 48
Autooxidation 92, 114

Bandowski's base 154
Benzoquinonediimine (BQDI) 103, 110, 112
Bisnitrone 112, 154
Bola-surfactant 53ff

CAC (critical alkyl group content) 17, 24, 25, 44
Carbon black 158
Catalysis 20, 40, 59
Chromophore 10, 20, 33
CMC (critical micelle concentration) 6, 17, 27, 28-35, 43, 44, 53-58
Conducting polymers, ^{15}N NMR 217
Conductometry 53, 56
Copolymer, alternating 16, 17, 25
-, block 3
-, graft 4, 7, 20, 29
-, random 8, 16, 20, 24, 25, 29, 30, 35, 40, 42

Copolymer, segmented block 5, 8, 25
-, star 4, 55, 56
Cosurfactant 31, 37
Counter ion 10-12, 28, 38, 53
Cross polarization 206, 207, 211, 223
Cyanate/cyanide based polymers, ^{15}N NMR 219, 221-224, 226-230

Decahydroquinoline 121
Decahydroquinoxaline 122
Decoupling, interrupted 233-235
Dendrimer 4, 48
1,2-Dihydro-2,2,4-trimethylquinolines (DHQ) 96
1,2-Dihydro-3-oxo-3H-indole 96, 117
9,10-Dihydroacridine 122
Dilution 25, 33, 43
Diphenylamines (DPA) 94, 102, 105, 110, 176
Diphenylphenazasiline 123
Dipolar rotational spin-echo 228-230
Discoloration, polymers 100, 113
Dissociation 12, 17, 25, 31, 36, 37
Double cross polarization 226-228
Drug delivery 20, 59

Electron microscopy 44, 48
Emulsifier 3, 39ff, 59
Energy transfer 20, 34, 42, 44, 46
ESR 20, 34, 40, 42
6-Ethoxy-2,2,4-trimethyl-1,2-dihydro-quinoline 118, 152

Flex-cracking 93
Flexibility 9, 19ff, 26, 44, 45
Fluorescence 20, 34ff, 42, 46, 55
Fluorocarbons 10, 11, 25
Functionalization 10, 20

Gemini surfactant 53ff
Geometry 6, 9, 13-17, 20, 26-56

Hindered amine stabilizers (HAS), appli-
    cation fields 125
-, basicity 125, 147
-, combination with phenolic AO 151,
    159
-, light stabilizers 124
-, radiolysis 146
-, salts with carboxylic acids 128, 146
-, thermal stabilizers 124, 159
Hydroperoxides 92, 136
Hydroxylamine, aliphatic 147
-, aromatic 101, 105, 115
-, heterocyclic 121, 134, 138

Ionene 17
Isotopic enrichment 192

Latex 26, 39, 59
Lifetime 34, 42, 43
Light scattering 46
Lipid 3, 16, 19
Liquid crystal 10, 16, 19, 49ff, 55, 59

Melamine-formaldehyde resins 211, 219,
    232, 233
Mesogen 10, 20, 22, 50
Micelle, polymeric 13, 14, 26, 34, 36,
    38, 42, 43ff, 58
Microviscosity 40, 41
Miscibility gap 50
Mobility 40-42, 55, 57
Monolayer (gas-water interface) 3, 15,
    27, 43, 55

Neutron scattering 46
Nitrogen, nonextractable 155
Nitrone 105, 121, 155
Nitroxide 128, 145, 155, 161
NMR 34, 36, 40, 41, 55
^{15}N-NMR, synthetic polymers 191-237
Nuclear Overhauser effect 193, 194

Oligomer 28, 32, 37, 39, 44, 53ff
Order parameter 40, 55
Ozonation 93, 112
Ozone scavenging 153

Peroxyacids 137
Phenolic resins 220, 221, 234
Phenylenediamines (PD), N,N-disubstitu-
    ted 96, 102, 109, 151, 176
Phenylnaphthylamines (PNA) 95, 102,
    105, 109, 176
Photo-oxidation 92
Photo-pollutants 91
Polyacrylonitriles, ^{15}N NMR 224
Polyaddition 6, 17
Polyamides 195, 207, 213-216, 230-232
-, crystallinity 230-233
Polyanilines, ^{15}N NMR 217
Polyazines, ^{15}N NMR 217
Polycondensation 6, 7, 17
Polyelectrolytes 12, 23-28, 46
-, ^{15}N NMR 204
Polymerization, degree 6-8, 44-48, 53-59
-, micellar 6, 14
Polypeptides, synthetic, ^{15}N NMR 197,
    206, 226
Polypyrroles 217, 235
Polysoap, natural 6, 39
Polyureas 198
Profile, dynamic 40
Pyrene 20, 34, 35, 36, 42

Quenchers, singlet oxygen 101, 140

Radiation-oxidation 150
Redox-activity 10, 22
Relaxation 42
Rigidity 20, 35, 42, 55

Saccharides 20, 31, 52
Salt effects 12, 25, 26, 31, 37
Solubilization site 34-40
Solvatochromism 34, 35
Spin-lattice relaxation time, ^{15}N NMR
    194, 205, 206, 211, 230-233

Stabilizers, oligomeric/polymeric   170
-, physical loss   166
-, polymer-bound   174
Staining of polymers   100
Sugars   12, 21
Surface activity   27ff, 37, 53-59
Surfactant, polymerizable   6-8, 14, 50,
   53, 59
- monomer   6, 12, 14, 17, 27, 36, 50, 58
Swelling   19, 50

1,2,3,4-Tetrahydro-2,2,4-trimethyl-
   quinoline   96, 121
Thickener   17, 24, 25

Thiosynergist   122, 148
TRFQ   46, 48

Urea-formaldehyde resins   198, 219
UV/Vis spectroscopy   20, 34

Viscosity   3, 5, 17, 23ff, 28, 46, 59

Wurster's ion-radicals   103, 111, 154

Zwitterion   10, 12, 25, 31

Springer-Verlag
and the Environment

We at Springer-Verlag firmly believe that an international science publisher has a special obligation to the environment, and our corporate policies consistently reflect this conviction.

We also expect our business partners – paper mills, printers, packaging manufacturers, etc. – to commit themselves to using environmentally friendly materials and production processes.

The paper in this book is made from low- or no-chlorine pulp and is acid free, in conformance with international standards for paper permanency.

Printing: Saladruck, Berlin
Binding: Buchbinderei Lüderitz & Bauer, Berlin